T0145218

Environmental Informatics

P. K. Paul · Amitava Choudhury · Arindam Biswas ·
Binod Kumar Singh
Editors

Environmental Informatics

Challenges and Solutions

 Springer

Editors
P. K. Paul
Department of Computer and Information
Sciences
Raiganj University
Raiganj, West Bengal, India

Arindam Biswas
Department of Computer Science
School of Mines and Metallurgy
Kazi Nazrul University
Berhampore, West Bengal, India

Amitava Choudhury
School of Computer Science
University of Petroleum and Energy Studies
Dehradun, Uttarakhand, India

Binod Kumar Singh
Department of Computer Science
and Engineering
National Institute of Technology
Jamshedpur
Jamshedpur, Jharkhand, India

ISBN 978-981-19-2085-1   ISBN 978-981-19-2083-7   (eBook)
https://doi.org/10.1007/978-981-19-2083-7

This Springer imprint is published by the registered company Springer Nature Singapore Pte Ltd.
The registered company address is: 152 Beach Road, #21-01/04 Gateway East, Singapore 189721,
Singapore

Preface

Environment is the aspect for everyone and important part of the society. Environmental Informatics is simply the utilizations and applications of Information Technology in Environment and Ecological Management. Environment is the major concerns not only for the individual but also for the organizations, institutions, and enterprises. Advanced tools and technologies are considered important in different environmental activities including environmental pollution. Today we all are getting benefits and support of the emerging technologies, and in this context, ICT applications in environment and ecology are considered as worthy. Technologists and scientists are engaged in healthy design and development along with implementation of environmental-friendly system for sustainable society. Environmental Informatics is a subfield of informatics which is dedicated in manual and computational environmental information systems. Informatics is board and interdisciplinary in nature and may be concentrated in different domain-based and functional areas. Domain-based Informatics may be restricted in the areas of biological science, social science, and pure science. Energy consumption is an important concern of Environment Informatics; therefore, energy informatics or environmental computing is also considered as similar area. Furthermore, Environmental Informatics may also known as

- Environmental information management
- Environmental information systems
- Sustainable computing
- Energy informatics.

In many countries, Environmental Informatics becomes a subject of study, research, and training and known as environmental information science. The latest applications of the computing, technologies, IT, and informatics principles in environmental aspects enriching Environmental Informatics for the support of environmental management, environmental systems and engineering, sustainable computing practice, etc. Environmental Informatics as a branch Informatics evolved first, and gradually, it is also known as environmental information technology and environmental information systems. Ecological informatics, forest informatics, irrigation informatics, geo-informatics, biodiversity informatics, etc., are also considered as valuable in healthy Environmental Informatics practice. Different sub-technologies of IT, viz. artificial intelligence, cloud computing, big data, Internet of things (IoT), edge and fog computing, robotics, and ML. Even some of the emerging information systems areas are also important in sophisticated environmental systems management. The effective and smarter ecological systems are worthy in advanced sustainable ecological practices using ICT. Speed, better monitoring, and management are considered as vital in promotion of the ICT in ecology and environmental practices. Governmental agencies, organizations, and scientific foundations are highly engaged in incorporation of environmental technological practices in recent past. This book is a comprehensive overview in environmental information science; therefore, it deals with the foundation topics, including basic and latest technologies applicable in environment and ecological systems. This research-based book, therefore, can be suitable for the readers interested in gathering knowledge of the following.

- Environmental Informatics with basics including applications using emerging technologies
- Role of ICCT underlying technologies in environmental and ecological management
- Green computing for business
- Green information centres and allied foundations
- Social and economic impact of artificial intelligence-based environmental
- E-waste in AI and digital healthcare context
- Smart energy conservation in irrigation management
- Artificial intelligence in agricultural system
- Advances and applications of bioremediation
- Internet of things (IoT) and drone applications in environmental systems
- Li-Fi-based energy-efficient traffic sensing and controlling system management
- Geospatial technologies and systems in water systems
- Automated geoprocessing model.

In summary, Environmental Informatics consists various technologies, tools, users, consumers, or data centre professionals. In better Environmental Informatics practice, some of the concerns may be important like cost effectiveness, energy and star rating, less power consumption and releases less carbon emission, less toxic material, and other harmful chemical, etc. Green informatics can also be suitable in general or manual documentation practice to current technology-based information systems management everywhere. Green information systems may be implemented. Environmental Informatics as a whole partially supported to the geo-informatics, forest management, and agriculture informatics to the enhanced, smarter, and sustainable ecological systems.

Raiganj, India   P. K. Paul
Gandhinagar, India   Amitava Choudhury
Asansol, India   Arindam Biswas
Jamshedpur, India   Binod Kumar Singh

Contents

About the Editors

Dr. P. K. Paul (Ph.D. IIEST Shibpur, PDFC-Environmental Informatics), working as Executive Director, MCIS and Assistant Professor, Department of Computer and Information Sciences, and also holds Information Scientist Position at Raiganj University, West Bengal, India. Additionally, he is also holding the Position of Honorary Professor, Logos University International, Louisiana, USA, and Chief Advisor (Innovative Program and Research Planning), Srinivas University, India. He has credited over 25 authored and edited books in diverse areas and focused on emerging and interdisciplinary information science. Further, he has about 200+ research/policy papers on cloud computing applications in information science, green computing or green IT in information field, I-schools aspects/usability engineering/information science educational aspects, etc. He is Chief Editor of IJASE, New Delhi, India, Chief Editor, IRA International Journal of Management and Social Sciences, and Chief Editor of IJISC, New Delhi, India. He is also involved as Editorial Board Member of more than 130 National and International Journals in diverse fields. He has completed many invited talk, viz. SRM University, Sikkim University, VIT University, Srinivas University, Mangalore University, TCG Digital, CloudNet, India. He has the track record of associating more than 200 international conferences as TPC, advisory committee, speakers, and in organizing body. Among his awards, few important are Best Researcher Award in Information Sciences (from IARA, Trichy), Best Faculty Award from IARA, David Clark Blair Young Scientist Award in Computer & Information Science (from BSS, India), International Young Scientist Award (from ISROSET) in Health Informatics, Citation Award (from Sri Sai University), Distinguished Young Information Science & Computing Academician in Asia from IRDP, Asian Information Science and Technology Ambassador Award-19, International Social Informatics and Digital Society Think Tank Award-2020, etc.

Dr. Amitava Choudhury working as Assistant Professor in the Department of Computer Science and Engineering, Pandit Deendayal Energy University Gandhinagar, Gujarat, India. He did his Ph.D. from India Institute of Engineering Science and Technology, Shibpur, India, and Master of Technology from Jadavpur University, Kolkata, India. He has more than nine years of experience in teaching and two years in research work. He serves as Reviewer of IEEE biomedical Transaction and Medical & Biological Engineering & Computing. He is Member of the IEEE, UP section, and Institute of Engineers, India. His area of research interest is computational geometry in the field of micromechanical modelling, pattern recognition, character recognition, and machine learning. Currently, he is associated with cluster computing, as Guest Editor. His areas of research interest are computational geometry in the field of micromechanical modelling, pattern recognition, character recognition, and machine learning.

Dr. Arindam Biswas received his M.Tech. degree in Radio Physics and Electronics from University of Calcutta, India, in 2010 and Ph.D. from NIT Durgapur in 2013. He was Postdoctoral Researcher at Pusan National University, South Korea, with prestigious BK21PLUS Fellowship, Republic of Korea. He got Visiting Professor at Research Institute of Electronics, Shizouka University, Japan. He has been selected for IE (I) Young Engineer Award: 2019-20 in Electronics and Telecommunication Engineering discipline, Institute of Engineers, India. He has 12 years of experience in teaching research and administration. Presently, he is working as Associate Professor in School of Mines and Metallurgy at KaziNazrul University, Asansol, WB, and India. He has 48 technical papers in different journals and 30 conference proceedings and six books, one edited volume and one chapter with international repute. He received research grant from Science and Engineering Research Board, Govt of India, under Early Career Research Scheme for research in Terahertz-based GaN Source. He also received Research Grant from Centre of Biomedical Engineering, Tokoyo Medical and Dental University in association with RIE, Shizouka University, Japan, for study of biomedical Thz imaging based on WBG semiconductor IMPATT Source. Presently, hc is serving as Associate Editor of Cluster Computing, Springer (SCI Indexed), and as Guest Editor of Nanoscience and Nanotechnology-Asia (Scopus Indexed), Recent Patent in Material Science (Scopus Indexed), Bentham science Publisher . He has produced four Ph.D. students in different topics of applied optics and high-frequency semiconductor device. He has organized and chaired difference international conferences in India and abroad. His research interest is in carrier transport in low-dimensional system and electronic device, nonlinear optical communication, and THz semiconductor source. He acted as Reviewer for reputed journals, Member of the Institute of Engineers (India), and Regular Fellow of Optical Society of India (India).

Dr. Binod Kumar Singh is Associate Professor in the Computer Science and Engineering Department of National Institute of Technology, Jamshedpur. He had completed his Ph.D. from Indian Institute of Technology Roorkee. He has a vast experience of 25 years in the research and education sector. He has a broad list of publications in SCI/SCIE indexed journals. His research interests lie in the field of image processing, network security, cloud computing, and distributed systems.

Chapter 1
Environmental Informatics: Basics, Nature, and Applications Using Emerging Technologies with Reference to Issues and Potentialities

P. K. Paul

Abstract Environmental Informatics is one of the emerging interdisciplinary knowledge domains which is also considered as a practicing field. This is the merger of domain Environmental Science with Information Science. Environmental Informatics is dedicated in information technology and computing applications in wide areas of Environment, Ecology, and Biodiversity Management. The role of Information Technology is emerging day by day and its results enhancement in Environment indirectly worthy for some of the other subjects such as Geology, Geography, Climatology, Oceanography, Agriculture, Forestry. According to the experts, the stakeholders of Environmental Informatics are IT, computing, and similar technologies from the technology side dedicated in effective environmental systems organization, management, and development. Environmental Informatics is associated with the management aspects and thus also worthy in environmental management. This chapter is conceptual in nature with the basics, features, and nature of the Environmental Informatics with the role in sustainable development practice. This chapter also illustrated the technologies involved for real practice with potential academic degrees and programs in this field.

Keywords Environmental Informatics · Computing · Environmental Information Science · Disaster management · Ecological development

1.1 Introduction

There are many subjects which are closely associated with the environmental-related subjects, like environment science, environment studies, environment engineering, environment management, etc., and therefore Environmental Informatics applicable on all these subjects using various Information Technology components like Software Technology, Web Technology, Database Technology, Networking Technology, and so on [1, 8]. IT is useful in several environment, ecological issues, and concerns and

P. K. Paul (✉)
Department of CIS, Raiganj University, Raiganj, India
e-mail: pkpaul.infotech@gmail.com

Fig. 1.1 Emerging IT components in Environmental Informatics required in environmental systems

right solution for solving the environmental aspects. Environmental Informatics in emerging scenario uses the sub-technologies like data analytics, cloud and virtualization computing, IoT systems, converged systems and network, usability systems and engineering, etc. [3, 20]. However, details of some other technologies are mentioned in Fig. 1.1.

Environmental Informatics in other words is also known as Environmental Information Science dedicated in ensuring eco-friendly information systems using state-of-the-art technologies. The society and community are modernizing and developing, at the same time, it is destroying the natural environment systems, and here use of Environmental Informatics-based systems would be suitable for improving the environment [7, 8, 28]. Several issues, challenges, and concerns of IT in environmental applications are emerging.

1.2 Objective

The chapter entitled 'Environmental Informatics: Basics, Nature and Applications using Emerging Technologies with reference to Issues and Potentialities' is conceptual and theoretical in nature and associated with deals with the following.

- To have an idea and concept of the subject Environmental Informatics including its evolution.
- To know about the characteristics, features, and nature (including components) of the Environmental Informatics.
- To get a picture of potential uses of the Environmental Informatics role and importance in developing environmental systems and development.
- To find out the core tools and technologies including environmental-related emerging technologies needed in environmental system management.
- To learn about the educational programs, potential programs in the areas Environmental Informatics and allied areas.

1.3   Methods

This chapter is theoretical in nature and conceptual too. The chapter is a review and having nature of environmental-related aspects lies on secondary and primary sources. To gather various attributes of Environmental Informatics like features and functions review of literature plays a leading role. Further, websites and web portals on environment and ecology are analyzed and mapped to get the current applications of IT in the environment and ecologies. In addition to this, various academic departmental websites reviewed and analyzed are associated with the Environmental Informatics education, training, and programs.

1.4   Environmental Informatics: Features, Role, and Stakeholders

The two main areas of Environmental Informatics are 'Environment' and 'Informatics'. Though *environment* does not mean it is only with the environmental science, it is also with other related areas such as environmental studies, ecology, agriculture, horticulture, and disaster management. *Informatics* includes the areas of IT, computing, as well as other similar technologies. The term Informatics was initially considered as practicing field, but gradually it has become a field of study and research [5, 30, 31]. Furthermore internationally in many universities, research centers Informatics as a branch widely started and practiced. The technologies of computing, technologies, and informatics are considered as worthy and increasing in almost all the sector due to its role in almost all the sectors. Environmental management today, in many ways supported by the sustainable computing, may sometimes also called as Environmental Informatics. Due to the interdisciplinary nature of the Environmental Informatics, the following features and characteristics may be considered as important.

- Applications and utilizations of the information technology and computing in environmental systems such as management and monitoring are possible with Environmental Informatics [2, 11].
- Environmental Informatics is supported by the environmental principles useful in information technology practice and development.
- The merging or integration of 'environment' and 'information technology'.

Environmental Informatics is purely interdisciplinary in nature and dedicated in modernizing environment and ecological information systems. It is required for various environment, disaster management, ecology, and waste management-related aspects.

- In the planning of energy, environmental systems tools and technologies supported by the IT and computing are considered as worthy. Environmental information technology therefore started in academics and in 'practice'.
- Regarding the simulation, optimization of environmental systems and monitoring Environmental Informatics is required [12, 22].
- In several areas and sectors, the increasing applications of GIS, remote sensing, and spatial IT are worthy and rising rapidly, and such technologies are part of Environmental Informatics.
- In the practice of environmental chemistry, biochemistry and allied activities also Environmental Informatics are valuable.
- As far as environmental management and monitoring are concerned, IT and computing systems are useful.
- In many environmental-related aspects, viz. atomic, molecular, and macromolecular scales, Environmental Informatics practice is emerging and urgent.
- In designing, developing, and modeling of biological environment-related processes, Environmental Informatics-based tools are highly required and supported.
- In the websites related to the environment and ecology role of Environmental Informatics practice is considered important in order to develop healthy Environmental Information Systems.
- In the modeling of biotechnological systems including pollution mitigation too, practice of Environmental Informatics is considered as worthy.
- In the managerial activities such as environmental statistics and environmental risk analysis including climate modeling and downscaling, the field Environmental Informatics is worthy [29].
- Regarding the impact and assessment of the adaptation planning, biological and disaster management systems promotion and development of Environmental Informatics can be considered as worthy and applicable [6, 13].

The applications of the animation and graphics include various visualization tools required in environmental decision support systems development; in this context, various emerging technologies may be considered as cloud computing, artificial intelligence, machine learning, deep learning, and so on are impacting. Initially,

Environmental Informatics nomenclature did not exist, but gradually due to its practice, it has become a field of study as well as research. Before the improvement of the Environmental Informatics, some of the domain-centric informatics include health informatics, geo-informatics, bio-informatics, medical informatics, and so on [9, 15]. Internationally in many universities and higher educational institutions educational, research and training programs have started in Environmental Informatics and allied areas. Many of them are offered as major degrees as Bachelors, Masters, and PhDs in Environmental Informatics and allied areas. There are some other nomenclatures too which are being used in different contexts such as

- Ecological informatics.
- Environmental computing.
- Ecological information science.
- Environmental information systems.
- Environmental information technology [8, 18, 24].

Environmental Informatics today has been evolved as an emerging interdisciplinary subject which comprises various environment and ecology-related subjects and also IT and computational sciences [19, 27]. Based on the analysis of the nature of Environmental Informatics, it may be considered as **stakeholders** comprising environment, content, and information, technologies, and people (i.e., User and HR). *Environment* is obviously an important component of environmental systems; as most of the living and non-living facets of the nature are part of environment. Owing the importance of ecology and environment, the field *Environmental Informatics* is treated as important for the real-time solutions. Because environment is a broad and interdisciplinary science, it has applications in diverse areas and allied area management like agriculture, forest management, oceanography, etc. Environmental Informatics dedicated in animals, forests, etc., [24, 25] and therefore in all the ecological and environmental-related aspects too is worthy and required. Environmental Informatics is indirectly applicable in various areas like

- Climatology sciences
- Oceanography systems
- Geological systems
- Geography and allied areas
- Agricultural systems
- Forestry, disaster management, etc.

All these stakeholders not only help in Environmental Informatics but also in allied areas, viz. agricultural informatics, forestry informatics, irrigation informatics, etc.

Content is treated as important stakeholder in Environmental Informatics subject and practice due to the role of information in developing environmental information systems and some other allied activities such as ecological information repositories, ecological and environmental database, and disaster environmental modeling [10, 14]. Environmental Informatics is purely based on content and similar contents [23,

26]. *Technologies* are another stakeholder in healthy Environmental Informatics practice, and it has uses in various basic information technology components, viz.

- Web technology.
- Networking technology.
- Database technology.
- Software technology.
- Multimedia technology, etc. [16, 17, 32].

In sophisticated Information Technology practice in environment and ecological systems, various subfields and parts of IT are important. Some of the subfields and emerging technologies are already mentioned in this work (Refer Fig. 1.2).

Users are obviously considered as important stakeholder in all type of informatics and engineering systems or computational systems. As far as Environmental Informatics is concerned, the users are importantly distributed in different areas, and apart from the basic users, the Human Resources are also considered as worthy in Environmental Informatics practice. Various types of Human Resources and skilled manpower are considered as crucial in such development.

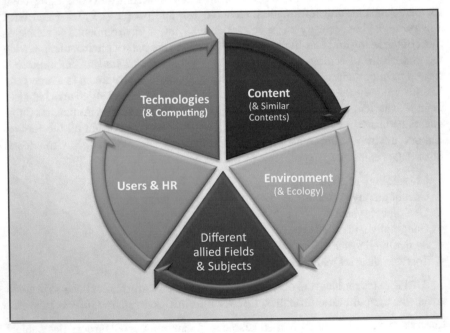

Fig. 1.2 Major stakeholders of developing Environmental Informatics

1.5 Issues and Challenges Concerning with the Environmental Informatics Practice

Environmental information science or Environmental Informatics practice is emerging rapidly throughout the world, particularly in developed countries. Though the analysis of current uses and concern of Environmental Informatics leads to various issues and challenges in the developing countries, some of them are as follows:

Technological Implementation—It is essential that proper, sufficient, and adequate technological implementation should be provided in order to avail proper benefits from the Environmental Informatics [17, 30].

Proper Awareness—The awareness of Environmental Informatics and its practice is need to care of urgently, including training and educational programs. Professionals and government personnel should take proper steps in environment and ecological systems development and monitoring [4].

Financing and Funding—Proper funding is important in developing Environmental Informatics practice. Financial arrangement is essential in order to take purchasing technology including implementation and development of the systems. Various environmental organizations, ministries, and departmental funding opportunities should be provided.

Management and Governance—Managing different technologies is an important issue in regard to environmental-related technologies. Here proper management and governance steps are essential to follow up.

Manpower and HR Development—It is a fact that there are many organizations, institutes, and higher educational institutions offering educational and training programs in this field and related areas; but there is a shortage of skilled manpower in designing and developing Environmental Informatics.

Interdisciplinary and Broad Cluster—Environmental Informatics is purely interdisciplinary and getting more broader day by day. Its requirement increasing gradually in all the areas of environment, agriculture, forestry, ecology, and other technological areas; therefore, the skill and knowledge development is a big and emerging concern [33].

1.6 Emerging Technologies, Environmental Informatics, and Environmental Management

Latest and emerging technologies of IT and computing dedicated in various activities and affairs of environment and ecological systems and among them important are big data management, cloud computing, robotics and artificial intelligence, Internet of things (IoT), usability engineering, and HCI, etc.

1.6.1 Big Data

Big data is an emerging component of information technology which is required in managing large number and complex data. The big data and analytics applications in the environment lead to several management problems such as global warming, climate change, satellite earth observation, and numerous data collection and its management. In forestry and forest management, in deforestation by counting trees too Environmental Informatics is applicable, in the creation of sustainable smart cities, in urbanization, in GPS sensors management too Environmental Informatics is important and required. Big data is useful in managing large amount of data management and therefore in pollution control in the cities, traffic flows, etc., and in all these allied areas, big data and analytics are effectively useful. As far as renewal energy management is concerned, big data and analytics tools are useful. In other power management-related aspects such as in wind and power management, big data is effectively useful by collecting the data and their proper analysis. Regarding hydroelectric power, in smart meters also big data analytics is being used. As far as geographic data management is concerned such as space, geospatial data management, cartographic naval navigation, regarding disaster, and emergency management also data analytics are useful and important. Satellite data is collected using GPS, remote sensing, and GIS; and in this context, big data management is worthy and important.

1.6.2 Cloud Computing

Cloud computing is the way and the platform of virtualization of different information technology systems. It helps in remote access of IT support from different place, and in this regard, it helps in environment-related aspects. The big data helps in generating and managing the data, and the same data is stored by cloud computing models and other cloud supported systems. In ecological monitoring, disaster management, forest management, wildlife management, and some other areas applications and utilizations of cloud computing are being used.

1.6.3 Robotics and Artificial Intelligence

Robotics and artificial intelligence are another two related aspects and sub-technology in the field of information technology and having impact in environment, ecology, and disaster management systems. Artificial intelligence is responsible in developing intelligent systems including the products and services. Therefore, it is required in the purpose of modeling and governance of environmental systems with intelligent mechanism. In identification of the tropical cyclone, weather forecasting also artificial intelligence and robotics are considered as worthy and important. In the natural disaster management, trees and forests are basically affected, and with the artificial intelligence and robotics supported systems, it is possible to identify the affected trees and forests. As far as intelligent environmental power management system is concerned, it is empowered with the robotics and artificial intelligence.

1.6.4 Internet of Things (IoT)

Internet of things (IoT) is the latest emerging technology in IT and computing and dedicated in the collecting data using Internet. Internet of things is helpful in managing different kind of modern services and products associated with the Internet. IoT sensors are dedicated in collecting data from different nodes and points and which are being used in environmental and ecological concerns. The sensors are dedicated in collecting the data and ultimately help in environmental decision making, ecological monitoring, forest management, agricultural management, and so on [5, 31].

In addition, the abovementioned emerging technologies in the environmental practice lead to the development of some other activities as mentioned (in Fig. 1.3) [9, 21, 26].

1.7 Educational Programs and Potentialities in Promoting Environmental Informatics

In addition to these, some of the universities have started offering joint or integrated or dual degree programs such as University of North Carolina at Chapel Hill, US offers BS-Environmental Science and MS-Information Science Dual Degree, and this is offered by College of Environment, Ecology and Energy with School of Information and Library Science. However, BSc Environmental Informatics and Business Information Systems (Dual Degree) is offered by The University of Applied Sciences, Germany.

Environmental Informatics has started popularity in some of the universities, and therefore, it has started as major program at Bachelor's Degree in some of the universities such as at Auburn University, Alabama, USA, Wuhan University, China, Northern Arizona University, USA, Virginia Polytechnic Institute and State University, USA, etc.

There are potentiality to offer the program of Environmental Informatics with some other allied nomenclature such as ecological informatics, disaster informatics, and natural and eco-informatics. The program can be offered in other allied branches as specialization such as computing, information technology, information science as proposed in Table 1.1.

Fig. 1.3 Different emerging technological applications in ecological informatics (top) and few tools live examples (below)

Table 1.1 Potential Environmental Informatics major in IT and computing-related programs

Environmental Informatics potentiality as a major of IT and computing-related subjects
Bachelor of Science/Master of Science (BS/MS/BSc/MSc) Information Technology (Environmental Informatics)
Bachelor of Science/Master of Science (BS/MS/BSc/MSc) Computing (Eco and Disaster Informatics)
Bachelor of Science/Master of Science (BS/MS/BSc/MSc) Computing (Ecological and Forest Informatics)

Table 1.2 Potential Environmental Informatics major in biological-related programs

Environmental Informatics potentiality as a major in biological and related subjects
Bachelor of Science/Master of Science (BS/MS/BSc/MSc) Biological Science (Environmental Informatics)
Bachelor of Science/Master of Science (BS/MS/BSc/MSc) Forest and Disaster Management (Environmental Informatics)
Bachelor of Science/Master of Science (BS/MS/BSc/MSc) Botany (Eco and Disaster Informatics)
Bachelor of Science/Master of Science (BS/MS/BSc/MSc) Environmental Science (Ecological and Forest Informatics)

Similar to IT and computing program Environmental Informatics can be offered in biological sciences as a major or specialization, as depicted in Table 1.2.

However, Environmental Informatics or allied programs may be offered at Master of Computer Application (MCA) or Bachelor of Computer Application (BCA) degrees as a specializations. The allied areas and nomenclature of Environmental Informatics may be offered such as agricultural informatics, ecological informatics, forest informatics, and disaster informatics. Furthermore, the branch Environmental Informatics may be offered as by research mode. And in this context, degrees may be all what proposed and depicted in Tables 1.1 and 1.2 may be offered like MS (by research) in information technology (Environmental Informatics), etc.

1.8 Conclusion

Environmental Informatics is scientifically and academically developed in the atmosphere of information technology for different ecological and environmental purposes including modeling, simulation, designing, development, and data analysis of environmental systems. In addition to the basic technologies such as geographical information systems (GIS), remote sensing, and GPS, various spatial technologies are being used in this context. The technological development in Environmental Informatics is noticeable and changing day by day. Government ministries, bodies and departments, scientific houses, and environmental organizations are engaged in diverse applications in Environmental Informatics and allied technologies. Different higher educational institutions and universities already started educational programs in developing proper manpower in this field. Though there is an urgent need of preparing skilled manpower, and thus short-term programs, certifications in the areas are highly important for complete sustainable development. Policymakers and the government bodies required proper steps in developing proper and effective Environmental Informatics practice. Robust benefits and complete solutions toward environment become possible from the field and for its final completion and further development.

References

1. Allen, T. F., Giampietro, M., & Little, A. M. (2003). Distinguishing ecological engineering from environmental engineering. *Ecological Engineering, 20*(5), 389–407.
2. Argent, R. M. (2004). An overview of model integration for environmental applications—Components, frameworks and semantics. *Environmental Modelling & Software, 19*(3), 219–234.
3. Borchardt, M., Wendt, M. H., Pereira, G. M., & Sellitto, M. A. (2011). Redesign of a component based on ecodesign practices: Environmental impact and cost reduction achievements. *Journal of Cleaner Production, 19*(1), 49–57.
4. Cruz, F. B. D. L., & Barlaz, M. A. (2010). Estimation of waste component-specific landfill decay rates using laboratory-scale decomposition data. *Environmental Science & Technology, 44*(12), 4722–4728.
5. Dmitrović, L. G., Dušak, V., & Milković, M. (2017). Involving environmental informatics in Croatian technical studies. *Technical Gazette, 24*(6), 1869–1875.
6. Frew, J. E., & Dozier, J. (2012). Environmental informatics. *Annual Review of Environment and Resources, 37*, 449–472.
7. Green, D. G., & Klomp, N. I. (1998). Environmental informatics-a new paradigm for coping with complexity in nature. *Complex Systems, 98*, 36–44.
8. Haupt, S. E., Pasini, A., & Marzban, C. (Eds.). (2008). *Artificial intelligence methods in the environmental sciences*. Springer Science & Business Media.
9. Hilty, L. M., Page, B., & Hřebíček, J. (2006). Environmental informatics. *Environmental Modelling & Software, 21*(11), 1517–1518.
10. Huang, G. H., & Chang, N. B. (2003). The perspectives of environmental informatics and systems analysis. *Journal of Environmental Informatics, 1*(1), 1–7.
11. Hunt, C. B., & Auster, E. R. (1990). Proactive environmental management: Avoiding the toxic trap. *MIT Sloan Management Review, 31*(2), 7–14.
12. Jones, D. A., Lelyveld, T. P., Mavrofidis, S. D., Kingman, S. W., & Miles, N. J. (2002). Microwave heating applications in environmental engineering—A review. *Resources, Conservation and Recycling, 34*(2), 75–90.
13. Laird, D. A., Yen, P. Y., Koskinen, W. C., Steinheimer, T. R., & Dowdy, R. H. (1994). Sorption of atrazine on soil clay components. *Environmental Science & Technology, 28*(6), 1054–1061.
14. Lu, J. W., Chang, N. B., & Liao, L. (2013). Environmental informatics for solid and hazardous waste management: Advances, challenges, and perspectives. *Critical Reviews in Environmental Science and Technology, 43*(15), 1557–1656.
15. Mayfield, C., Joliat, M., & Cowan, D. (2001). The roles of community networks in environmental monitoring and environmental informatics. *Advances in Environmental Research, 5*(4), 385–393.
16. Maxim, L., & van der Sluijs, J. P. (2011). Quality in environmental science for policy: Assessing uncertainty as a component of policy analysis. *Environmental Science & Policy, 14*(4), 482–492.

17. Mykrä, H., Heino, J., & Muotka, T. (2007). Scale-related patterns in the spatial and environmental components of stream macro invertebrate assemblage variation. *Global Ecology and Biogeography, 16*(2), 149–159.
18. Page, B. (1996). Environmental informatics towards a new discipline in applied computer science for environmental protection and research. *Environmental Software Systems*, 3–22.
19. Page, B., & Wohlgemuth, V. (2010). Advances in environmental informatics: Integration of discrete event simulation methodology with ecological material flow analysis for modelling eco-efficient systems. *Procedia Environmental Sciences, 2*, 696–705.
20. Paul, P. K. (2013). Environment and sustainable development with cloud based green computing: A case study. *Scholars Academic Journal of Biosciences (SAJB), 1*(6), 337–341.
21. Paul, P. K., Ganguly, J., & Sinha, A. (2014). Environmental information processing for eco friendly social and business development—A short communicationin. In *National conference on corporate social responsibility: Expectations and challenges* (pp. 258–264).
22. Paul, P. K., Kumar, K., Das, P., Karn, B., & Poovammal, E. (2016). Environmental information systems: Tools, technologies and other facets better green world: A brief conceptual study. *International Journal of Recent Researches in Science, Engineering & Technology, 4*(05), 142–147.
23. Paul, P. K., Bhuimali, A., Ghose, M., & Chatterjee, D. (2016). Eco informatics and green IT as an interdisciplinary environmental-computing-management domain: With a case study of united kingdom programs. *Palgo Journal of Education Research, 4*(6), 225–229
24. Paul, P. K., Bhuimali, A., Ghose, M., Ganguly, J., & Ghosh, M. (2017). Information technology and green-eco environment: The aspects in interdisciplinary scenario. *International Journal of Scientific Research and Modern Education (IJSRME), 2*(2), 27–30.
25. Paul, P. K., Aithal, P. S., Bhuimali, A., & Kalishankar, T. (2020). Environmental informatics vis-à-vis big data analytics: The geo-spatial & sustainable solutions. *International Journal of Applied Engineering and Management Letters (IJAEML), 4*(2), 31–40.
26. Paul, P. K., Bhuimali, A., Aithal, P. S., & Kalishankar, T. (2020). Environmental informatics: Educational opportunities at bachelors level—International context and Indian potentialities. *International Journal of Applied Engineering and Management Letters (IJAEML), 4*(1), 243–256.
27. Paul, P. K., Aithal, P. S., & Bhuimali, A. (2020). Environmental informatics and educational opportunities in post graduate level—Indian potentialities based on international scenario. *IRA-International Journal of Management & Social Sciences, 16*(2), 45–58.
28. Paul, P. K., Saavedra, M. R., Aithal, P. S., Aremu, P. S. B., & Baby, P. (2020). Environmental informatics: Potentialities in iSchools and information science & technology programs—An analysis. *International Journal of Management, Technology, and Social Sciences (IJMTS), 5*(1), 238–251.
29. Paul, P. K., Bhuimali, A., & Aithal, P. S. (2020). Environmental informatics: The foundation, allied & related branches—Analytical study. *International Journal of Social Sciences, 9*(1), 01–07.
30. Sandler, S. I. (1996). Infinite dilution activity coefficients in chemical, environmental and biochemical engineering. *Fluid Phase Equilibria, 116*(1–2), 343–353.
31. Thompson, S., Treweek, J. R., & Thurling, D. (1997). The ecological component of environmental impact assessment: A critical review of British environmental statements. *Journal of environmental Planning and Management, 40*(2), 157–172.
32. Tochtermann, K., & Maurer, H. A. (2000). Knowledge management and environmental informatics. *Journal of Universal Computer Science, 6*(5), 517–536.
33. Yakhou, M., & Dorweiler, V. P. (2004). Environmental accounting: An essential component of business strategy. *Business Strategy and the Environment, 13*(2), 65–77.

Chapter 2
Exploring the Role of ICCT Underlying Technologies in Environmental and Ecological Management

P. S. Aithal ⓘ **and Shubhrajyotsna Aithal** ⓘ

Abstract Technology is an enabler of the implementation of various strategies in solving environmental problems. Information communication and computation technology (ICCT) and nanotechnology (NT) are two new emerging general-purpose technologies that have the capabilities to solve many problems of society in an innovative and effective manner. These technologies have potentiality to manage the natural environment and ecology of the earth to support sustainable living creatures. In this chapter, we made a detailed analysis of the role of ICCT underlying technologies in environmental and ecological management for maintaining sustainable living systems on earth. The chapter discusses the technology interventions and management of natural environmental and ecology and identifies the role of ICCT underlying technologies in environmental and ecological management. The chapter also discusses implementation strategies of use of ICCT underlying technologies in environmental and ecological management and analysis of the possible role of ICCT in ecological management using a qualitative ABCD analysis framework.

Keywords Environmental management · Ecological management · ICCT underlying technologies · ICCT in environmental management · ABCD analysis for ICCT in ecological management

2.1 Introduction

Managing the physical, social, and economical environment is an important and essential responsibility of everyone in society for sustainable living on the earth. The physical environment also called the natural environment comprises of everything within and around every living creature. In a literature sense, environment means

P. S. Aithal (✉)
College of Management and Commerce, Srinivas University, Mangalore 575 001, India
e-mail: psaithal@srinivasgroup.com

S. Aithal
College of Engineering and Technology, Srinivas University, Mangalore, India

© The Author(s), under exclusive license to Springer Nature Singapore Pte Ltd. 2022   15
P. K. Paul et al. (eds.), *Environmental Informatics*,
https://doi.org/10.1007/978-981-19-2083-7_2

everything surrounding an individual, object, element, or system and their relationship and interaction. For a sustainable environment, there should be a cordial and long-lasting interrelationship between individuals, objects, elements, or systems with their surroundings [1]. Though the concept of environment is complex in nature, they can be simplified using a system model that is based on the interaction between input, output, processes, and surroundings of a system. Systematic study of the relationship between the components of any system in terms of its physical, social, and economical concepts constitutes environmental science. It is the branch of environmental science that focuses on the interaction between various organisms and their environment. It is interesting and appropriate to know the possible impact of technology and its management for maintaining a sustainable environment and ecology of living and nonliving systems including individuals and organizations. Any system is stable if and only if its environment supports it, i.e., the characteristics, performance, and productivity of every system depend on its environment [2]. Management of the environment and ecology of the earth using innovative technologies for the sustainability of living beings is finding primary importance and should be given utmost preference for the survival of living beings. The current scenario of environmental degradation due to natural and man-made disasters is alarming and threatens the continuation of living beings on mother earth. Sustainable environment and ecology are important for human well-being and prosperity. Ecology enriches the interdependence of human beings and nature by maintaining clean air and water, nutritious food, and sustaining biodiversity in a changing climate. In this chapter, the effect and role of ICCT underlying technologies on environmental management and ecology management are discussed and analyzed.

2.2 ICCT Underlying Technologies

There are many debates on the interaction between technology and the environment [3]. Technology, being an application of science, is an enabler and driver of all industries to progress and prosper. Technology-based various industry generations are identified [4], and currently, industry 4.0 is active that is driven mainly by the Internet of Thigs (IoT) to ubiquitously connect various cyber-physical systems. Recently, two technologies are identified as universal technologies that have capabilities to solve many problems in all industry sectors including primary industry sector, secondary industry sector, tertiary industry sector, and quaternary industry sector [5]. These two universal technologies are information communication and computation technology (ICCT) and nanotechnologies, which have capabilities to support intangible and tangible products and services in all industry sectors. Information communication and computation technology (ICCT) is growing like an umbrella and contributing through many emerging new sub-technologies called ICCT underlying technologies [6]. Table 2.1 lists twelve ICCT underlying technologies along

Table 2.1 Twelve ICCT underlying technologies with their objectives and potentialities

S. No.	ICCT underlying technology	Objectives	Potentialities
1	Artificial intelligence and robotics technology	To create intelligent machines	Machines can think and make decisions better than human beings
2	Big data and business intelligence technology	To analyze continuously generated data in business processes and find the pattern to predict the future	Use of big data indications in the form of business intelligence for making current decisions based on future predictions
3	Blockchain technology	Connecting digital information in the form of digital ledger of transactions in such a way that it is vulnerable to change, hack, or cheat	To allow digital information to be recorded and distributed in such a way that it cannot be edited at any one point during transaction
4	Cloud computing technology	To access and use any third-party electronic device through Internet so that the concept of rental usage of third-party hardware and software resources instead of owning them	Organizations and individuals can use rented digital devices for processing and storage of information ubiquitously to reduce the expenditure and to optimally utilize available resources
5	Cybersecurity and forensic technology	To stop digital crimes by identifying and gathering proofs of crime in an exceedingly forensically sound manner with required evidence	Possibility of providing perfect security for digital information during processing, transmission, and storage by effectively handling the unauthorized intrusion and disclosing the crime
6	Digital business and marketing technology	Reaching every potential customer and providing information and awareness about products/services	Ability to improve business models toward ideal business model
7	3D-printing technology	Printing physical objects of any size layer by layer by mixing and processing required raw materials systematically	Ubiquitous printing of physical objects anytime, anywhere. This avoids logistics of physical products between countries
8	Internet of Things (IoT) technology	Connecting cyber-physical devices through Internet	Automated controlling of Internet-connected devices ubiquitously

(continued)

Table 2.1 (continued)

S. No.	ICCT underlying technology	Objectives	Potentialities
9	Information storage technology	To enhance digital information storage ability of devices along with decrease in their size and cost	Storing huge amount of data and digital information for future tech-world
10	Optical and quantum computer technology	Increasing the speed and capability of computers beyond electronic computers capability	High-speed computers which can cater the need of every one of this world through cloud computing platform
11	Online education technology	Education to every one through effective online ubiquitous models	Providing education to every one irrespective of their geographical and economic background, at any age and any time
12	Virtual and augmented reality technology	To provide real-world experiences through virtual environment, i.e., mimicking reality through virtual setup	Creating artificial environment with scenes and objects that mimics the reality

with their objectives and potentialities. In this chapter, we made a detailed analysis of the role of ICCT underlying technologies in environmental and ecological management for maintaining sustainable living systems on earth.

In this chapter, how ICCT underlying technologies are useful in managing a sustainable environment and ecology is discussed and analyzed.

2.3 Objectives

(1) To discuss the technology interventions and management of natural environmental and ecology.
(2) Identify and analyze the role of ICCT underlying technologies in environmental and ecological management.
(3) Implementation strategies of use of ICCT underlying technologies in environmental and ecological management.
(4) Analysis of the possible role of ICCT in ecological management using qualitative ABCD analysis framework.

2.4 Role of ICCT Underlying Technologies in Environmental Management

(1) **Role of Artificial Intelligence and Robotics in Environment and Ecology Management**

    (i)   Artificial intelligence technology helps environmental management, especially planning, monitoring, and controlling a sustainable environment. It includes analyzing the current environment, forecasting future changes, monitoring, and controlling the environmental changes using various AI-supported monitoring and decision supporting systems [7].

    (ii)   AI and robotics technology is also useful for modeling environmental systems using case-based reasoning, rule-based models, artificial neural networking, fuzzy logic models, multi-agent systems, genetic algorithms, machine learning models, cellular automata, swarm intelligent models, and hybrid modeling [8].

    (iii)   Predicting and managing environmental data in environmental management systems, sustainable green human resource management, and water resource management for sustainable environmental planning, atmospheric forecasting and management, and e-waste management for environmental planning, using the principles of artificial intelligence [9].

    (iv)   Artificial intelligence-based expert systems are helpful in sophisticated environmental management models of materials with reduce, reuse, recycle, and recover slogans [10].

    (v)   Artificial intelligent technology is used in ecological management based on modeling and simulation, integration of qualitative and quantitative knowledge, theoretical aspects of ecological modeling, and natural resource management and policy analysis, etc [11]. Artificial intelligence-based expert systems are expected to provide useful tools for ecological research and ecological knowledge management applications [12].

    (vi)   Robotics technology has attractive applications in environmental engineering, environmental monitoring, and environmental management of the planet and its environmental processes. Robots are used to explore deep oceans, track harmful algal blooms and to control the spread of pollution, and monitor remote volcanoes [13].

    (vii)   Areal robotics are used for forest management and seeding, Autonomous robotics are used in identification and management of invasive aquatic plant species, pest control in agriculture, ecological management of agricultural weed, plant eco-phenotyping, etc [14].

(2) **Role of Big Data and Analytics Technology in Environment and Ecology Management**

    (i) Big data generally represent the mass volume of data generated using video detectors (CCD cameras) and continuously monitor the changes in any system and cannot be processed using ordinary data processing tools and practices. In environmental management systems, big data are used to monitor and control the continuously changing environment due to various environment degradation activities.

    (ii) Big data-based analytics are used to describe a situation, predict a situation, and prescribe a solution to control a situation in environmental management.

    (iii) Descriptive analytics of environmental information describes what is already happened based on analysis of environmental and ecological data.

    (iv) Predictive analytics of environmental information predicts what could happen in the future based on analysis of environmental and ecological data [15].

    (v) Prescriptive analytics of environmental information prescribes what should happen and how variations can be controlled based on analysis of environmental and ecological data.

    (vi) Big data technology helps effective analysis of agricultural and rural ecological management systems [16].

    (vii) Big data and analytics support societal development and environmental sustainability. It also supports urban ecological environmental management through the upgraded geographical management system.

    (viii) Study of environmental impact on the earth surface, marine, and atmosphere using big data and analytics and intervention on disaster resilience through big data for environmental sustainability [17].

(3) **Role of Blockchain Technology in Environment and Ecology Management**

    (i) Blockchain technology also called distributed ledger technology is helpful for maintaining undistracted environmental data for long period for continuous analysis.

    (ii) Blockchain technology can be used for the management of effective treatment of industrial wastewater and safely discharge to water bodies [18].

    (iii) Blockchain technology, with its decentralized property, has the capability to protect and sustain the global environment at various levels including life on earth, life below the earth's surface, and climate changes. Blockchain technology is used to monitor the climate change, biodiversity, conservation of healthy water bodies to manage the ecological threats [19].

(iv) Blockchain technology is used in securing environmental data due to its unique property of non-modifying data feature at any one stage. Hence, it can be used in industrial pollution data security, weather monitoring and forecasting data security, marine data security, etc [20].

(v) Blockchain technology has potential applications in designing a smart environment and smart mobility by planning the use of renewable energy sources and creating awareness regarding environmental and energy sustainability [21].

(vi) Blockchain technology enables individuals and organizations to manage their carbon emission footprints, the social and environmental costs, and environment management policies and strategies [22].

(vii) Blockchain technology can be used to monitor and control unauthorized looting of natural resources including mines, forests, sand from mafias and hence protect natural resources for environmental and ecological sustainability.

(viii) Blockchain technology can be used effectively or to improve the efficiency of waste management including solid, liquid, and gases wastes from various economic activities in society to maintain a sustainable environment and ecosystem.

(ix) Blockchain supported e-agriculture and animal husbandry to include public participation and support for basic resources like water and feed management.

(4) **Role of Cloud Computing in Environment and Ecology Management**

Cloud computing technology allows individuals and organizations to use ubiquitous computing resources (both software and hardware) for digital information processing and storage. The major roles of cloud computing technology are as follows:

(i) Use of cloud computing platform to decrease the cost of using computing processes related to environmental planning, monitoring, and controlling.

(ii) Cloud computing platforms can be used for urban ecological environment investigation and management using computer neural network algorithm [23].

(iii) Cloud computing can be also used for reducing the operational cost of regional environmental monitoring and management systems and agencies in every country [24].

(iv) Cloud computing platforms can be also used to develop healthy and sustainable development environmental systems [25].

(v) A new model of cloud computing platform based on modern eco-agriculture is possible [26].

(vi) Cloud computing platform can be used to evaluate ecological environment quality in tourist areas of smart cities.

(vii) In modern forest fire management, green construction management, and power consumption management systems, cloud computing platforms can be extensively used for better cost and time management.

(5) **Role of Cybersecurity and Forensics in Environment and Ecology Management**

(i) Cybersecurity and forensics have applications in environmental data protection during such data transactions and storage.

(ii) Cybersecurity and forensics technology have advantages in protecting the data of primary industry sector including agriculture, forestry, fisheries, and mining.

(iii) Minimizing crimes in various industry sectors that affect the environment and natural ecology are possible using cybersecurity and forensics technology.

(iv) Cybersecurity technology supports protecting research data on environmental monitoring and natural ecology management on the earth's surface [27].

(6) **Role of 3D Printing Technology in Environment and Ecology Management**

Three-dimensional (3D) printing is an additive manufacturing process of physical objects layer by layer from a digital design. Ideal 3D printing should manufacture any physical object of any size using the combination of required input raw material made by any element and compounds. 3D printing technology is expected to play a special role in future environmental and ecological management:

(i) 3D printing process can use reusable/recycled materials thereby can contribute to control environmental degradation [28].

(ii) 3D printing technology supports effective waste management due to the fact that product/object printing takes place in a location of demand and waste materials after recycling can be used as input [29, 30].

(iii) 3D printing has many positive and negative implications on the environment and natural ecology managing them by proper planning and precautions [31].

(iv) The negative environmental impacts of manufacturing industries can be minimized using 3D printing technology due to its systematic layer-by-layer manufacturing process model [32].

(v) 3D printing enables improving the processes related to evolution and ecology in terms of better experimental techniques, greater flexibility, reduced cost, and ease of production [33].

(vi) 3D printing models provide an opportunity to conduct research using more life-like models with more appropriate responses from organisms involved without really affecting the living ecology of nature.

(vii)   Raid and low-cost printing of various ecological models for museums to educate and create awareness of the environment and its effective management is possible using 3D printing technology.

(7)   **Role of Digital Business and marketing technology in Environment and Ecology Management**

Digital business models which are different versions of traditional brick and mortar models have changed the way of doing business and provided many advantages and benefits to both customers and business organizations. The digital business technology using the Internet allows ubiquitous sales of both tangible and intangible commodities. E-business in any form including mobile business (m-business) [34] allows to improve the business models to improve toward ideal business model [35]. Similarly, online marketing models using Internet technology allow individuals and organizations to reach every potential customer across the world. Digital business and digital marketing technologies contribute substantially to environmental and ecological management by decreasing many physical business processes in almost all industries.

(i)   Digital business and marketing models decreased environmental pollution due to ubiquitous online business opportunities.

(ii)   Many business processes including transportation of human and other resources, both for tangible and intangible commodities.

(iii) This technology provided many features for effective business leading to customers' satisfaction and delight.

(iv)   E-business and e-marketing strategies allowed to plan the business processes in such a way to decrease various industrial waste, effective reuse, and hence decrease the cost of doing business [36].

(8)   **Role of Internet of Things (IoT) in Environment and Ecology Management**

Internet of Things (IoT) describes a group of physical objects interconnected through the Internet called things, embedded with sensors, software, and other features for the exchange of data and instructions. IoT contains cyber-physical systems connected through the Internet in such a way that the connected things can be controlled by each other without human intervention [37]. IoT has many potential applications in environmental and ecological management:

(i)   IoT systems can be used to track the data of various detectors connected in an environmental monitoring system and analyze them to study the changes in environmental patterns.

(ii)   IoT has the potential ability to be used in climate change and environmental monitoring and management. IoT supports functions like storing, organizing, processing, and sharing data and information and their applications in environmental monitoring and management [24].

(iii)   IoT can be used effectively for automatic waste management through monitoring and controlling connected devices in smart cities [38].

(iv)   IoT has many technological advantages for ecological research and monitoring wild animals, especially using its networked sensor technology to measure environmental parameters related to accurate, real time, and comprehensive data for monitoring conservation of wildlife and animal ecology [39].

(v)   IoT has the potential ability to support forest eco-management systems and stability. IoT can be used in the conservation of habitats and species, prevention of degradation of forest soil, forest fire prediction and control, timber production and management, etc. [40].

(9)   **Role Information Storage Technology in Environment and Ecology Management**

Information storage and management for long-time information ecology so that information storage systems can be leveraged to achieve eco-efficiency, eco-equity, and eco-effectiveness [41].

(i)   Information storage and management systems support such information storage that relates to the development of a sustained, robust, persistent infrastructure for data collection in environmental science research [42].

(ii)   Information storage management technology helps to process huge wildlife location data, tracking systems that have large, continuous, and high-frequency datasets of wildlife behavior, to be processed using Global Positioning Systems (GPSs), and other animal attached sensor devices [43].

(10)   **Role of Quantum Computing in Environment and Ecology Management**

Quantum computers and optical computers are high-speed computers that use quantum optical principles for computation unlike electronic signals in currently used electronic computers. Most widely used model of quantum computer uses quantum circuits that function based on quantum bits or qubits. Qubits can be either 0 or 1, or both (i.e., in a superposition of 0 and 1). In quantum computing, the information is encoded in qubits. The speed of quantum computers is several million times of the speed of supercomputers (supercomputers operate at a speed of the order of several hundred petaflops). The role of quantum computers in environmental and ecological management includes:

(i)   Quantum computers are potential machines for innovative climate change solutions through the simulation of quantum-level atomic interactions, could pave the way to discovering a new catalyst for carbon capture, heralding a new era of scrubbing carbon directly out of the air [44].

(ii)   Quantum computing may help to optimize the design of carbon-intensive materials to help to reduce carbon emissions from buildings, transportation, or any other heavy industry products [45].

(iii)   Quantum computers have the potential to predict extreme weather situations based on numerical climate and weather prediction models.

(11) **Role of Online Education Technology in Environment and Ecology Management**

Online education technologies are useful to educate people by providing environmental science education in online mode. This will create responsibility among the citizens to maintain a clean and green environment. Through online education technology, governments and NGOs can create awareness among the public about water, air, and soil pollution and precautions to be taken to maintain a quality environment and ecosystem.

(12) **Role of Virtual and Augmented Reality in Environment and Ecology Management**

Virtual reality is a technology used to mimic reality in a synthetic or virtual mode. Augmented reality enriches the real world by superimposing computer-generated effects on it.

(i)   Virtual reality is used for environmental management training [46].

(ii)   Virtual reality technology is used for visualizing ecological data [47].

(iii)   Virtual reality technology is used in visualizing natural environments from data [48].

(iv)   Virtual reality technology is used to visualize forests under climate change [49].

(v)   Virtual reality technology is used to promote awareness creation on climate change [50].

(vi)   Improving video games with players-augmented reality.

(vii)   Augmented reality technology has potentiality in providing impressive environmental education.

(viii) Effective teaching on climate change using augmented reality technology.

(ix)   Use of mobile-augmented technology for environmental monitoring [51].

2.5 Implementation Strategies of Use of ICCT Underlying Technologies in Environmental and Ecological Management

Any technology management framework involves six stages that include (i) identification, (ii) analysis, (iii) selection, (iv) acquisition, (v) exploitation, and (vi) protection of technology to be used to fulfill the individual or organizational objectives.

(i)   Identification stage involves both identifications of need/want to change called problem and identification of possible technologies that drive the change.
(ii)   Analysis of technologies in connection to the identified problem using a suitable analysis framework which may include SWOC analysis, ABCD analysis, PESTLE analysis, and six thinking hats analysis.
(iii)   Based on analysis, find a suitable technology as optimum technology which provides incremental/radical/architectural/disruptive innovation and depends on objective adopting technology.
(iv)   The chosen technology should be adopted in the business processes, through systematic acquisition. This includes in-house development of solutions using chosen open-access technology.
(v)   Exploitation of acquired technology by utilizing other resources of the organization to increase the performance and to get maximum benefits is essential for getting benefits of new technology used. Failure to recognize and manage new technologies by organizations results in inefficiencies and frustration.
(vi)   Protection of technology copying by others, especially by competitors through certain strategies like having patent/trademark/copyrights are essential and desirable.

ICCT underlying technologies have the capability to innovate and provide new solutions to existing problems in society and hence are having their applications in all industry sectors in one or the other way. These technologies can be potentially used for environmental protection. Due to liberalization and subsequent globalization, economic development without environmental considerations causes the environmental crisis. As a result, organizations are cautious about environmental degradation and hence call for diligent management of the environment which in turn is indispensable for sustainable development. ICCT underlying technologies are capable to solve many environmental and ecological problems and are considered as general-purpose technologies of the twenty-first century.

2.6   Analysis of Possible Role of ICCT Underlying Technologies in Environmental and Ecological Management Using Qualitative ABCD Analysis

ABCD analysis framework developed by our group [1, 52] has two models as simple ABCD listing or quantitative factor and elemental analysis. In this section, the ABCD listing [53] of the use of ICCT underlying technologies in the environment and ecological management is considered:

Advantages

(1)   Effectiveness of managing the natural environment and its ecology.
(2)   Cost-efficiency of managing the natural environment.
(3)   Saves time of intervention in managing the natural environment.

(4)   Better awareness creation among people to manage natural environment.
(5)   Systematic environmental monitoring to detect and manage changes.
(6)   Visualizing natural environments from data.
(7)   Efficient waste management through recycling and pollution management.

Benefits

(1)   Maintaining a sustainable natural environment and its ecology.
(2)   Low-cost players in managing sustainable natural environment and its ecology.
(3)   Corrective action can be taken in time to manage the natural environment.
(4)   Awareness about the importance of a natural and sustainable environment leads to an involvement of everyone in minimizing environmental pollution.
(5)   Systematic environmental monitoring using ICCT underlying technologies leads to proper planning to control and manage changes.
(6)   Based on data analysis using virtual reality and simulation, corrective measures shall be taken if the deviation is found.
(7)   Effective management of waste leads to balancing ecology on earth.

Constraints

(1)   Identification, procuring, and implementation challenges.
(2)   New technologies are prohibitively costly and complex.
(3)   Increased skill gaps between different tech-generations of people in society.
(4)   New technology-based disruptive changes are difficult to accept by stakeholders.
(5)   Adequate training is required for existing employees to utilize the new technologies.

Disadvantages

(1)   Job loss due to automation.
(2)   Risk of social isolation and addiction of technology.
(3)   Risks of failure of management of technology.
(4)   Ethical and legal concerns.
(5)   Extreme dependability.

2.7   Conclusion

ICCT underlying technologies, due to their general-purpose technology characteristics, have potential applications to innovate efficient natural environmental management and natural ecological management in society. The role of ICCT underlying technologies in environmental and ecological management is identified and analyzed. Various implementation strategies of the use of ICCT underlying technologies in environmental and ecological management are discussed. Analysis of the use of ICCT underlying technologies in natural environment management and ecological management using qualitative ABCD framework is depicted. It is found that ICCT underlying technologies have potential advantages and benefits in managing the natural environment and ecology in order to sustain living beings for a longer period.

References

1. Aithal, P. S. (2016). Review on various ideal system models used to improve the characteristics of practical systems. *International Journal of Applied and Advanced Scientific Research, 1*(1), 47–56.
2. Järvilehto, T. (1998). The theory of the organism-environment system: I. Description of the theory. *Integrative Physiological and Behavioral Science, 33*(4), 321–334.
3. Foray, D., & Grübler, A. (1996). Technology and the environment: An overview. *Technological Forecasting and Social Change, 53*(1), 3–13.
4. Aithal, P. S., & Aithal, S. (2020). Conceptual analysis on higher education strategies for various tech-generations. *International Journal of Management, Technology, and Social Sciences (IJMTS), 5*(1), 335–351.
5. Aithal, P. S. (2019). Information communication & computation technology (ICCT) as a strategic tool for industry sectors. *International Journal of Applied Engineering and Management Letters (IJAEML), 3*(2), 65–80.
6. Aithal, P. S., & Aithal, S. (2019). Management of ICCT underlying technologies used for digital service innovation. *International Journal of Management, Technology, and Social Sciences (IJMTS), 4*(2), 110–136.
7. Cortès, U., Sànchez-Marrè, M., Ceccaroni, L., R-Roda, I., & Poch, M. (2000). Artificial intelligence and environmental decision support systems. *Applied Intelligence, 13*(1), 77–91.
8. Chen, S. H., Jakeman, A. J., & Norton, J. P. (2008). Artificial intelligence techniques: An introduction to their use for modelling environmental systems. *Mathematics and Computers in Simulation, 78*(2–3), 379–400.
9. Chen, J., Huang, S., BalaMurugan, S., & Tamizharasi, G. S. (2021). Artificial intelligence based e-waste management for environmental planning. *Environmental Impact Assessment Review, 87*(1), 106498.
10. Yu, K. H., Zhang, Y., Li, D., Montenegro-Marin, C. E., & Kumar, P. M. (2021). Environmental planning based on reduce, reuse, recycle and recover using artificial intelligence. *Environmental Impact Assessment Review, 86*(1), 106492.
11. Loehle, C. (1987). Applying artificial intelligence techniques to ecological modeling. *Ecological Modelling, 38*(3–4), 191–212.
12. Rykiel, E. J., Jr. (1989). Artificial intelligence and expert systems in ecology and natural resource management. *Ecological Modelling, 46*(1–2), 3–8.
13. Dunbabin, M., & Marques, L. (2012). Robots for environmental monitoring: Significant advancements and applications. *IEEE Robotics & Automation Magazine, 19*(1), 24–39.
14. Donhauser, J., van Wynsberghe, A., & Bearden, A. (2021). Steps toward an ethics of environmental robotics. *Philosophy & Technology, 34*(3), 507–524.
15. Hampton, S. E., Strasser, C. A., Tewksbury, J. J., Gram, W. K., Budden, A. E., Batcheller, A. L., Duke, C. S., & Porter, J. H. (2013). Big data and the future of ecology. *Frontiers in Ecology and the Environment, 11*(3), 156–162.
16. Shin, D. H., & Choi, M. J. (2015). Ecological views of big data: Perspectives and issues. *Telematics and Informatics, 32*(2), 311–320.
17. Ma, H., Xiong, Y., Hou, X., & Shu, Q. (2020, February). Application of big data in water ecological environment monitoring. In *IOP Conference series: Materials science and engineering* (Vol. 750, No. 1, p. 012044). IOP Publishing.
18. Hakak, S., Khan, W. Z., Gilkar, G. A., Haider, N., Imran, M., & Alkatheiri, M. S. (2020). Industrial wastewater management using blockchain technology: Architecture, requirements, and future directions. *IEEE Internet of Things Magazine, 3*(2), 38–43.

19. Sivarethinamohan, R., & Sujatha, S. (2021). Unraveling the potential of artificial intelligence-driven blockchain technology in environment management. In *Advances in mechanical engineering* (pp. 693–700). Springer, Singapore.
20. Yang, Z., Xie, W., Huang, L., & Wei, Z. (2018, March). Marine data security based on blockchain technology. In *IOP conference series: Materials science and engineering* (Vol. 322, No. 5, p. 052028). IOP Publishing.
21. Orecchini, F., Santiangeli, A., Zuccari, F., Pieroni, A., & Suppa, T. (2018, October). Blockchain technology in smart city: A new opportunity for smart environment and smart mobility. In *International conference on intelligent computing & optimization* (pp. 346–354). Springer, Cham.
22. Howson, P. (2019). Tackling climate change with blockchain. *Nature Climate Change, 9*(9), 644–645.
23. Liu, T. (2021). Urban ecological environment investigation based on a cloud computing platform and optimization of computer neural network algorithm. *Arabian Journal of Geosciences, 14*(15), 1–15.
24. Fang, S., Da Xu, L., Zhu, Y., Ahati, J., Pei, H., Yan, J., & Liu, Z. (2014). An integrated system for regional environmental monitoring and management based on internet of things. *IEEE Transactions on Industrial Informatics, 10*(2), 1596–1605.
25. Paul, P., Bhuimali, A., Aithal, P. S., Kalishankar, T., & Saavedra M. R. (2020). Artificial intelligence & cloud computing in environmental systems-towards healthy & sustainable development. *International Journal of Inclusive Development, 6*(1), 01–08.
26. Liu, S., Guo, L., Webb, H., Ya, X., & Chang, X. (2019). Internet of things monitoring system of modern eco-agriculture based on cloud computing. *IEEE Access, 7*(1), 37050–37058.
27. Noussia, K. (2019). Cybersecurity and environmental impact: insurance as a better protection mechanism for liability from incidents in oil and gas. *InsurTech: A Legal and Regulatory View, 1*, 231–239.
28. Nadagouda, M. N., Ginn, M., & Rastogi, V. (2020). A review of 3D printing techniques for environmental applications. *Current Opinion in Chemical Engineering, 28*, 173–178.
29. Sharma, A., Mondal, S., Mondal, A. K., Baksi, S., Patel, R. K., Chu, W. S., & Pandey, J. K. (2017). 3D printing: It's microfluidic functions and environmental impacts. *International Journal of Precision Engineering and Manufacturing-Green Technology, 4*(3), 323–334.
30. Zgodavová, K., Lengyelová, K., Bober, P., Eguren, J. A., & Moreno, A. (2021). 3D printing optimization for environmental sustainability: Experimenting with materials of protective face shield frames. *Materials, 14*(21), 6595.
31. Maric, J., Rodhain, F., & Barlette, Y. (2016). 3D printing trends and discussing societal, environmental and ethical implications. *Management des technologies organisationnelles, 6*, 126–138.
32. Kreiger, M., & Pearce, J. M. (2013). Environmental impacts of distributed manufacturing from 3-D printing of polymer components and products. *MRS Online Proceedings Library (OPL), 1492*, 85–90.
33. Walker, M., & Humphries, S. (2019). 3D printing: Applications in evolution and ecology. *Ecology and Evolution, 9*(7), 4289–4301.
34. Aithal, P. S. (2015). Mobile business as an optimum model for ideal business. *International Journal of Management, IT and Engineering, 5*(7), 146–159.
35. Aithal, P. S. (2015). Concept of ideal business & its realization using e-business model. *International Journal of Science and Research (IJSR), 4*(3), 1267–1274.
36. Yi, L., & Thomas, H. R. (2007). A review of research on the environmental impact of e-business and ICT. *Environment International, 33*(6), 841–849.
37. Sachin Kumar, S., Dube, D., & Aithal, P. S. (2020). Emerging concept of tech-business-analytics an intersection of IoT & data analytics and its applications on predictive business decisions. *International Journal of Applied Engineering and Management Letters (IJAEML), 4*(2), 200–210.
38. Dipak, G., & Aithal, P. S. (2021). Smart city waste management through ICT and IoT driven solution. *International Journal of Applied Engineering and Management Letters (IJAEML), 5*(1), 51–65.

39. Guo, S., Qiang, M., Luan, X., Xu, P., He, G., Yin, X., Xi, L., Jin, X., Shao, J., Chen, X., Fang, D., & Li, B. (2015). The application of the internet of things to animal ecology. *Integrative Zoology, 10*(6), 572–578.
40. Salam, A. (2020). Internet of things for sustainable forestry. In *Internet of things for sustainable community development* (pp. 147–181). Springer, Cham.
41. Chen, A. J., Boudreau, M. C., & Watson, R. T. (2008). Information systems and ecological sustainability. *Journal of Systems and Information Technology, 10*(3), 186–201.
42. Baker, K. S., & Bowker, G. C. (2007). Information ecology: Open system environment for data, memories, and knowing. *Journal of Intelligent Information Systems, 29*(1), 127–144.
43. Urbano, F., Cagnacci, F., Calenge, C., Dettki, H., Cameron, A., & Neteler, M. (2010). Wildlife tracking data management: A new vision. *Philosophical Transactions of the Royal Society B: Biological Sciences, 365*(1550), 2177–2185.
44. Berger, C., Di Paolo, A., Forrest, T., Hadfield, S., Sawaya, N., Stęchły, M., & Thibault, K. (2021). Quantum technologies for climate change: Preliminary assessment. arXiv:2107.05362
45. Chen, S., & Ji, S. (2018). The application of quantum computing and quantum information in ecology. *Ekoloji, 27*(106), 651–658.
46. Markowitz, D. M., & Bailenson, J. N. (2021). Virtual reality and the psychology of climate change. *Current Opinion in Psychology, 42*(1), 60–65.
47. Huang, J., Lucash, M. S., Scheller, R. M., & Klippel, A. (2019, March). Visualizing ecological data in virtual reality. In *2019 IEEE conference on virtual reality and 3D user interfaces (VR)* (pp. 1311–1312). IEEE.
48. Huang, J., Lucash, M. S., Simpson, M. B., Helgeson, C., & Klippel, A. (2019, March). Visualizing natural environments from data in virtual reality: Combining realism and uncertainty. In *2019 IEEE conference on virtual reality and 3D user interfaces (VR)* (pp. 1485–1488). IEEE.
49. Huang, J., Lucash, M. S., Scheller, R. M., & Klippel, A. (2021). Walking through the forests of the future: Using data-driven virtual reality to visualize forests under climate change. *International Journal of Geographical Information Science, 35*(6), 1155–1178.
50. Fauville, G., Queiroz, A. C. M., & Bailenson, J. N. (2020). Virtual reality as a promising tool to promote climate change awareness. *Technology and Health*, 91–108.
51. Veas, E., Grasset, R., Ferencik, I., Grünewald, T., & Schmalstieg, D. (2013). Mobile augmented reality for environmental monitoring. *Personal and Ubiquitous Computing, 17*(7), 1515–1531.
52. Aithal, P. S. (2016). Study on ABCD analysis technique for business models, business strategies, operating concepts & business systems. *International Journal in Management and Social Science, 4*(1), 95–115.
53. Aithal, P. S. (2021). Analysis of systems & technology using ABCD framework. In Chapter 8: *Ideal systems, ideal technology, and their realization opportunities using ICCT & nanotechnology* (pp. 345–385). Srinivas Publication, India (April 2021). ISBN: 978-81-949961-3-2.

Chapter 3
The Practice of Green Computing for Businesses

Abdool Qaiyum Mohabuth

Abstract Organisations are more than ever reliable on the extensive use of computing resources for their businesses. This has become more prominent with the pandemic situation of COVID-19. Many enterprises have been compelled to re-engineer their ways of doing business. They needed to turn towards information and communication technology to maintain their businesses. These have caused the use of computing resources to increase significantly among businesses. Many of them are looking forward to use computing devices in order to improve in terms of productivity and quality of products and services they offered. As a means of avoiding crowds, many of them are proposing online services. Others are making use of appointment booking systems so that customers do not have to queue for services at their offices. This is causing a significant increase in the use of computing resources, and this trend will surely continue to brighten the businesses. On the other hand, very little attention is being paid to the impact of computing on the environment. Lead, chromium and mercury in computer devices produce harmful chemicals that contaminate air and water in the environment. Computer accessories such as laser printers comprise toner particles that can affect the lungs. Prolonged use of computers promotes global warming. Computer parts are not biodegradable and are not recycled with the absence of disposal techniques. As there is much concern regarding the carbon footprint generated by computing resources, green initiatives must be encouraged at the level of the businesses. Green computing focuses on the effective design, efficient development, usage and disposal of computers together with its associated systems with minimal impact to the environment. The scope of green computing covers energy efficiency to the use of hazard-free systems. Almost every organisation has multiple computing devices. These devices are heavy consumptions of amounts of megawatts. Promoting the use of computers, servers and tablets that are energy efficient is now critical. New computing devices come on the market almost every day. Temptation to acquire the latest technology is high, and organisations must get rid of unwanted computing and electronic devices. There is a crucial need to devise appropriate disposal and recycling procedures for these unwanted devices.

A. Q. Mohabuth (✉)
Department of Software and Information Systems, Faculty of Information Communication and Digital Technologies, University of Mauritius, Reduit, Mauritius
e-mail: a.mohabuth@uom.ac.mu

Green computing is a responsible use of computers and their resources in the environment. Green computing practices include the deployment of appropriate green knowledge to employees, efficient energy consumption computers, the use of energy harvesting technologies in data centres, dematerialisation with reduction of the use of paper-based documents together with enhanced disposal and recycling methods. This chapter focuses on the practice that organisations should put forward to achieve green computing initiatives.

Keywords Green knowledge · Green use · Energy efficiency · E-waste · Recycling

3.1 Origin of Green Computing

The term green was coined in 1991 by the US Environmental Protection Agency (EPA) which set-up a 'Green Lights program' with the aim to encourage power-efficient lighting besides improving lighting quality (Fig. 3.1). The EPA probed further by initiating a regulatory programme known as 'Energy Star' in 1992 with the aim to look at the power efficiency of products through the use of labels to inform consumers. The implementation of the Energy Star programme became successful in the US and helped businesses and families to reduce carbon footprint by 3.5 billion metric tons. The term 'green computing' emerged shortly after the Energy Star programme. Green computing is defined as the study and practice of using computers and their related resources in an environmentally sensitive manner [2]. One of the first application of the Energy Star programme to green computing was the sleep mode function in PCs that placed a computer on standby mode for a set duration of time when being in an idle state. Alongside, the Swedish organisation TCO development (Tjänstemännens Centralorganisation) initiated the TCO certification programme to promote low magnetic and electrical emissions from CRT-based monitors [3]. With advancement in technology, green computing continues to gain momentum. It starts

Fig. 3.1 Green computing logo [1]

to include energy cost accounting, virtualisation practices, thin client solutions and e-waste.

Essentially, the whole green aspect emerged in the 1990s when there was news that the environment was becoming a non-renewable resource. People were really shocked, and they started realising that every individual had to do something to protect the environment. There was the establishment of the 'triple bottom line' which consisted of people, planet and profit, an expanded spectrum of values and standards for measuring societal and organisational success. This was considered vital with regards to green computing. The triple bottom line strived to achieve three goals: social responsibility, economic viability and the repercussion on the environment. The objective was to make the whole process of computing friendlier to the environment, economical and societal. This implies manufacturers aim at producing computing equipment in a manner that projects the triple bottom line with optimism. Once computers were sold to businesses and people, they needed to be used in a green way by reducing electricity consumption and discarding them appropriately or reutilising them. The goal is to make computers a green product throughout the processes from its creation to its end [4].

3.2 Evolution of Green Computing Standard and Regulations

Advancement in green computing initiatives led to the establishment of many regulatory bodies which set-up standards and regulations that were used by many organisations to reduce the energy and resources. In order to ease and impose the application of green computing practices many countries worldwide have made use of these standards. Reference [5] found that in many countries there is low level of awareness about policies, laws and recycling initiatives among the citizens. Indeed, legislations and policies are essential to make way for the adoption of green initiatives. Government has a major role to play in taking up green policy decisions. 'The government has to play a crucial part to decrease the e-waste by comprehensive law and regulations. The awareness should be fuelled among the computer users to access the channels to dispose of undesirable computer systems safely and responsibly. The government–public initiatives would make consumers realise the effect of e-waste in nature' [6]. The following green entities emerged:

(1) EPEAT: Electronic Product Environmental Assessment Tool, an evolution tool initiated by a non-profitable organisation called Green Electronics Council (GEC) founded in Portland, Oregon, USA to achieve an environment in which only sustainable IT products are designed, produced and acquired. EPEAT was a user-friendly resource for producers, purchasers, resellers and others to point out environmentally sound equipment. It was an internationally recognised grading system for green electronics (Fig. 3.2).

Fig. 3.2 The new colour in
computer technology [7]

This evaluation tool of GEC was evaluating electronic devices based on 51 criteria
where 23 are core requirement criteria and the remaining 28 are optional. These
criteria are now reorganised into the following performance categories aiming at:
Decreasing and abolishing materials that are hazardous to the environment,
Choosing materials,
Forecasting for the end of life of products (such as reuse and recycling),
Product lifetime,
Energy preservation,
End-of-life cycle management,
Corporate performance,
Packing and covering.

These categorisations measure a product's efficiency and sustainability features.

(2) RoHS: Restriction of hazardous substances instilled by the European Union
appropriated in 2003 with primarily objective to shun the use of six perilous
materials—lead, mercury, cadmium, chromium, hexavalent or flame retardants
within rational limits in equipment (Fig. 3.3).

(3) ACPI: Advanced Configuration and Power Interface which enabled the
management of different power-saving aspects of basic hardware by an operating system. Intel-Microsoft standard called advanced power management
took over as the predecessor of ACPI which enabled the Basic Input–Output
System (BIOS) of a computer to manage the power management functions [9]
(Fig. 3.4).

(4) Energy Star 7.1 Standard: This standard became effective in November 2018
which stated that products shall be manufactured with the provision of facility

Fig. 3.3 RoHS logo [8]

Fig. 3.4 Advanced
configuration and power
interface representation [10]

for disassembling and recycling. Electronic subparts, chassis, external parts and sub-enclosures are easily dismantled by hand, by the use of ordinary tools, or by an automated recycling process [11] (Fig. 3.5).

(5)   W.E.E.E Law: The European Waste Electrical and Electronic Equipment Directive was in force in 2003. According to the law the equipment manufacturers had the responsibility to take back the electrical and electronic waste without charge (Fig. 3.6).

The objective was to reduce waste from electronic and electrical devices and provide encouragement for designing devices that improved environmental performance throughout the life cycle [14]. Manufacturers were compelled to join a compliance scheme and registered in each EU country. Violations against the law were actionable and prosecutable [15].

Fig. 3.5 Energy Star logo
[12]

Fig. 3.6 WEEE logo [13]

3.3 Green Computing Industry Association

The following industry association promoted advancement in green computing initiatives and practices.

(1) Green Grid: It is a voluntary universal non-profit-making organisation where its principal objective is to establish standards to assess data centre efficiency comprising both the facilities and the devices inside. Partner companies give out information about technologies and processes that can facilitate data centres improve performance against some metrics. The latter consists of Intel, Microsoft, IBM, HP, AMD, APC, Dell, EMC and Sun.

(2) The Climate Savers Computing Initiative: A member company dedicated to acquire energy-efficient desktops and servers and to devise power management strategies widely. By openly proving their support for this effort, companies declared their commitment to the greening of computing and collaborate with other industry-leading firms and organisations which are involved in sustainable information technology and corporate social responsibility. Its board members included Dell, HP, Google, Microsoft, Intel, Lenovo and the World Wildlife Fund.

(3) The Uptime Institute: It offers educational and consulting services for organisations keen to boost up data centres uptime and sustainable IT. The Institute has developed industry-standard metrics which evaluate data centre availability. The Institute encourages learning among its partners and holds events such as conferences and site tours. It delivers knowledge about best practices, benchmarking and addresses issues about abnormal incidents. Besides these, it provides certification in terms of tier levels and site robustness for data centres.

3.4 Green Computing Initiatives in Organisations

Green computing is the practice of eco-friendly procedures in the design, production, use and disposal of computing resources without compromise on overall computing performance [16]. Green computing started in 1992 when the US Environment Protection Agency put in place a programme called the Energy Star which aimed at saving energy. The widespread use of computers has been found to have an adverse effect on the environment, since computer parts are not biodegradable and are rarely recyclable. Toxic chemicals used in the manufacturing process also have harmful effect on the environment when computers are disposed of. Absence of appropriate disposal techniques contributed to the negative impact on the environment. The raising concern about the environment of not being a renewable source populated green computing movement. [17] devised the pathway for achieving green computing which he classified in terms of 'green manufacturing' where computer parts are to be produced with minimal impact on the environment; 'green design' where the focus is on energy efficiency; 'green use' where computing resources have to be used in an eco-friendly sound manner with reduction in the energy consumption; 'green disposal' where obsolete computers and electronic parts are recycled. However, lack of knowledge on green computing has made this pathway very difficult to be applied in practice. A number of surveys have shown that the biggest barrier that hinders green computing practices is lack of knowledge [18]. For instance, a survey carried out in the UK revealed that only 18% out of the 120 IT decision makers took into account the carbon footprint of computer systems on acquisition and more than 50% did not dare to consider any environmental impact when using their computing resources. Many were unaware about requirements for green systems in their organisations. Even though the concept of green computing was initiated in the early 1990s, it is only now that it is gaining more popularity due to the reality of climate change [19]. The widespread use of computing in all spheres of life impacted heavily on the environment. Green computing has a crucial role in improving the processes and efficiency towards the reduction of energy consumption and carbon footprint [20].

The prime objective of greening computer usage was to allow organisations to use computing resources more efficiently while keeping or improving overall performance. The first wave of bringing along green endeavours was to focus mainly on reducing power consumption for datacentres and technical equipment (such as desktops and projectors). The pros of greening computers in terms of lessening power usage and corporate carbon footprints are direct and comparatively fast to achieve. Nevertheless, to advance beyond internally focused green computing initiatives into the field of competitive advantage and corporate sustainability, greater emphasis is being focused on how a second wave of green computing practices will fit with and allow corporate sustainability strategy. A related research was held in the year 2014 in Saudi Arabia where a set of 500 staff members from 272 public and private firms were inquired with a view to find the impact level of 15 different factors on their decision to launch green computing compliant systems in their organisations. The research showed that exempting government regulations on carbon emissions, and their related taxes was the main motivator for transitioning towards green computing, while elements such as environmental consciousness and corporate social responsibility did not lead to a serious consequence [21]. While computer solutions are being driven in organisations to enhance business productivity, there is need to look at sustaining green initiatives beyond energy efficiency.

3.5   Importance to the Environment

Green computing aims at having durable computing to the environment. It has similar objectives as green chemistry, which looks at extending the lifetime of products with more energy efficiency. It advances to care for disposing products which are more easily recycled and biodegradable. The use of the less hazardous substances to the environment health is not forgotten. The issue of escalating electronic waste has become a conundrum. Many adverse consequences are gradually appearing towards the environment. Due to the quick bottleneck of electronics, this culminated towards an awful 70% of all hazardous wastes. A lot of poisonous materials, such as heavy waste and flame-resistant plastics, have excessive computer waste, which effortlessly appears in groundwater and bio-accumulation. Moreover, the production of electronic chips in huge amounts cause more lethal gases and chemicals to be released in the environment. In the US, around 24 million computers become outdated every year. About 14% (3.3 million) of these will be reused or given away. The rest representing more than 20 million computers are considered for dumping, destroying or otherwise disposed of in temporary storage [22]. In reality, they are either dismantled in landfills or sent to emerging countries. The e-waste trade chain is growing up in emerging nations for importing outdated electronics for the business of reuse or extraction of useful components. Unfortunately, these practices are generally not regulated and there is no supervision from the part of the authorities. It is, therefore, vital to create sustainable business processes, leverage environmental initiatives and mushroom up green awareness to reduce energy consumption and overall carbon emission. Additionally, it is important to reduce dangerous material usage in computers as well as hazardous disposing methods. It is equally important to look at optimising energy efficiency, promote recycling, preserve and effectively utilise natural resources and encourage device efficient disposal methods. Most importantly, there is need to educate and motivate people to initiate green computing practices in terms of promoting reduce, reuse and recycle activities.

3.6   Greening for IT in Organisations

Enterprises dispose a million of computing equipment every year. Integrated components in these equipments contain many toxic substances and rare materials as stated above, and it is necessary to recycle them to create suitable channels and well distributed. Recovery of waste should become a culture in organisations. Efforts must, therefore, be made to reduce the negative impact on the environment. This may be achieved in industries principally through:

- Intelligent reduction of their consumption energy,
- Recycling and recovery of computer waste,
- Reduction of greenhouse gas emissions throughout the life cycle of product: conception, design, implementation, use, transport and recycling.

The above scopes which are applied in the first instance in organisations to develop a culture of reducing the environmental impact of ICT are commonly known as 'green for IT'.

3.6.1 Eco-responsibility of Enterprises

Green for IT is about the application of strictly IT measures on the use of software, hardware, services and processes that reduce the impact of IT on the environment through an eco-responsible approach in organisations which involve eco-design, energy saving and waste management. The methods used to achieve eco-responsible computing in enterprises are mainly focused on optimising existing technical processes and on the energy savings of different equipments. Green IT is a cross-cutting exercise that affects both the infrastructure and the post held by the workers. Today, 60% of computers in organisations remain on 24 h a day. One of the green IT principles is to optimise the use of the equipments and reduce their usage costs. To win in the long run, it is necessary first of all to invest, and that requires often to negotiate with his own management.

3.7 Application of Basic Solutions to Integrate Green Computing

Figure 3.7 shows an architecture with simple solutions making it possible for enterprises to quickly initiate steps towards the practice of green computing.

(1) On procuring new equipment
      When renewing or acquiring new computing resources

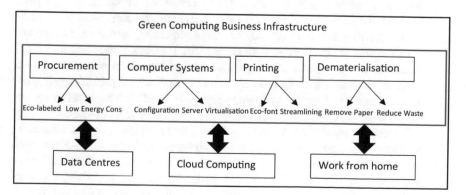

Fig. 3.7 Greening organisations' business infrastructure

- Choose eco-labelled equipments which offer less pollutant output, controlled electricity consumption and end-of-life product recycling integrated from the design phase of these products.
- Choose computers with processors with low power consumption (ULV) and with the right processor speed: simply willing to systematically acquire the most powerful computers available from the market when it is known that these equipments will be used only for some using some office packages does not make sense.
- Replace CRT screens which are high-energy consumers, through screens LCD or LED ones which are much less energy consuming.

(2) At the level of computer systems used

- Configure the computers to go straight to the sleep mode when being inactive for some point of time. Refrain to use screen savers as they do not lead to energy savings.
- Use software agents and tools management system to automate standby, shutdown and restart remote computers based on predetermined time schedules. For example, a particular IT department with eight men days may invest on such a solution and this may result in savings of $30 to $40 per day at each workstation. Annually, this may turn over $100,000 savings considering the whole IT infrastructure in the organisation.
- Use software with incorporated specialised modules to monitor energy consumption. Address the impact of renewing the computer system by using Energy Star, Dotgreen and VisionSoft logistics.
- Consider the use of new computers generation, such as thin clients, or use third-party software to modulate the available power of computers
- Go for server virtualisation. This process uses software application to divide a physical server into multiple virtual servers. Each server can then execute on their own independently, the benefit of server virtualisation is the cost savings that can be realised through the provision of web hosting services and the efficient use of existing IT infrastructure resources. The processing capability of servers is compromised when there is no server virtualisation. Only a small fraction of the processing power is used. As a result, servers remain idle because the workload is distributed to only a small portion of the servers on the network. Data Centres are overloaded with underutilised servers, preventing power and resources to be used at their optimum level. When each physical server is divided into multiple virtual servers, server virtualisation enables each of these virtual servers to function as a single physical system. Each virtual server can run its own applications and operating system. This process increases resource utilisation by making each virtual server act as a physical server and increases the capacity of each physical machine. Organisation may achieve server virtualisation through complete virtualisation, paravirtualisation or virtualisation at the level of systems. Complete virtualisation uses a hypervisor, a type of software that communicates directly with the disc space and CPU of a physical server. The hypervisor monitors physical server resources and keeps each virtual server independent and in the dark of other virtual servers. It also relays

physical server resources to the appropriate virtual server when applications run. The biggest limitation of full virtualisation is that a hypervisor has its own processing needs. This may slow down applications and impact server performance; unlike full virtualisation, paravirtualisation requires the entire network to work together, as one coherent unit. As all virtual server operating systems know about each others with paravirtualisation, the hypervisor does not need to use that much processing power to manage systems; with virtualisation at the level of systems, a hypervisor is not used. Instead, the virtualisation feature, which is part of the physical server operating system, performs all of the tasks of a hypervisor. However, with this method of server virtualisation, all virtual servers must run the same operating system.

(3)   At the level of the users

At the user level, it is necessary to raise awareness among all staff to comply with certain simple guidelines

- Turn off PCs every evening or at least the weekend;
- Place orders for equipment purchases adapted to the needs of users, agreement with the IT department, do not follow advertisements extolling the merits of the latest computer in fashion.
- Print documents only when necessary. Today, a company can spend up to 5% of turnover in printing stuff (material, ink, paper, etc.). In fact, 16% of pages printed by users are directly thrown away without being used, 20% of printed pages have a lifetime of a few seconds. A company generally spent 60% of its annual budget on consumables (cartridge, ink and paper), the remainder for the purchase of the hardware, maintenance and electrical consumption.
- Reduce the number of printers by sharing the facilities: switch from printers located in offices, to a printer multifunction by floor or department.
- Replace ageing printers with new, eco-labelled printers, and equipped with the latest saving devices in terms of energy and remote management. The acquisition cost will be quickly offset by reduced costs to use.
- Configure printers for printing on both sides (recto–verso) by default, and print multiple pages on single sheet (e.g. for PowerPoint presentation).
- Print on eco-labelled recycled paper. The use of recycled paper certified to low weight saves on average 18% wood, 14% water and 20% energy.
- Introduce policies that enable staff to be aware of the proper use of printer, to avoid printing personal, unnecessary or unclaimed data.

Organisations may also consider to outsource their printing task to a service provider if it is more viable.

At the level of software, consider the use of specific applications such as *PrintFriendly, GreenPrint or Print What* You Like, for e.g that allow to remove unnecessary parts of the document before printing, such as advertising inserts when printing web pages.

- Encourage staff to use more economical print characters such as Century Gothic font, 30% thinner than Arial or the Ecofont font, would result in an immediate 30% reduction on ink used to print text.

- Reducing the number of pixels to print means a reduction in the amount of ink or toner to use. In fact, this has been tried and tested at the University of Green Bay in USA where the academics and students were requested to use simple fonts in their office application instead of the bold thick ones. The results were an immediate reduction in ink consumption using the simple font instead of the default ones used in Windows.
- Streamlining the printing policy is an important source of savings and improvement productivity within any organisation, but it requires governance at the highest level of management, because the implementation of this policy is a project that expands across the organisation.

(4) Dematerialisation in organisations

Dematerialisation refers to all techniques that remove paper in favour of electronics. Well-known examples are the dematerialisation of incoming mail, the dematerialisation of invoices, the dematerialisation of cash flows and cheques and the dematerialisation of administrative procedures (tax declarations, vat returns and calls for tenders). Dematerialisation has several advantages, including reduced costs and processing times, secure flows, homogenised document processes and reduced waste such as packaging waste, non-recyclable paper and ink cartridges, reduced company carbon footprint through reduced transport and paper. Dematerialisation reduces energy consumption and associated greenhouse gas emissions. It represents a saving potential of 7800 million tons of CO_2 in 2020, or nearly 30% of total CO_2 emissions [23] and thus contributes to efforts in the process of combating global warming. Dematerialisation can proceed in organisations except that today, some legality cheque are still insured on paper. And so it is for the transmission of documents to public services that still require re-materialisation of important files on paper. Although efforts are underway, all public organisations will have to, however, be fully equipped to be able to move to zero paper in the long run. However, it is certain that the trend on dematerialisation of administrative documents is going in the right direction in terms of sustainable development in enterprises.

3.7.1 Green Computing at Data Centres

Data centres are hungry consumers of computing resources. They undertake the processing of activities for public organisations such as the management of consumers' financial flows, store emails, photos or videos and other users exchange over the Internet. These data centres use hundreds of computers working together 24 h a day and are highly strategic for many sectors in the economy. They made use of powerful servers with their attached equipment in terms of UPSs, storage and network equipment as well as air conditioning equipment to maintain a suitable temperature. The composition of a data centre varies by servers up to thousands. The consequence of the explosion in demand Internet supplemented by the needs of businesses that are growing strongly, is that data centres have become essential elements of business life, and they concentrate a major part of the energy expenditure. They need electrical power and air conditioning system permanently. Data centres have increasing cost due to the cooling system that need to be applied. Data centres need to optimise the use of equipment through the virtualisation of servers and storage that will enable reducing the number of physical machines and therefore, the total consumption of data processing. Indeed, green technologies are more expensive at the current price of energy, but the return on investment is, however, fast.

3.7.2   Migration Towards Cloud Computing

Companies may also move towards cloud computing to reduce power usage and budget. Cloud computing provides businesses with practical models for accessing software platforms, computing infrastructure and app offerings on a pay-per-use basis. With cloud computing, companies are able to re-engineer their businesses, avoid IT maintenance, free up capital and adopt better business models, enhance security measures and apply more flexibility into their services and solutions, helping customers in new ways, and grow their business in the ever-changing market conditions. They may go for the following models:

- Public Cloud: A public cloud is an environment with many tenants, i.e. it is designed to be shared by a certain number enterprises and is located in the premises of the cloud service provider. When an organisation decides to opt for this type of cloud model, the logistics in terms of platform, hardware and software are provided and managed by the service provider. The organisation accesses its accounts by the use of web browsers. This model fits well for small and medium enterprises.
- Private Cloud: It refers to the exclusive use of cloud resources by a single organisation. A private cloud can be located in the organisation's own data centre or hosted in the cloud by a service provider specifically for use by the organisation. This model is suitable for large enterprises.
- Hybrid Cloud: In this model, data and application are shared by using a combination of the properties associated with public and private clouds. Hybrid clouds connect existing resources that are not in the cloud to the infrastructure and applications are cloud based. This approach is more flexible to organisations, as it enables interoperability of applications and allows easy data management intended for public use and those intended for use internally.

Cloud computing services provide convenient pay-per-use models that eliminate maintenance cost as well as other expenses. Cloud vendors own a wide range of infrastructure, platforms and application that they can host on-site and lease, giving companies the flexibility to adapt cloud services as per their needs. Companies have choices among the following services:

- Infrastructure as a Service (IaaS): It is where cloud computing resources are provided through the net within the virtual environment. It presents the main architecture of the cloud where instead of the business organisation owing the computing resources in terms of platform, storage, network connections, IP addresses, servers, etc., these are hosted by the service providers. IaaS providers also take responsibility for performing backup tasks and system maintenance. Typically, the computing resources are scattered across many data centres through a number of servers and networks which are managed and maintained by the service providers. The customer has access to virtualised components to develop their own IT platform. With IaaS, businesses do not need to purchase, manage and maintain the underlying infrastructure, yet they can choose when and how to manage their workloads. IaaS makes the infrastructure quickly operational to enterprises using a pay-per-use model.
- Platform as a Service (PaaS): PaaS is based on the IaaS model, but is generally specific to application development tools. It is a cloud computing service that enables organisations to outsource the hosting of software and hardware tools for application development. Businesses may develop their own applications by accessing the logistics in terms of hardware and software provided by the PaaS providers. These logistics are not hosted locally but at the provider's site. Thus, companies do not need to make internal installation of hardware and software to launch or develop new applications. Key features offered by PaaS vendors include programming platform, database management system, operating system, server software, storage, network access, design and development tools and support. It is preferable for enterprises to consider all offers on a case-by-case basis.
- Software as a Service (SaaS): In this model, cloud service providers host and manage an entire infrastructure, including end-user applications. With SaaS, there is no dedicated hardware or software to purchase, install, maintain or update: the process is transparent to the enterprises, who simply have to worry about their daily tasks. When businesses choose a SaaS model, there is nothing to install at their local sites. They simply log in and start using the service provider application running on their infrastructure immediately. SaaS is generally available by subscription.

Using the above services means that all employees, including mobile salespeople, can access companies' IT system from any Internet access point. Examples of simple and popular applications include Google Docs and Microsoft Office 365, two services that allow users to work and collaborate, wherever they are, on files hosted by the software provider's servers. These resulted in considerable savings and simplify tasks for businesses. Better yet, the service provider takes care of all the technical issues that companies cannot handle due to time and lack of expertise. In addition:

- The provider protects and backs up companies' data.
- Systems will not go offline due to software problems.
- Companies do not waste time or money updating a system, only to find out two years later that it is inadequate or obsolete.
- Companies do not have to spend a lot of money to hire IT staff in-house.

Cloud providers offer an increasingly comprehensive list of services that include the latest enterprise resource planning systems, remote data access and secure data storage among others. This can lead to considerable savings. In general, an organisation may save about 65% when it uses a cloud-based ERP system, rather than using the traditional method of buying software and then installing it on its computers.

Another major cost-saving issue relates to maintenance. Businesses tend not to have a sufficient budget to maintain new systems and do not have the expertise to do the work in-house. As a result, the failure of many systems is due to a lack of maintenance, process discipline, training and user loyalty. Cloud computing eliminates the maintenance issue.

For organisations to take full advantage of cloud computing, they need to train their staff appropriately. A cloud-ready company will need a workforce trained to easily use and exploit new cloud services. The maintenance of the traditional IT system will compromise the agility, innovation and economic benefits of cloud computing. Employees will need to be able to recognise new models of the cloud and develop new mindset and skills to exploit services offered in a dynamic cloud computing market in order to use effectively and continuously promote an approach that prioritises cloud computing. In addition, when recruiting IT employees, their experience in cloud computing need to be taken into account. IT staff trained in cloud computing should be able to perform the following tasks:

- Design cloud-optimised solutions, leverage the benefits of the cloud in terms of scalability, multiple users and high availability taking into account the sensitivities of the cloud in terms of latency, performance and security.
- Manage cloud computing solutions efficiently to minimise costs.
- Integrate cloud services and other internal systems using modern Web technologies through APIs.
- Use data flow technologies or modular cloud service platforms to customise cloud services to deliver new functionality.
- Work with users and provider communities for continuous capacity building of the enterprise.

Despite that the cloud revolution holds great promise for growth flexibility and cost savings for businesses, adoption of uncoordinated cloud services can easily deprive the business of the benefits of cloud computing. The one-time purchase of different types of services of cloud computing across multiple vendors may create a highly heterogeneous and fragmented structure in organisations. To prevent this from happening enterprises will have to adapt and reinvent themselves. They need to strengthen their efforts to migrate to cloud computing in a coordinated manner with comprehensive and structured ways together with clear plan for the future.

3.7.3 The Concept of Work from Home

Businesses may also introduce the concept of work from home which may considerably reduce the load of the work setting computing infrastructure (Fig. 3.8). Work from home means any form of work which was generally performed by an employee on the companies' premises, is now performed outside most specifically at his residence using information and communication technologies. The concept of work from home has become more prominent with the COVID-19 pandemic situation. However, in many countries, it is still under test stages and it has not yet been integrated in their labour law or code of practice. Up to now, there are particular criteria or conditions for determining the possibility or desirability of implementing work from home in an undertaking for the benefit of employees. In theory, work from home applies to all occupational categories of workers. However, this is not a right for the employee. Within the same company, it is possible that the profile of one position prevents the implementation of work from home, while other positions may benefit from it. For example, employers may decide to implement work from home only for certain categories of staff.

Major Advantages of Working from Home

For the enterprises

- Increase production
- Achieve economies of scale on premises and on current expenses
- Improve the quality of life of employees and consequently increase their motivation and involvement
- Reduce absenteeism
- Reduce energy consumption, less load on the IT infrastructure of companies.

  For the employees

- Time savings, especially in travelling to and from the work settings
- Better management of the working time

Fig. 3.8 Work from home initiative [24]

- Greater autonomy in task management
- Better concentration leading to better productivity.

Work from home surely saves time, efficiency and concentration, which is often compromised when working at the work settings: an employee is generally interrupted on average 140 times a day! By eliminating commuting to the office, teleworking is also ideal for reconciling personal and professional life. For example, it allows employees to free up time for their family life or to invest in sports or associative activities. However, work from home can sometimes be perceived as a loss of professional identity: isolation, reduced interaction with colleagues and decreased sense of belonging to a company. Employees need to feel morally capable of working remotely or in isolation. In addition, informal practices have been seen to be developed, with employees using their smartphones, tablets or laptops for example, reading professional emails in transport or finishing a file late at night in a quiet home. Employees working from home may tend to allow themselves little time off and perform tasks even beyond the required hours. This may lead to deterioration in their physical and mental health, including exhaustion and depression. This can affect worker performance and increase the risk of resignation, which can ultimately harm companies. The risks of burnout have also been noticed. If employees are having trouble completing their tasks, they need to talk to their hierarchy to find other solutions (closed office, fewer meetings and other working methods). These hazards can be overcome if companies put in place the necessary structure to support the employees during the process.

How Enterprises Can Easily Adopt the Work from Home Concept

In many cases, it is not necessary to amend the employment contract to allow an employee to work from home. To implement work from home, there are generally three possibilities:

(1)   A simple agreement between the employer and the employee, by any means (verbal agreement, email, letter, etc.)
(2)   A collective agreement
(3)   A charter drawn up by the employer, after consulting the respective labour authority.

In the context of a simple agreement with the employee (without collective agreement or charter), the latter formalises this agreement with the employer.

In the case of a collective agreement or charter, the following points are specified:

- Conditions for moving to work from home and conditions for returning to contract performance without work from home
- The terms and conditions for employee acceptance of work from home concept
- The arrangements for controlling working time or regulating working load
- The determination of the time slots during which the employer can usually contact the employee.

3.8 Recycling of Obsolete Equipment in Organisations

Several studies and surveys in the field pointed out that much of the computer equipments used in businesses are not treated adequately once they reach the end of their life. In fact, very often they are left behind at the landfill site. These materials are incinerated, thus releasing into the environment substances that are highly harmful to the planet, to humans, to wildlife, but also to flora. Various problems, including lung disease, cancer, the destruction of the groundwater reserve and the extinction of certain animal species, are emerging as a result of this mistreatment of electronic waste. The electronics used in many computer products are generally safe in normal use. However, when these products are discarded, the substances in a computer can become toxic due to lack of stability. Thus, a CRT monitor is made of lead glass and PVC that produces toxins when incinerated. Each monitor contains up to 4 kg of lead in the funnel and its electronic components. A monitor also contains barium and phosphorus, elements that are toxic to the environment and human health. The CPU is also made up of products such as beryllium, lead and hexavalent chromium. That is why electronic wastes should not be thrown away. They should be recycled. Obsolete computers, old-fashioned printers, hard drives and other consumables that come to the end of their life in enterprises need to be disposed of properly. The equipment can, in theory, be returned to the manufacturer (not the seller) of the products. They are required to collect it. But the reality is otherwise.

Computers and computer hardware follow a specific lifecycle and upgradeability. They cannot be dropped off on the street or thrown away with household waste. Devices running on mains, batteries or batteries must be cleaned before recycling. After that, they will begin a second life, useful and respectful of the environment. It should be noted that almost all IT equipment can be recycled: processor, hard drives, computer screens, printers, laptops, motherboards, scanners, keyboards, mice, USB sticks, cables, speakers, headphones, tablets, etc. The average computer composition is found to be as follows:

- 37% plastic that comes primarily from the casing of a computer
- 11% non-ferrous metals and 13% ferrous metals of cables and various components
- 16% electronic cards
- 20% controlled substances, i.e. capacitors, batteries, etc.

The obvious and necessary general rule in an organisation is to hire a computer recycling professional. Indeed, it would be difficult for companies to take on this responsibility within their structure. IT recycling is organised as follows: sorting, collecting, repackaging and processing. The entire process may be facilitated with the help of a selected provider.

Sorting

It is done in several stages. Firstly, separate computers that are still usable from those that will end up in a specialised dump. There are two categories:

- Computer equipment that runs on battery, or with mains power. They are recycled with small electrical and electronic devices. They will join old toasters and hair dryers.
- Equipment with a screen. They are recycled separately because they must be cleaned before treatment. Such is the case for notebooks and tablets.

Collection

The professional offers a collection service that will depend on the organisation, but also on the quantity and condition of the equipments. It is recommended to deal in advance with the different providers.

Reconditioning

CPUs, laptops, etc. The professional re-uses entire devices or only components. Some products may be tested, upgraded and reused. Others will be dismantled to extract the metals and components still usable and may, if necessary, be attributed to other equipments. All will be sent again to the functional equipment market and under warranty.

The Treatment

In this case, the collected waste is disassembled in order to recover the material. The metals and components still usable are recovered and then sorted according to precise classification standards. They are, then dispatched to recycling companies.

In all computer equipment, whether it be in the casing, electronic cards (motherboard, memory card, graphics card, etc.) or in the cables, there are plastic, ferrous metals such as steel and non-ferrous metals such as aluminium or copper. All these materials may be recycled in their specialised sectors or resold like the metals that will go to foundries. Screens on their sides are cleaned, recycled and reclaimed.

Batteries and hard drives still remain. Contrary to popular belief, batteries cannot be reused as such. They have a limited lifespan and go directly to recycling. Hard drives are crushed to destroy confidential data. The destruction of hard drives is an important step for both the recycling company and the owner of the devices to be recycled. In addition to containing data that is sometimes confidential and non-divulgable, the hard drive is made up of elements that can cause numerous harms to the environment and human health. Ethics and security rules must, therefore, apply to its destruction so that it can proceed as smoothly as possible. Only about 17% then remains, this may be treated by incineration, storage centres, etc.

Donation to Charitable Institutions

Old computers can be used to equip low-income households, schools, institutions or be sent to developing countries. There are associations, some of which are supervised by the United Nations, which are responsible for upgrading old equipment for this purpose. Emmaus also takes care of the reconditioning. It is, therefore, always a good idea for enterprises to inquire with local associations who could recover their old computer equipments.

Promote the Prolonged Use of Computer Equipment

Avoid replacing digital equipments in organisations on a whim, or following a promotional offer. Try to increase the use of digital devices such as tablets and computers from 2 to 4 years. This will improve its environmental record by 50%. Organisations should have proper maintenance contracts for their IT resources. In addition, they should be having protective security plan to safeguards their IT assets. Installation of anti-virus and anti-malware protection as well as appropriate firewalling system will enhance the security infrastructure and avoid outages and save money.

3.9 Summary

This chapter presents green computing initiatives for businesses. The origin and the different terminologies in the arena of green computing are explained clearly which allows employees in organisations to build up their knowledge about the subject. The importance of moving towards the green environment is enlighted and the hazards and harm caused by carbon footprint due to the increasing use of computing are discussed. The application of green practices that enterprises can put forward are exposed and these spread from acquisition of computing resources in terms of looking for eco-label with low energy consumption to the use of computer systems in the offices with proper configuration and appropriate server virtualisation. The printing task which includes a large volume of data is also considered, aiming at streamlining the procedures and using of the eco fonts. Dematerialisation is also discussed which looks at withdrawing the use of paper-based documents and reducing waste. Besides, the ways in which data centres can become more efficient are considered and the benefits to organisations to migrate towards cloud computing are illustrated. In addition, the concept of work from home is explained as means to reduce the load on the computing infrastructure of enterprises. Finally, the ways of disposing of obsolete computers are discussed and consideration for recycling is elucidated.

References

1. Green Computing logo. (2022). Available from http://sites.google.com/site/greencomputingb ypaige/green-computing. Last Accessed February, 2022
2. Harris, J. (2008). *Green computing and green it: Best practices on regulations and industry initiatives, virtualization, power management*. Emereo Publishing.
3. Miraz, H., Excell, P., & Sobayel, K. (2021). Evaluation of green alternatives for blockchain proof-of-work (PoW) approach. *Annals of Emerging Technologies in Computing, 5*(4), 121–128.

4. Jindal, G., & Gupta, M. (2012). Green computing future of computers. *International Journal of Emerging Research in Management &Technology*, 14–18.
5. Ramzan, S., Liu, C., Munir, H., Xu, Y. (2019). Assessing young consumers' awareness and participation in sustainable e-waste management practices: A survey study in Northwest China. *Environmental Science and Pollution Control Serves, 26*(19), 20003–20013.
6. Chandar, K. (2009). Computing practices: Going green. *SCMS Journal Of Indian Management, 6*(4), 58–64.
7. The new colour in computer technology. (2022). Available from http://www.ijcsms.com. Last Accessed on: February, 2022.
8. RoHS logo. (2022). Available from www.edding.com. Last Accessed on: February, 2022.
9. Shinde, S., Nalawade, S., & Nalawade, A. (2013). Green computing: Go green and save energy. *International Journal of Advanced Research in Computer Science and Software Engineering, 3*(7), 1033–1037.
10. Advanced Configuration and Power Interface representation. (2022). Available from http://www.Computerhope.com. Last Accessed on: February, 2022.
11. Harmon, R., & Moolenkamp, N. (2012). Sustainable IT services: Developing a strategy framework. *International Journal of Innovation and Technology Management, 9*(2), 112–118.
12. Energy Star logo. (2022). Available from http://www.energystar.gov. Last Accessed on: February, 2022.
13. WEEE logo. (2022). Available from http://www.stssapp.com. Last Accessed on: February, 2022.
14. Roy, S. (2021). Green computing: An eco-friendly and energy-efficient computing to minimize E-waste generation. In: *Go Green for Environmental Sustainability*. CRC Press, pp. 87–100.
15. Shittu, O., Williams, I., & Shaw, P. (2021). Global E-waste management: Can WEEE make a difference? A review of e-waste trends, legislation, contemporary issues and future challenges. *Waste Management, 120*, 549–563.
16. Saha, B. (2018). Green computing: Current research trends. *International Journal of Computer Sciences and Engineering, 6*(3), 467–469.
17. Murugesan, S. (2008). Harnessing green IT: Principles and practices. *IEEE Computer Society, 10*(1), 22–33.
18. Adimoolam, M, John, A., Balamurugan, M., Kumar, T. (2021). Green ICT communication, networking and data processing. In: *Green Computing in Smart Cities: Simulation and Techniques*. Springer, Cham., pp. 95–124 (2021).
19. John A, Kumar T, Adimoolam M, Blessy A (2021) Energy management and monitoring using IoT with CupCarbon platform. In: *Green Computing in Smart Cities: Simulation and Techniques*. Springer, Cham., pp. 189–206 (2021).
20. Saxena, S., Khan, M., & Singh, R. (2021). Green computing: An era of energy saving computing of cloud resources. *International Journal of Mathematic Sciences and Computing, 2*, 42–48.
21. Khan, S., Ahamed, M., & Ravinath, D. (2014). A study on green IT enablers for Saudi Arabian consumer purchasing behaviour using structural equation modelling Middle East. *Journal of Business, 9*(3), 9–21.
22. Needhidasan, S., Samuel, M., & Chidambaram, R. (2021). Electronic waste—An emerging threat to the environment of urban India. *Journal of Environmental Health Science and Engineering, 12*, 1–9.
23. ARCEP. Nos publications chiffres. (2022). Available from https://www.arcep.fr/cartes-et-don nees/nos-publications-chiffrees.html. Last Accessed on: February, 2022.
24. Work from home initiative. (2022). Available from http://www.fotolia.com. Last Accessed on: February, 2022.

Chapter 4
Green Information Centres and Allied Foundations: *The Concern of Environmental Information and Documentation Practice*

Sarmistha Chowdhury, K. S. Tiwary, and Jayati Lahiri Dey

Abstract Green Information Centre is an emerging concept of the information centre, and it is required for the modern information system practice. It is supported with the green and eco-friendly approach and dedicated in the designing, development, management and evaluation of the information systems. Eco-friendly tools, techniques and principles are considered worthy in Green Information Centre over the traditional information centre operation. Green Information Centre is not only the institutions having information solutions but also the applications of the green principles to the allied institutes where information practice is considered as worthy. Once environment-friendly green principal and systems are practised in the traditional documentation, then the establishment is known as Green Information Centre. Such types of information centres are called as sustainable information centre with integrated approach of intelligent concept in environmental management. In addition to the computing and IT, in Green Information Centre the aspects of management science, information studies, information management, etc. are highly associated with green knowledge management. The libraries and similar knowledge resource centres are also these days adopting eco-friendly principles, and this chapter discusses various aspects of Green Information Foundations in brief manner. Here, basic features and functions of the green and eco-friendly information-related foundations are discussed with emphasis on Green Libraries.

Keywords Green information systems · Eco-friendly information management · Green libraries · Sustainable information infrastructure

S. Chowdhury (✉) · J. Lahiri Dey
Department of Computer and Information Science, Raiganj University, Raiganj, India
e-mail: sarmistha.gini14@gmail.com

K. S. Tiwary
Deanship of Faculty of Science and Management, Raiganj University, Raiganj, India

4.1 Introduction

The government bodies and associations are moving towards greener and eco-friendly policies and strategic adaptation in different ways. Green information mechanism is the need of hour, and therefore, institutions are putting efforts in developing green and eco-friendly procedure to build organizations greener including eco-friendly and greener information systems [1, 19]. There are multiple ways in eco-friendliness in the organizations, and as far as Green Information Centres and Foundations are concerned, it is dedicated in proper designing, developing and managing green information systems [6, 7]. The applications of computing and information technologies are important in designing and developing green information systems, and in this context, various allied technologies are being used such as—

- Software engineering/technology.
- Communication technology.
- Database technology.
- Web technology.
- Networking technologies.
- Security technologies, etc.

According to the green and eco-friendly approaches in all the green information-related aspects and facets, the principals and procedures of Energy and Environmental Management should be provided. In Green Information Centre and Allied Foundations, some of the concepts and establishments are purely applicable, and here, green strategies are important. As the information centres are of two types, i.e. traditional and computational information management, in all types of information centres and foundations, green principles are applicable [3, 20].

4.2 Objective

The present chapter entitled 'Green Information Centres and Allied Foundations: *The Concerns of Environmental Information & Documentation Practice*' is theoretical in nature and consists with following aim and objectives—

- To know about the basic of the Green Information centres and foundations dedicated in information-related activities.
- To gather knowledge about the Green Information Centre and Green Information Systems emphasizing basic nature, areas and functions.
- To get concepts and knowledge regarding contemporary scenario of Green Data Centre with Green Information Analysis Centre.
- To know about the Green Libraries including its basic features and functions and emerging characteristics.
- To get a detailed overview on latest principles, strategies, policies and development in Green Libraries.

4.3 Information Dealing Foundations with Green and Environmental Approaches: The Basics

Environment is a vital concern for all of us and directly and indirectly associated with us. In different sectors, proper development and sustainability are essential to increase [29, 30]. Sustainability is the capacity for biosphere and human civilization to co-exist sustainability and environment are co related. Like other sectors in the information sector too interaction with environment and sustainability considered as worthy and important [9]. Sustainability is based on three pillars, and these are as follows—

- Economics (this is required in profit).
- Environment (also called as planet).
- Social (also known as people).

Information sector is broad and increasing day by day, and furthermore, it consists with different units and foundations such as data centres, documentation centres, information centres, information analysis centre, libraries and knowledge centre [4, 5, 32]. All these as a whole is called information foundation. Today, in all the areas and sectors environment including sustainability is considered as worthy because better interaction and applications are very much important. It is worthy to mention that Green Information Foundations may consist of the following—

- Green Information Centre (GIC)
- Green Information Systems (GIS)
- Green Information and Knowledge Network (GIKN)
- Green Libraries (GL)
- Green Information Resource Centre (GIRC)
- Green Data Centre
- Green Information Analysis Centre, etc.

The concept of environment, ecology and sustainability are rising rapidly. These information foundations may be called as eco-friendly information centres [10, 13]. Here, Fig. 4.1 depicted various Green Information Foundations dedicated in building of eco-friendly information infrastructure.

4.3.1 Green Information Centre: Basics

Green Information Centre is a kind of information centre supported by and with environment-friendly and ecological principle. Furthermore, such Green Information Centres are dedicated in promotion of green and eco-friendly culture and activities of green principle and management in the information activities such as in collection and selection, use of green strategy in processing of information. In addition to these in Green Information Centres, green method in dissemination of information is applicable. Information centre is dedicated in information-related affairs, and such

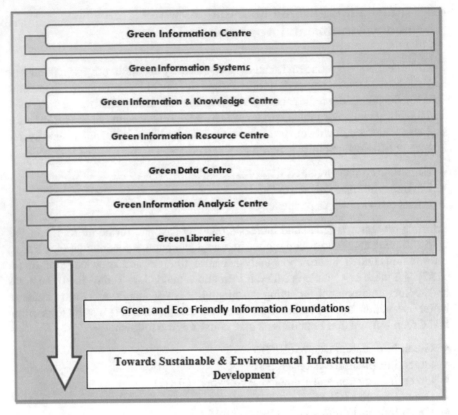

Fig. 4.1 Greener information infrastructure powered by various eco-friendly information centres

activities may be via manual and computational systems. Therefore, green principles, green strategies and methods are applicable in both types of information systems [16, 25].

4.3.2 Green Information Systems: Foundations

Information system is broader concept than information centre, and it is considered as the merger or combination of different types of information centre or similar type of establishments. Information system is a kind of integrated functioning body dedicated in information-related and specific works. As information is required in all the areas and sectors such as health care, business, transportation, education, manufacturing, weather and climatology and so on, information systems and its role are crucial for complete information infrastructure development. Adopting green technology and green computing principles in the information systems leads healthy information infrastructure which is sustainable and healthy [8, 24]. Information system is responsible in the proper functioning with the connected information centre. Information system can be established based on the territory such as regional, national and international. Moreover, based on the way of delivery information system could be manual and also technological in nature. In information system, various principles of environmental systems, green technologies, energy informatics and other concerns are applicable.

4.3.3 Green Information and Knowledge Centre: Basics

The concepts of Information Resource Centre are emerging but before moving to such area first gather information about the Green Information and Knowledge Centre. Such kinds of centres and establishments are like information centre; however, these establishments also additionally deal with knowledge and resource managements. The Green Information and Knowledge Centre is followed by green principles. The green libraries or green computing and technology principles and procedure are adopted into this kind of Green Information and Knowledge Centre. The nature of electronic information and knowledge systems in such establishments is worthy and increasing [11, 22].

4.3.4 Green Information Resource Centre: Overview

Green Information Resource Centre is a kind of establishment dedicated in information-related affairs. Similar to 'Green Information Centre' and 'Green Information and Knowledge Centre', this kind of centre heavily concerns about the information resource collection and its management. Among the resources important are primary sources, secondary sources and tertiary sources. Green Information Resource Centre consists with knowledge materials like documents, books, journals, encyclopaedias and so on. In addition to information resources and collections these days, such type of establishment concerned about the computing facilities and service. Green Information Resource Centre is dedicated in technological green and eco-management and also eco-friendly traditional document management [12, 21]. Such kind of organization is a part of an organization, or it may be a stand-alone institution. Similar to the Green library-based principles or green technologies, Green Information Resource Centre also depends on such principles and methods. Furthermore, energy management designing and development principles are also applicable in Green Information Resource Centre including knowledge network.

4.3.5   Green Data Centre: Basics

Green Data Centre in another word can be called as Eco or Environmental Data Centre. Today, most of the data centre is computational or electronic or computationally supported. In some context, such type of data centre is also known as cloud centre. However, in respect of traditional information studies, data centre may be manual document based which are connected or merged with the following establishments (and allied foundations and institutions) such as—

- Information centre,
- Information systems,
- Information and knowledge network,
- Information resource centre,
- Documentation centre,
- Information analysis centre, etc.

Such types of Green Data Centre are rising internationally and merging with technological and tools oriented towards environment-friendly approaches [2, 14, 15].

4.3.6   Green Documentation Centre: Basics

Documentation centre is a kind of information-related foundations dedicated purely on documentation activities. Here, the major documentation procedures are primary, secondary and tertiary sources of information. Documentation centre may be established in an organization or company to serve the organization by offering information and documentation need or may be established on different subject/s. Documentation centre directly and indirectly helps in other information centre or libraries. As documentation centre deals with the information activities, it is supported by the manual procedure and also computing and information technologies. Here, green principles and environmental strategies are essential to follow up.

4.3.7   Green Information Analysis Centre: Basics

Green Information Analysis Centre is the kind of information foundation dedicated in analysing data and information using various knowledge organization tools and supporting information technologies. Here, collection of required data is important with proper analysis of information based on various feature and criteria. The 'Green' Information Analysis can be computational or manual [13, 17]. There are different methods and procedure adopted in information analysis, and all these can be followed in Green Information Analysis Centre as depicted in Fig. 4.2.

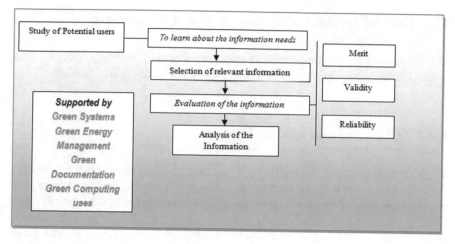

Fig. 4.2 Backbone of green information analysis centre

4.3.8 *Green Library: Basics*

In the field of library and similar foundations, the concept 'Green Library' is considering and emerging rapidly [10, 23]. Green libraries are dedicated in designing, development of environment and green institution responsible for sustainable library development including energy management (electrical). In the Green Libraries, the natural resources are being used in the building constructions including the activities of the—

- Library shelves,
- Library equipment,
- Documentation tools,
- Tools and products used in the libraries,
- Waste management and disposal (i.e. recycling, etc.).

In recent past, the IT and computing systems and its applications in the libraries are emerged, and in this context, it is recommended to use the green and environment-friendly approaches and systems into the libraries; this is also required to make libraries more greener as well as energy efficient. There are different opinions available on Green Libraries but among these one of the important opinions is provided by Oxford English Dictionary, *'green/sustainable libraries as a library designed to minimize negative impact on the natural environment and maximize indoor environmental quality by means of careful site selection, use of natural construction materials and biodegradable products, conservation of resources like water, energy, chapter, and responsible waste disposal recycling, etc'*.

Within Green Information Foundations, the aspects of Green Libraries are being popular and interestingly rising, and in this context, following parameters are important.

4.4 Green Libraries with Concerns and Parameters

Various parameters are considered as important concern in designing and developing Green Libraries, and among them, few important are mentioned as follows.

4.4.1 Electrical and Similar Systems

The *electrical system* is the backbone of the modern libraries. Various sections and divisions of the libraries are also electronic technologies and system integrated. In the libraries, various electronic and electrical products are being used such as lights, fans, speakers and air-conditioning systems. Thus, during non-use it is essential to off all the lights, fans and other electronic products. Furthermore, low power consumption lights such as CFL and LED and fans are essential use having green and environmental factor. It is also essential to follow up the aspects of maintenances.

4.4.2 Building Material and Constructions

Green Library is also normally established in a construction settings, and in this context, *building material* should be considered important regarding the use of sustainable material. According to the experts, the proposed green building is expected to be eco-friendly material supported or featured. The construction of the building should be supported by the quicker and renewal materials, and in this context, bamboo can be used and wood can be avoided. Further, it is important to note that the material used in the building recycling facilities or not [27, 31].

4.4.3 Indoor Air Quality

Inside of the library also, it is essential to establish the environmental friendly systems; furthermore, library building should be more and properly ventilated. Air-conditioning machines are being used these days but it is essential to use minimum as much as possible. Different and proper healthy techniques, tools and designing become mandatory in entire library and also for the individual room or section. There-fore, green building concept is useful here in respect of different types of contents and paints, carpeting elements, etc. In a library, there are different sections impor-tant; among these important are reference sections and information kiosks; in such places too, following green principles is essential; and it should be followed inside and outside of the building. The green (both natural and technological) inside can

Fig. 4.3 Green libraries and look (a library in USA in left and Anna University Library India in right)

be seen in many international libraries; here, a sample library of USA is depicted in Fig. 4.3 left, whereas right side is an example of Indian green library.

4.4.4   Control and Proper Population

In big and multistored libraries, lot of divisions, and units are common these days, and therefore, proper transportation could be considered as important keeping in mind the increasing population [26, 28]. Thus, in modern library systems nearby parking zone is essential to plan, so that user can use the services effectively and exactly as per the need and demand.

4.4.5   Use of Contents and Documents

Eco-friendly contents and documents like books, journals and periodicals are considered as important regarding developing green libraries. Furthermore, as far as electronic content is concerned such as e-book, e-journal and e-document, the energy management system should be followed. The digitalization of existing products and documents also supports eco-friendliness and green principles.

4.4.6   Green Design

Green libraries are considered as worthy by adopting proper and effective designing or sustainability, and in this context, different models and principles are available to

Table 4.1 Brown suggested green libraries factors	Brown suggested keys for green libraries
	Community collaboration
	Enhanced daily life
	Green material
	Green roofs
	Raised floor system
	Energy efficiency
	Natural ventilation
	Green power and reveal energy
	Indoor environmental quality

be followed. However, brown suggested models and keys in the year 2003 considered as important as depicted in Table 4.1.

4.4.7 Recycling of the Products

Recycling is being considered as most important and timely these days, and therefore, proper steps are essential in different areas like data, documents, plastic products and other managements. Recycling of the documents and books and scanning of manuscript are worthy moves to think in respect of environmental friendliness.

4.4.8 Technological Systems

Regarding information technology and computing, proper and scientific procedure and policies are essential to follow-up, and in this context, the green computing and green information policy are considered as important to adopt. Here, green computing and ICT practice may lead to proper environment friendliness, and in this regard, following can be adopted.

- Products longevity
- Data centre design
- Software and deployment optimization
- Power management
- Material recycling
- Cloud computing and technology
- Edged computing
- Telecomputing, etc. could be used [18, 32].

4.4.9 Social Responsibility and Greenery

Social responsibility is being considered as important in the green designing. And here, it is essential to follow the social responsibility so that libraries can be green and eco-friendly. Emerging trend is environmentalism today, and therefore, social responsibility must be followed up.

4.4.9.1 Good Sanitization System

Good sanitization systems are considered as important to adopt in managing good green libraries including its designing and development. Natural resources are essential to use in different areas like washroom, drinking water etc. Here, reuse of waste including proper water harvesting should be considered as worthy. In proper rainwater systems and management, the approach of the green designing and development is noticeable. Some of the essentials or prerequisite is need to be followed up in this regard.

- In the libraries, polythene and unnecessary plastics are essential to avoid or reduce.
- It is worthy to look as much as natural wind facilities are available with support of green building designing as well as development procedure.
- It is essential to follow more and more eco-friendly pesticide regarding pest control and management.
- Regarding the aspects and functions of stacking, documentation, computerized information centre, use of the network printer are concerned eco-friendliness can be noted. Further here, sharing of the printer could be treated as green library management.
- Refilling of the printing devises is important rather purchasing of the new printing materials.

Gradually, the concepts of green libraries are rising and various nations are involving with various initiative for its proper development. Regarding some notable worldwide libraries, few are depicted in Fig. 4.4 in which green library model is adopted effectively.

Different concerns are rising in developing Green Libraries, and in this regard, various strategies and methods are essential among the users and service providers. In developing green libraries and similar institutions, proper awareness are also solicited.

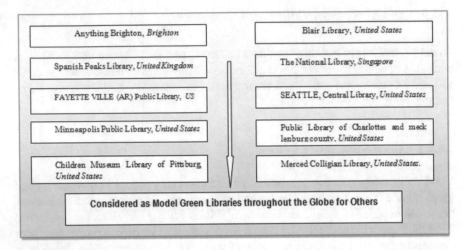

Fig. 4.4 Some of the model green libraries globally

4.5 Conclusion

The concepts of the Green Information Foundations are emerging, and in various ways, green and eco-friendly approaches are rising. Different organizations and institutions are internationally putting efforts in conceptualizing and developing green technology, green computing and Green Information Foundation concepts. Here, technologies, tools, procedure, strategies and mechanism are considered as important regarding green and environment friendliness. Further, proper awareness from different levels is required. Different governmental initiatives, associations and organization's efforts are required in building Green Information Institutions. In most of the corporate, companies and organizations, information centre or libraries become integral part these days; therefore, adopting green technologies, tools, procedure and strategies are considered as worthy. Educational training, awareness and initiatives are important in proper green information mechanism and better environmental informatics practice.

References

1. Akman, I., & Mishra, A. (2014). Green information technology practices among IT professionals: Theory of planned behavior perspective. *PROBLEMY EKOROZWOJU–Problems of Sustainable Development, 9*(2), 47–54.
2. Akman, I., & Mishra, A. (2015). Sector diversity in green information technology practices: Technology acceptance model perspective. *Computers in Human Behavior, 49*, 477–486.
3. Aulisio, G. J. (2013). Green libraries are more than just buildings. *Electronic Green Journal, 1*(35).

4. Bai, C., & Sarkis, J. (2013). Green information technology strategic justification and evaluation. *Information Systems Frontiers, 15*(5), 831–847.
5. Dalvi-Esfahani, M., Alaedini, Z., Nilashi, M., Samad, S., Asadi, S., & Mohammadi, M. (2020). Students' green information technology behavior: Beliefs and personality traits. *Journal of Cleaner Production, 257*, 120406.
6. Dedrick, J. (2010). Green IS: Concepts and issues for information systems research. *Communications of the Association for Information Systems, 27*(1), 11.
7. Dias, S. M. (2017). Environmental sustainability for public libraries in Portugal: A first approach. *Electronic Green Journal, 1*(40).
8. Jankowska, M. A. (2014). Practicing sustainable environmental solutions: A call for green policy in academic libraries. *Against the Grain, 22*(6), 12.
9. Jr, B. A., Majid, M. A., & Romli, A. (2017). Green information technology system practice for sustainable collaborative enterprise: A structural literature review. *International Journal of Sustainable Society, 9*(3), 242–272.
10. Junior, B. A., Majid, M. A., & Romli, A. (2018). Green information technology for sustainability elicitation in government-based organisations: An exploratory case study. *International Journal of Sustainable Society, 10*(1), 20–41.
11. Kaushal, C. (2015). Green initiatives for libraries: an environment for next generation. *International Journal of Tropical Agriculture, 33*(2 (Part IV)), 1893–1897.
12. Kurbanoğlu, S., & Boustany, J. (2014, October). From green libraries to green information literacy. In: *European Conference on Information Literacy* (pp. 47–58). Springer, Cham.
13. Meher, P., & Parabhoi, L. (2017). Green Library: An overview, issues with special reference to Indian libraries. *International Journal of Digital Library Services, 7*(2), 62–69.
14. Mishra, D., Akman, I., & Mishra, A. (2014). Theory of reasoned action application for green information technology acceptance. *Computers in Human Behavior, 36*, 29–40.
15. Noh, Y. (2015). A study on developing the evaluation items for the green libraries certification. *Journal of the Korean Society for information Management, 32*(3), 99–130.
16. Noh, Y., & Ahn, I. J. (2018). Evaluation Indicators for green libraries and library eco-friendliness. *International Journal of Knowledge Content Development & Technology, 8*(1), 51–77.
17. Pangail, R. K. (2015). Green libraries: Meaning, standards and practices. *Episteme, 4*(3), 1–9.
18. Paul, P. K. (2016). Green information science: Information science and its interaction with green computing and technology for eco friendly information infrastructure. *International Journal of Information Dissemination and Technology, 3*(4), 292–296.
19. Paul, P. K., Bhuimali, A., Ghose, M., Ganguly, J., & Ghosh, M. (2017). Information technology and green-eco environment: The aspects in Interdisciplinary scenario. *International Journal of Scientific Research and Modern Education (IJSRME), 2*(2), 27–30.
20. Paul, P. K., Aithal, P. S., Bhuimali, A., & Kalishankar, T. (2020). Environmental informatics vis-à-vis big data analytics: The geo-spatial and sustainable solutions. *International Journal of Applied Engineering and Management Letters (IJAEML), 4*(2), 31–40.
21. Paul, P. K., Bhuimali, A., Aithal, P. S., & Kalishankar, T. (2020). Environmental informatics: Educational opportunities at bachelors level—International context and Indian potentialities. *International Journal of Applied Engineering and Management Letters (IJAEML), 4*(1), 243–256.
22. Przychodzen, W., Gómez-Bezares, F., & Przychodzen, J. (2018). Green information technologies practices and financial performance—The empirical evidence from German publicly traded companies. *Journal of Cleaner Production, 201*, 570–579.
23. Qing, H. G. (2019). Green information technology government regulation components: Improving Indonesia green information technology. *Journal of Theoretical and Applied Information Technology, 97*(16), 4467–4477.
24. Saha, P., & Padhan, H. (2019). Green libraries effect to the academic institutions: A special study on US based libraries. *Library Philosophy and Practice*, 1–9.
25. Santhanam, A., & Keller, C. (2018). The role of data centres in advancing green IT: A literature review. *Journal of Soft Computing and Decision Support Systems, 5*(1), 9–26.

26. Sarkis, J., Koo, C., & Watson, R. T. (2013). Green information systems and technologies—This generation and beyond: Introduction to the special issue. *Information Systems Frontiers, 15*(5), 695–704.
27. Sengan, S., Priya, V., & Dadheech, P. (2020). Energy and green IT resource management analysis and formation in geographically distributed environmental cloud data centre. *Energy, 29*(6), 4144–4155.
28. Sornasundari, R., & Sara, C. (2016). Green library: A study. *International Journal of Research Instinct, 3*(2), 616–621.
29. Uddin, M., Shah, A., & Memon, J. (2014). Energy efficiency and environmental considerations for green data centres. *International Journal of Green Economics, 8*(2), 144–157.
30. vom Brocke, J., Watson, R. T., Dwyer, C., Elliot, S., & Melville, N. (2013). Green information systems: Directives for the IS discipline. *Communications of the Association for Information Systems, 33*(1), 30.
31. Wong, J. K. W., & Zhou, J. (2015). Enhancing environmental sustainability over building life cycles through green BIM: A review. *Automation in Construction, 57*, 156–165.
32. Zhang, N., & Xie, H. (2015). Toward green IT: Modeling sustainable production characteristics for Chinese electronic information industry, 1980–2012. *Technological Forecasting and Social Change, 96*, 62–70.

Chapter 5
A Study on the Social and Economic Impact of Artificial Intelligence-Based Environmental Forecasts

Paramita Bhattacharjee, Ajitesh Moy Ghosh, and Pabak Indu

Abstract An ever-evolving environment is a constant on our planet Earth. Our economic and social well-being is influenced by the environment. People have been attempting to manipulate the environment since the dawn of technology in a variety of ways. When we try to control the environment, it just sails away. Humans have come to realize that regulating the environment is impossible since nature always finds a way to get its own way. Humans, on the other hand, have evolved ways for predicting the environment's behavior by monitoring abrupt changes in the paraments and making essential calculations. Computers have taken over the majority of the computations, and artificial Inelegance algorithms have taken over the prediction systems, over time. However, the quality of the information utilized to make an informed decision is critical to protecting our environment. In today's environmental protection, computers play an important role in duties including monitoring, data processing, and communication. AI approaches have had a profound effect on environmental decision support systems' conception and development. Many articles were analyzed and categorized in order to uncover answers that already exist related to the impact of the environmental changes. Access to dynamic environmental information may be gained using GeoWeb APIs. In order to bring processing to the data, cloud computing and array-friendly databases would be helpful. Geotagging and location sensing will help convert citizen scientists into environmental data collectors. The usage of a combined cycle steam power plant for electricity generation has been observed. When looking at the environmental effect, a thorough investigation is necessary. Acidification potential (AP) was determined to have the greatest influence, with NOx being the most significant cause. On-site energy consumption data were gathered, which included the amount of power used by each piece of equipment and each structure. This study examines the link between the organizational context and the development of dynamic capacities in the setting of environmental unpredictability. Economic tools from poultry farms to the production of consumer goods all rely heavily on environmental monitoring. Additionally, IoT and blockchain have a significant impact on environmental monitoring and prediction. Open-data sources

P. Bhattacharjee · A. Moy Ghosh · P. Indu (✉)
Department of Computer Science and Engineering, Adamas University, Kolkata, West Bengal, India
e-mail: pabakindu@yahoo.co.in

© The Author(s), under exclusive license to Springer Nature Singapore Pte Ltd. 2022
P. K. Paul et al. (eds.), *Environmental Informatics*,
https://doi.org/10.1007/978-981-19-2083-7_5

that contain a wide range of environmental information may be utilized for a number of environmental analytical objectives, which in turn can have a positive impact on human well-being in the long run. An example of this can be the system that replicates the water quality in Moscow's rivers and canals. A neural network was trained to classify a condition as either good or bad based on the data it received from the indicators. Using AI framework, policymakers, researchers, and innovators may make educated decisions on the potential and problems posed by recent changes in the environment. Authors believe that AI-based prediction systems can help humans achieve steady economic and societal goals.

Keywords Economic and social well-being · Environmental decision support systems · GeoWeb APIs · Geotagging and location sensing · Acidification potential · IoT and blockchain

5.1 Introduction

We will never fully appreciate the value of our surroundings. However, we may estimate a portion of the importance our surroundings possess, which will aid us in appreciating it. Human health and economy are directly related to the environment. On the one hand, it provides humans with oxygen, food, and other essential ingredients for survival, while on the other side, it gives economic resources. Humanity's whole life support system is contingent upon the health of all environmental factors. Every species is dependent on its surroundings for the energy and resources necessary to maintain existence. Agriculture, water treatment, and hygiene advances have had a far higher influence on human health than any other medical technology, while also contributing to economic stability.

As with any other element, unanticipated environmental changes might result in catastrophic failure, resulting in loss of health and economy. To avoid such disasters, we need to have enough environmental warning mechanisms in place, such as understanding the likelihood and strength of hail storms so that we may take appropriate precautions before they strike. In the event of a landslide, traffic can be suspended temporarily. Forecasting large waves can assist fishermen and sailors in determining when to begin their expedition and when to seek refuge.

To gain a better understanding of how the environment behaves, environmental monitoring has been in great demand during the last few years. IoT devices in conjunction with an artificial intelligence-based decision support system may be used to collect and evaluate environmental data. These analytical findings will aid us in making a conclusion. Geotagging technology can assist in discovering locations with comparable mineral content of soil, planting needs, and so forth. The agricultural industry is rapidly adopting AI-based technology to assist farmers in producing healthier crops, controlling pests, monitoring soil and plant growth conditions, organizing data for farmers, assisting with workload, and improving a variety of agriculture-related activities throughout the food supply chain [1].

With the ever-changing global environment and fast-growing pollution, farmers are having difficulties determining the optimal time to begin plantings. In these settings, weather forecasting powered by artificial intelligence assists farmers in determining the type of seeds to sow.

Farmers may take preventative actions to safeguard the safety and health of their crops by utilizing AI-enabled soil and crop health monitoring systems [2].

Artificial intelligence firms are creating robots capable of performing a wide variety of activities in agricultural environments. This sort of robot is designed to manage and harvest crops more quickly and efficiently. These robots are programmed to evaluate crop quality and to harvest and package crops. These gadgets are capable of determining the product's quality and, in certain cases, the price band [3].

Apart from being a saving grace for agriculture, AI and IoT-based technologies may also assist in keeping farm animals healthy, resulting in environmental and economic growth [4]. Farmers can now follow their livestock without having to accompany them wherever they go, thanks to the development of battery-powered sensors and smart equipment. Ranchers may use IoT livestock monitoring to keep an eye on their animals in real time. It assists in resolving a variety of challenges that arise in the livestock business. The Internet of Things-enabled livestock monitoring system has the potential to benefit livestock farmers. The Internet of Things-enabled livestock management devices give data on several aspects of cow health. A sensor-equipped wearable collar or tag monitors an animal's position, temperature, blood pressure, and heart rate in near-real time and wirelessly delivers the data to farmers' devices. Each farm animal has a customizable tracker. Because the tracker is compact and lightweight, it does not protrude from the animal's body. The GPS tracker and sensors are integrated into the device, allowing for round-the-clock monitoring of the animals' movements and health seven days a week. The gadget is designed to monitor animal behavior. IoT-enabled devices can also aid in lowering livestock mortality rates, since when farm animals become ill, the disease may be identified and treated before it is too late [5].

Yet another technological advance blockchain technology is already being employed in a number of industries, several governments have included it into their environmental sustainability strategies [6]. Without a doubt, blockchain technology will enable new methods of green manufacturing, the monitoring, and storage of data-related activities that contribute to pollution and environmental harm, and the real-time collecting and analysis of green or low-carbon data for speedy decision-making. Additionally, blockchain technology can assist in the construction of a green supply chain.

With very minor changes to the environment, we should expect a slew of problems to human wellness. The majority of climate forecasting research, on the other hand, is presently concentrated on mechanistic, bottom-up methodologies such as physics-based general circulation models and Earth system models. We examine the performance of a phenomenological, top-down model created using a neural network, and large quantities of global monthly mean temperature data in this study. The neural network algorithm correctly forecasts temperature rises and declines for the next 10 years by creating graphical representations of monthly temperature data for the previous 30 years. When using LeNet for the convolutional neural network, the best global model has a 97.0% accuracy [7].

Given the efficacy of IoT devices, AI, and Ml-based technologies in achieving diverse goals, the authors feel that these technologies have the potential to be a game changer for environmental maintenance and enhancement. Economic growth is inextricably linked to environmental growth, and with the aid of all of these technologies, it will soon be feasible to achieve all of the economic and societal goals that humanity has set for the future [8].

5.2   Some Existing Solutions

Over the years, researchers have attempted to develop a viable weather forecasting method based on either tools-based forecasting or mechanism-based forecasting models. The following are some of the technologies and mechanisms now utilized to forecast weather.

5.2.1   Tools

Nearly two decades ago, AI was included into the development of the dynamic-integrated forecasting (DICast) [9] system, which uses meteorological data to generate automated and accurate forecasts.

The dynamic integrated forecast system is responsible for ingesting meteorological data (observations, models, statistical data, and climate data, among others) and producing meteorological forecasts for forecast locations and lead times specified by the user. To do this, DICast employs a variety of forecasting algorithms to provide independent estimates for each data source. At each user-defined forecast site, a single consensus forecast is constructed from a succession of individual predictions using a processing method that takes into account the recent skill of each forecast module. Leading commercial weather service providers have also incorporated the technique [3]. Figure 5.1 shows the weather forecast pictures captured by the DICAST.

NCAR scientists have been analyzing storm characteristics such as temperatures at various altitudes, wind direction and speed, updraft, and humidity using a neural network as part of a meteorological model. The computer then searches for patterns indicative of the presence of a storm capable of generating hail, which can travel at rates of up to 120 miles per hour and cost between $8 and $10 billion in damage yearly in the United States alone [10].

Computer vision and other forms of artificial intelligence have made substantial contributions to the agriculture economy. Crop growth monitoring has always been reliant on manual evaluation, which is neither timely nor reliable. Computer vision [11] enables continuous real-time monitoring of plant development as well as the

Fig. 5.1 DICAST weather
forecast pictures image
courtesy [9]

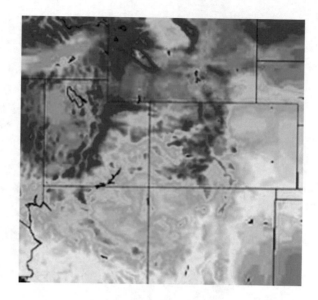

identification of crop alterations caused by malnutrition or disease. Precision agriculture is largely reliant on reliable, high-resolution maps to track crop development.
These maps were created using data from satellite-borne radars and optical sensors,
as well as airborne and drone-borne optical sensors. There is a unique method for
constructing growth deficit maps with a 5 cm precision and a spatial resolution of
1 m using differential synthetic aperture radar interferometry (DInSAR). Figure 5.2
shows DInSAR-based crop growth monitoring tools-based images [12].

Human labor is both costly and inefficient when compared to automated solutions.
Additionally, conventional weeding practices include the use of pesticides, which
may pollute neighboring healthy plants, water, or animals.

Intelligent weed detection and eradication using robots such as Econobox is
enabled by computer vision, cutting costs, and improving productivity [13].

Odd Bot [14], another similar robot, was on show at the consumer electronics show
(CES) in Las Vegas earlier this month with its information booth and weed-plucking
equipment. According to Martijn Lukaart, Founder & CEO, Odd Bot is presently
being developed for use in organic agricultural zones to simplify the weed-pulling
process for vast farms that now perform all labor manually.

Numerous large-scale farmers have previously invested in a platform that enables
workers to sleep on a bed while traveling between rows of crops. As a consequence,
workers will be able to weed more effectively and enjoyably. Figure 5.3 shows the
image of Odd Bot and agricultural robot.

Farmers may secure their crops and avoid damage by discovering insect pests
early, but this can be extremely hard and time-consuming, whereas camera-based
agricultural monitoring systems can identify, categorize, and count insects that pose
a hazard to crops. Automated insect pest management is cost-effective and contributes

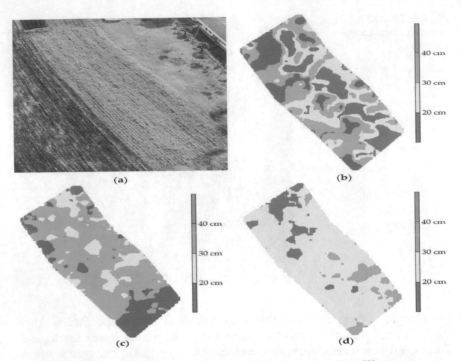

Fig. 5.2 Crop growth monitoring with drone-borne DInSAR image courtesy [9]

Fig. 5.3 Odd Bot, an agricultural robot image courtesy [15]

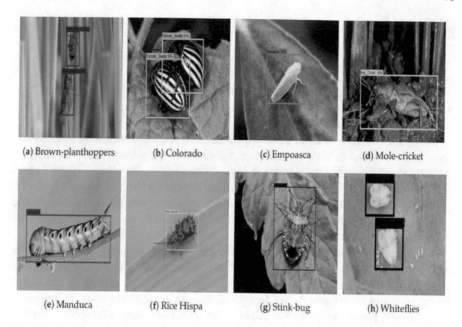

(a) Brown-planthoppers   (b) Colorado   (c) Empoasca   (d) Mole-cricket

(e) Manduca   (f) Rice Hispa   (g) Stink-bug   (h) Whiteflies

Fig. 5.4 Farm field and garden identification of insects through AI-based systems image courtesy [17]

to pesticide reduction [16]. Figure 5.4 shows an AI-based system having ability to identify insect pests from images.

Remote and real-time monitoring of animals and their behaviors is possible using computer vision systems. It is particularly handy for counting animals, recognizing illness or abnormal behavior, and recording births.

Additionally, image or video data may be combined with other sensor-collected environmental data (such as temperature or ventilation), enabling AI-powered systems to provide farmers with actionable insights regarding animal health and availability to food and water. For example, "All flex Livestock Intelligence" monitors dairy cows [18].

MSD Animal Health Intelligence health monitoring software, for example, assists dairy producers in identifying growing health problems before to output declines or clinical symptoms. Additionally, they enable farmers to analyze the success of veterinary treatments in real time, enabling them to optimize recovery and treatment costs. Dairy farmers may use the apps to make educated and early treatment and intervention decisions, allowing them to take preventative measures to avert deterioration, cut treatment costs, limit or avoid impact on milk production, and decrease death rates.

MSD Animal Health Intelligence's (All flex Livestock Intelligence) reproduction monitoring software minimizes the ambiguity and inconsistency associated with analyzing the reproductive condition of each individual heifer and cow. Dairy producers may utilize the applications to enhance conception rates while saving money and time on expert labor [19].

The reproduction monitoring programmers employ behavior monitoring of activity, rumination, eating, and other unique significant cow behaviors to provide unmatched heat detection accuracy, including detection of weak heat indicators, while minimizing false positives. They deliver actionable knowledge in real time and precise insemination timing directions.

Dairy producers may utilize the applications to reduce the calving interval and reduce or eliminate their dependency on reproductive hormones. Farmers can boost their herd's milk production and genetic quality by limiting the number of days open.

Group Monitoring: MSD Animal Health Intelligence's (Allflex Livestock Intelligence) [20] group monitoring tools provide dairy farmers with timely and actionable information that enables them to make better herd health, welfare, and production choices.

Farmers may optimize their nutrition strategy by rapidly evaluating how their cows react to ration changes, such as new batches, new suppliers, a toxin or feed issue, and ratio modifications. Benefits of ration changes can be evident in as short as a few hours, allowing for informed nutrition decisions that increase herd health, reproduction, and milk productivity. By visualizing long-term trends spanning several months, nutritionists can gain a new perspective on diets and adjustments. Environmental issues such as severe heat, a lack of water, damp bedding, overcrowding, wind, and dogs, as well as changes in grouping, equipment, and humans, can all have an effect on cows' health. This enables informed decision-making about current concerns, the activation and fine-tuning of heat abatement systems, and the creation of policies to prevent or mitigate future problems.

Figure 5.5 shows the sensor-based cattle health monitoring device.

Geotagging is the process of adding geographical information, such as latitude and longitude, to various media, such as a photograph or video. Users may obtain a range of location-specific information from their devices via geotagging. It informs customers of the location of the content of an image. Several assets are generated in

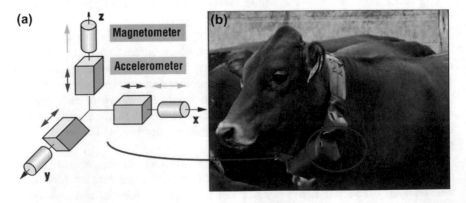

Fig. 5.5 Cattel health monitoring sensors image courtesy [21]

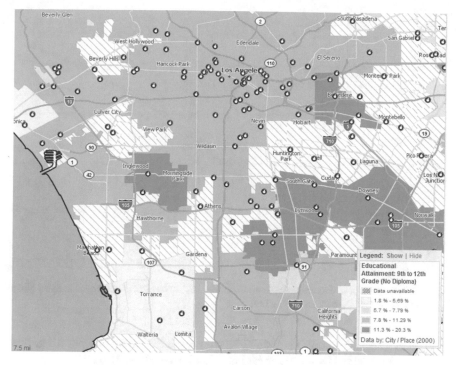

Fig. 5.6 Geotagging image courtesy [23]

states as a consequence of numerous Ministry of Agriculture programmers. Additionally, RKVY funds [22] have been used to construct infrastructure and assets in the agriculture and allied sectors, including soil testing laboratories, pesticide testing laboratories, bio-fertilizer manufacturing units, custom hiring centers, vaccine manufacturing units, veterinary diagnosis laboratories, dispensaries, milk collection centers, fish production units, godowns, cold storage, and shade nets. States and the Indian government must monitor large-scale firms closely in order to analyze financial flows, inventory assets, encourage transparency, plan future assets, and tell farmers about available resources [23]. Figure 5.6 shows the geotagging image.

5.2.2 Mechanism

According to the paper [24], artificial intelligence for environmental applications might contribute up to $5.2 trillion USD to the global economy in 2030, representing a 4.4% increase over current business practices. Simultaneously, AI levers have the potential to reduce global greenhouse gas (GHG) emissions by 4% in 2030, equivalent to 2.4 Gt CO2e, or the combined annual emissions of Australia, Canada, and Japan

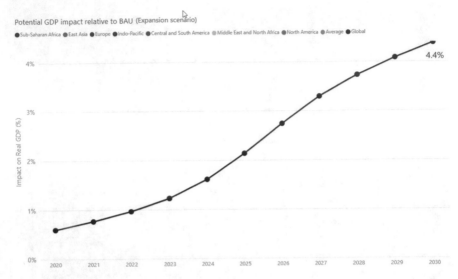

Fig. 5.7 Using AI for environment applications in four key sectors has the potential to boost global GDP by 4.4% by 2030, image courtesy [24]

in 2030 [25]. As a result of this change, AI may result in the creation of 38.2 million net new jobs worldwide, mostly in higher-skilled occupations (Fig. 5.7).

5.3 How AI Impacts Sustainable Development Goals

By detecting energy emission reductions, CO_2 removal, assisting in the development of greener transportation networks, monitoring deforestation, and forecasting extreme weather events, AI has the ability to expedite global efforts to safeguard the environment and save resources [26].

Within buildings, smart sensors and meters may be placed to gather data to monitor, analyze, and optimize energy consumption.

Predictive analytics driven by AI, in conjunction with drones, advanced sensor platforms, and similar technologies, can monitor for tremors, floods, windstorms, sea-level fluctuations, and other natural catastrophes. AI-enabled technology for promoting a circular economy and establishing resource-efficient smart cities.

The advancement of mobile solutions and their use (Smartphones equipped with GPS-type location capabilities, Wi-Fi, and 3G) adds a new dimension to the GeoWeb, significantly improving users' mobility habits. Figure 5.8 shows impact of AI on society, economy, and environment [27].

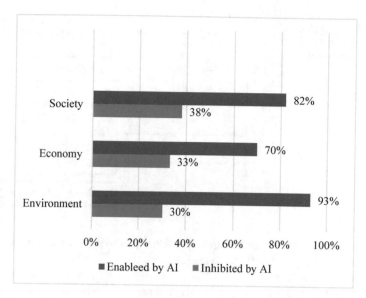

Fig. 5.8 Impact of AI on society, economy, and environment

5.4 Addressing Environmental Issues Using AI

While user-generated geospatial data are a more informal source of local knowledge, it has the potential to be so rich that it may currently serve as an important addition to official data.

We have seen the emergence of new spatial practices (like collaborative mapping and georeferencing material), as well as changes in the roles of experts and amateurs in the field (such network vectorization and collective place certification). Indeed, there is a pressing need for timely access to accurate and up-to-date information in crisis and disaster management, which necessitates the mobilization and use of donated geographic information. Figure 5.9 describes the priority areas for environmental challenges which can be addressed by AI systems.

5.4.1 Climate Change

Numerous recent events involving political crisis, emergency situations, or natural disasters have resulted in the creation and spontaneous coordination of map mashups, as well as the usage of instant communication platforms (hurricane Katrina, H1N1 flu, Haiti or New-Zealand earthquake, etc.). By combining authoritative and non-authoritative data, this novel technology-based method significantly alters the usual information chain utilized in crisis management (Goodchild 2009). Due to the advancement of telecommunications networks, the ability of ICT and geospatial

Fig. 5.9 Priority action areas for environmental challenge image courtesy [24]

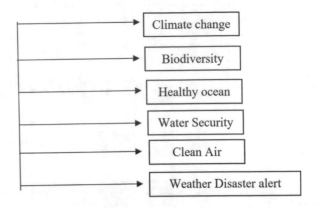

technologies to efficiently acquire and distribute information during times of crisis has increased (Internet, mobile phones) [28].

Although information is critical for preparing an emergency and giving life-saving aid to victims during times of crisis, communication networks and associated technologies are actual lifelines, and crisis management methods are dependent on geographic information and associated technology [29].

5.4.2 Biodiversity

Every living and non-living thing contributes significantly to the advancement and modification of the surrounding climate. Thus, conserving biodiversity is a necessary precursor to environmental change. AI can detect changes in land use, vegetation, forest cover, and the aftermath of natural disasters when paired with satellite data. Using the technologies described above, invasive species may be watched, detected, and tracked; recognizing and tracking their presence, as well as removing them, is all accomplished through the use of machine learning and computer vision. Blue River Technology is utilizing artificial intelligence to detect the presence of invasive species and other changes in biodiversity. Anti-poaching troops are using predictive algorithms to organize their patrol routes.

5.4.3 Healthy Ocean

An artificial intelligence system may be used to collect information and generate predictions regarding the health of an ocean, as well as the pH (Acidic potentiality) factor and other aspects of its environment [30].

5.4.4 Water Security

Artificial intelligence (AI) may be used to minimize contaminants in water, which in turn reduces water contamination and lack of safe drinking water. The use of artificial intelligence to detect the number and composition of dangerous contaminants is possible since AI operates on the basis of optics, which can improve the efficiency of waste management systems [31].

5.4.5 Clean Air

AI models are capable of forecasting changes in ozone layer. As opposed to this, scientists "train" the model by providing historical air quality data, which the computer then uses to learn how ozone reacts under a variety of different climatic conditions [32].

5.4.6 Weather Disaster Alert

Artificial intelligence-based weather forecasting analyzes and predicts changes in meteorological conditions with the help of high-performance computers and enormous amounts of data. It is becoming increasingly popular. It is considered to as a predictive technology since it has the potential to alter the method in which we avoid natural disasters such as tornadoes and hurricanes from occurring [33]. They are frequently employed in mid- to long-term strategic planning situations. Qualitative forecasting methods include the Delphi methodology, market research, and historical life-cycle comparison, to name a few. Statistical models, such as quantitative forecasting models, are used to predict future data as a function of past data.

According to a study [34], artificial intelligence (AI) has the potential to contribute to the achievement of 79% of the sustainable development goals (SDGs). The application of artificial intelligence in urban mobility allows for the prediction of traffic bottlenecks as well as the suggestion of alternate routes. This system estimates vehicle demand by zone and time in the context of shared mobility [35].

5.5 Achieving the Sustainable Development Goals with AI Technology

The Global Education 2030 Agenda agency for education has been entrusted with the responsibility of leading and educating, which is critical to achieving all of these goals of "provide inclusive and equitable quality education." The Education 2030

Framework for Action makes the following recommendations: The United Nations Educational, Scientific, and Cultural Organization (UNESCO) published this book in 2019. (UNESCO). The activities and collaborative actions aimed at leveraging artificial intelligence (AI) to accomplish [36].

The contribution of AI to sustainability is in the area of traffic management. By incorporating artificial intelligence into urban mobility, it is possible to forecast traffic congestion and recommend alternate routes. This technique forecasts vehicle demand by zone and time zone for shared mobility. This means that businesses can arrange for citizens to have access to vehicles based on their needs. This approach not only improves mobility but also helps the environment [37].

To this end, the emergence and widespread usage of mobile solutions (such as smartphones equipped with GPS-type location capabilities, Wi-Fi, and 3G) has added a new dimension to the GeoWeb, on which users' mobility habits are founded, as a complement [38].

Communication networks and associated technologies are true lifelines in a crisis, despite the fact that information is a critical component of emergency preparation and the provision of life-saving aid to victims [39].

Crisis management relies on geospatial data and other related technology. There has been an increase in the use of new information and communication technology such as mobile phones, the Internet, and social networks [40].

Many recent occurrences, whether they were related to a political crisis, an emergency scenario, or a natural disaster, have resulted in the deployment and spontaneous arrangement of map mashups, as well as the use of instantaneous communication tools to convey information (hurricane Katrina, H1N1 flu, Haiti or New-Zealand earthquake, etc.) [41].

By combining authoritative and non-authoritative data, this innovative technology-based strategy makes significant changes to the conventional information chain utilized in crisis management, which is currently in use.

On the technological and application side, the ability of ICT and GIS technologies to gather and disseminate information efficiently in times of crisis has increased as a result of the expansion of telecommunications networks, which has made it easier to gather and distribute information (Internet, mobile phones) [42]. Figure 5.10 represents a fire map from KPBS radio.

Precursors of acid rain are substances that have a high acidification potential. Other substances include sulfur dioxide (SO_2), nitrogen oxides (NOx), and nitrogen monoxide (NO). SO_2-equivalence is a common way to gauge acidification potential [43]. Figures. 5.11 and 5.12 represent the accuracy of prediction and various applications of the prediction models. And Fig. 5.13 depicts the block diagram of the predictive model.

Figure 5.14 shows various application fields of ML in IoT, and Fig. 5.15 shows various algorithms of ML used for environmental prediction. Tables 5.1 and 5.2 show different optimizing components and features of IoT applications.

Fig. 5.10 Fire map (KPBS radio) image courtesy [42]

Impact Category	No. of Testing data points	Mean Squared Error	R Squared
Global Warming Potential	196	0.28	81%
Photochemical Ozone Creation Potential	196	0.07	70%
Acidification Potential	196	1.12	68%
Eutrophication Potential	196	0.02	50%
Ozone Depletion Potential	196	0.01	54%
Abiotic Depletion Potential for non-fossil resources	196	1.40	62%
Abiotic Depletion Potential for fossil resources	196	0.01	54%

Fig. 5.11 Accuracy of the prediction

5.6   IoT Device-Based Monitoring Systems

The use of machine learning (ML) is critical in the improvement of quality of service (QoS) and connectivity between IoT end-to-end (E2E) devices for the provisioning of smart applications. Indeed, machine learning (ML) is unquestionably the principal vessel for AI research into its applications across a wide range of fields, including robotics, smart devices, smart industries, computer vision, smart processing, Internet of Things device connectivity, and autonomous systems. Furthermore, although the

Environmental Impact Indicators	Original Values	Units	Predicted Values
Global Warming Potential	0.6834	Kg CO$_2$ eq.	0.564
Photochemical Ozone Creation Potential	0.000266	Kg Ethane eq.	0.00019152
Acidification Potential	0.001282	Kg SO$_2$ eq.	0.00071792
Eutrophication Potential	0.0001744	Kg Phosphate eq.	0.000097664
Ozone Depletion Potential	2.085E-14	Kg RI I eq.	1.272E-14
Abiotic Depletion Potential for non-fossil resources	2.302E-7	Kg Sb eq.	1.3812E-7
Abiotic Depletion Potential for fossil resources	7.627	Mj	6.102

Fig. 5.12 Application of the prediction model

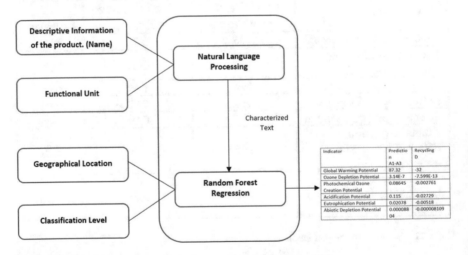

Fig. 5.13 Block diagram of the prediction model image courtesy [42]

connection and data gathering of IoT devices are extremely qualitative, the authors did not include a quantitative description of the ML for IoT connections in their article.

Rain mX [45] is a meteorological and environmental monitoring system that makes advantage of the Internet of Things to collect and analyze weather and environmental data. It is available for purchase separately or as a component of our LIQUA-Level and INFIL-Tracker systems. A self-contained rain gauge, the Rain

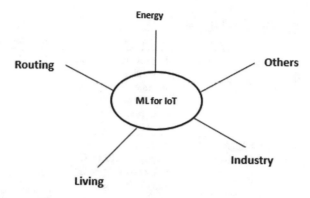

Fig. 5.14 Use of ML in IoT

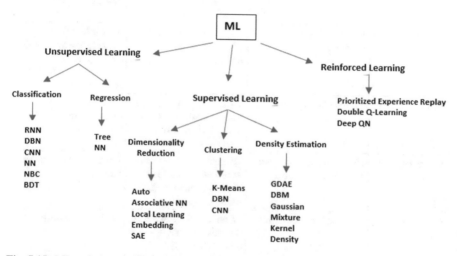

Fig. 5.15 ML techniques exploited for IoT issues image courtesy [44]

mX can also monitor temperature, pressure, and humidity with the addition of an optional converter. Rain mX has proven to be a popular add-on device to our INFIL-Tracker and LIQUA-Level [46] systems since it gives valuable information for determining stormwater catchment and groundwater infiltration levels, as well as for analyzing groundwater infiltration levels. Anyone who wishes to measure rainfall and environmental data on their own will be able to do it with the help of the device.

For the most part, using a weather app on your phone or connecting to your favorite browser-based weather watching site is the most frequent method of staying on top of the current weather conditions. In the majority of cases, the weather conditions you receive are for a location other than your own. Rain mX provides you with the opportunity to obtain weather information whenever you require it and to monitor it from any location you choose. Weather records for entire days, hours, months,

Table 5.1 ML for enhancing for IoT QoS

ML techniques	Optimized components	Features	IoT applications
R-learning	QoS metrics	To provide the ability to respond to IoT and be suitable for wireless communication operations with enhancing QoS	QoS optimizations
Bayesian	Battery charge and transmission power	To obtain energy-efficient IoT using adaptive ECG signal filtering	
KLMS	Missing Web service QoS value prediction	To predict the missing QoS values for the industrial IoT-based KLMS technique	Web service QoS of industrial Internet of Things
SOM and NN	Delay and energy consumption	To obtain an accurate way to route data through the network	Sensor intelligence routing, energy-aware routing, and IoT monitoring and activity recognition
ML	Traffic classification	To realize adaptive the traffic classification framework of different kinds of networks	QoS-aware traffic classification
ML review	Heavy operational load, packet lengths, delay	To classify the IP traffic in IP networks and also to classify unknown traffic using ML	IP traffic classification techniques
DL	Reducing process time	To apply ML techniques for management in smart cities where IoT constitutes information resources	QoS in smart cities

years, or weeks can be collected using Rain mX, including rainfall, relative humidity, temperature, and barometric pressure. Rain mX is free to download and use [44]. Figure 5.16 is a picture of Rain mX in practical use.

The ATSTORM storm detection system, which uses dual-sensor technology and an Internet of Things connection, may be able to improve performance over time by tweaking its algorithms and calculation formulae. The sensors are continually connected to the centralized computation system of Applications Technological S.A., which analyzes their signals, monitors their operation, and delivers notifications to the users [47].

ATSTORM allows for the correct identification of all phases of a thunderstorm as well as the early detection of hazard escalation in the event of a thunderstorm. Lightning strike avoidance devices can be properly controlled both when they are installed

Table 5.2 ML-based identification and connectivity of IoT devices

ML techniques	Features	Highlights
Classification	Device identification based on network traffic Distinguishing between IoT and non-IoT devices connected to the network	Identifying different kinds of nodes connected to the network
DNN *BPNN* *DNN and FNN* *ANN*	Failure detection Human activity classification Tracking accuracy	Classifying the images collected by the drone Detecting sensor failure in IoT network Improving IoT network tracking efficiency
CNN	The accuracy of detecting images	Enhancing the image detection accuracy of the IoT network
Bayesian	IoT device classifications	Classifying visual size of everyday objects Identifying smartphones
KNN/SVM	Network traffic	Traffic generated by applications' background activities
Random forestry	Network traffic	Identifying device types

Fig. 5.16 Rain mX image courtesy of [45]

Fig. 5.17 Thunderstorm detector—ATSTORM image courtesy [48]

and when they are switched off as a result of these measures. This system is distinguished by its centralized computation system, Internet of Things connectivity, and dual-detection technology (using a fully electronic electrostatic field sensor with no moving parts and an electromagnetic field-based sensor). It is entirely self-sufficient in terms of connectivity and energy because it is equipped with a 2G/3G communication module, a solar panel, and a battery. Because of ATSTORM's characteristics, it is straightforward to set up and has low client needs. Furthermore, both systems may be redundant in order to provide the best possible protection [48]. Figure 5.17 shows a prototype model of thunderstorm detector by ATSTORM.

The logics of the situation using our IoT-based monitoring platform, the PowerAMR automated weather monitoring station (WMS) [49] automatically monitors and transmits weather sensor data in real time to our PowerAMR automated weather monitoring station [49]. An entirely computerized, digital, and self-contained power source system, it consists of several sensors set on a tripod platform, as well as a solar panel for charging the battery, and a sealed water-proof container for the data logger, the solar charger, and battery.

The all-in-one weather station lowers the costs of installation, support, and maintenance while increasing the robustness and manageability of the PV plant monitoring system, smart farming system, and environmental monitoring system [50].

The following are some of the advantages of this platform:

Tracks the system's real-time and historical weather conditions, allowing for a complete analysis of the system's overall performance and effectiveness.

A wide range of climatic factors are measured, including solar radiation, wind speed and direction, humidity, air and module temperature, pressure, and a variety of other variables.

Solar irradiance, power, and temperature are represented by curves.

Calculate the PR, the expected vs. actual power output, the module efficiency, the total system efficiency, and a variety of other variables.

Any flaws in the system should be identified and corrected as quickly as possible.

Both the forecast of production losses and the receipt of maintenance warnings are made easier with this tool. Figure 5.18 shows a weather monitoring station and system by PowerAMR.

The Internet of Things has significantly reduced the amount of human work required in practically all areas. A large IoT ecosystem will be formed by billions of machine-type gadgets such as linked cars and wearable sensors, as well as everyday things, all of which will be connected to the Internet. The Internet of Things (IoT) allows objects to communicate with one another across wireless networks and gather and process data in real time. Smart homes, smart grids, smart agriculture, smart health care, and smart streets and parking are just a few of the Internet of Things applications in smart cities. However, in order to fully realize smart city applications, the Internet of Things (IoT) must overcome a number of hurdles, including processing, transmission capability, local data analysis in real time, end-to-end (E2E) latency, huge device connectivity, and privacy. Furthermore, the resources and energy accessible to Internet of Things devices are restricted [51].

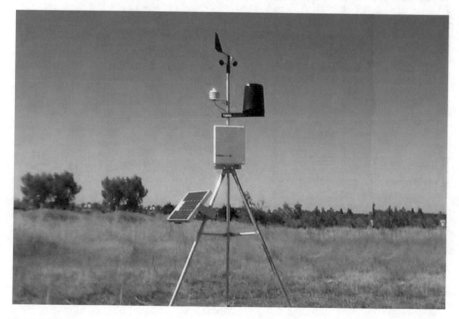

Fig. 5.18 Weather monitoring station and system, logics PowerAMR image courtesy of [48]

SEM has received a plethora of contributions that are divided into two categories: those that serve a purpose and those that employ various methods. As a result, the related research has been divided into three main subsections: those that investigate smart agriculture monitoring systems (SAMs), smart water pollution monitoring systems (SWPMs), and smart air quality monitoring systems (SAQMS). By providing a comprehensive analysis of various SEM application fields, such as soil monitoring (SM), ocean environment monitoring (OEM), marine environment monitoring (MEM), air quality monitoring (QAM), water quality monitoring (WQM), and radiation monitoring (RM), a wide range of SEM applications have been covered [52]. Figures 5.19 and 5.20 are the AI-based system to monitor water quality and waste management system.

In the course of studying the existing literature on SEM methods, particularly on advancements in IoT and sensor technologies for SEM systems, the current study on advancements in IoT and sensor technologies used for SEM systems provides insight to the scientists, policymakers, and researchers in developing a framework of appropriate methods for monitoring the environment, which is facing challenges primarily due to poor air quality, water pollution, and radioactive contamination. Agricultural production, which is the backbone of any developed or developing economy, is similarly affected by these factors; consequently, the development of AI-enabled smart agricultural monitoring (SAM) has been a revelation in the AI-enabled environment. SAM takes into account the following variables: SM, OEM, MEM,

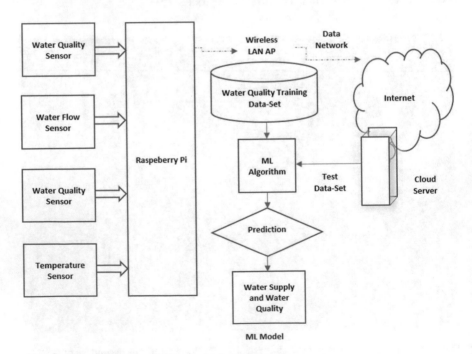

Fig. 5.19 AI-enabled water quality monitoring system

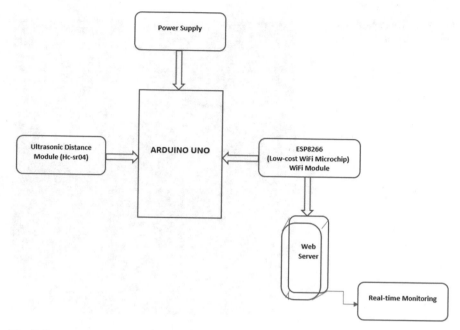

Fig. 5.20 Block diagram of IoT-based waste management system

AQM, WQM, and RM. It has been observed that greenhouse gas emissions have had an impact on soil monitoring methodologies. Sensors, wireless sensor network (WSN), and the Internet of Things (IoT) have all been used to build ocean and marine SEM systems; however, these approaches have primarily been hampered by cost, coverage, and installation concerns. The use of a mobile sensor network, wireless sensors, and Internet of Things devices that run on artificial intelligence and machine learning has been proposed for air pollution control and air quality management. We can see in Table 5.1 that the many types of SEM systems are planned and executed for a variety of diverse reasons, and that there is no robust way for dealing with any of the environmental difficulties that exist [53]. Figures 5.21 and 5.22 are the pictures of windspeed detector and rain gauge.

Integrated sensors and software for determining the concentrations of ambient pollutants such as PM2.5, PM10, CO, NOx, SOx, and O3 in an urban environment. Environmental sensors aboard poll drone measures meteorological characteristics including as noise, temperature, humidity, ambient pressure, rainfall, and flooding. It transfers real-time data to a cloud platform using a wireless communication protocol. The equipment is entirely solar powered and has a lengthy battery backup. Its IP65 enclosure protects it from extreme environmental conditions. Poll drone data are accessible through our air monitoring software, which visualizes and analyzes it in a variety of ways, including reports, alarms, heatmaps, and trend analysis [54]. Figure 5.23 shows a block diagram for air quality monitoring for a smart city.

Fig. 5.21 Windspeed and direction sensor

Fig. 5.22 Rain gauge temperature and humidity sensor

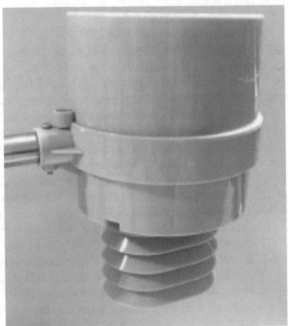

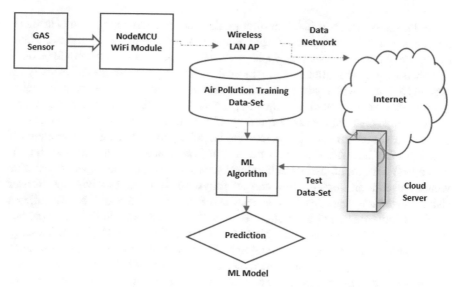

Fig. 5.23 Air quality monitoring solution architecture for smart city image courtesy [54]

5.7 Conclusion

For a long time, AI, ML, and IoT have had a significant impact on our daily lives, including our environment and all living things.

Artificial intelligence has the potential to speed up global efforts to protect the environment and conserve resources by detecting reductions in energy emissions, removal of CO_2, assistance in the development of greener transportation networks, monitoring of deforestation, and forecasting of extreme weather conditions. Some of the world's most pressing environmental problems could be addressed with artificial intelligence, as illustrated by these real-world instances. Improved grid systems with greater predictability and efficiency, as well as the use of renewable energy, can all help combat climate change by using machine learning to optimize energy output and demand real time. Installing smart sensors and meters in a building is a viable option for gathering data, analyzing it, and optimizing energy usage. Machine learning algorithms are already being utilized in smart transportation, such as Google Maps and Waze, to improve navigation and increase safety, thanks to AI. Buildings can be equipped with smart sensors and meters that collect data to monitor, analyze and optimize energy use, safety and the security of its occupants. AI can detect changes in land use, vegetation, forest cover, and the effects of natural disasters when paired with satellite data. The technology detailed above can be used to detect, identify, and track invasive species. Automated systems are employed to identify and remove them, using machine learning techniques and computer vision. Biodiversity changes and invasive species are detected using artificial intelligence by Blue River Technology. AI can help safeguard species and habitats by gathering data from remote ocean locations that are difficult or impossible to access. The application of artificial intelligence in the fight against illegal fishing may be possible in the future. Monitoring ocean conditions such as pollution levels, temperature, and pH could be done by robots driven by artificial intelligence (AI). An AI-powered predictive analytics system combined with drones, advanced sensor platforms, and other analogous technologies can keep track on earthquakes, floods, hurricanes, sea-level changes, and other potential natural disasters in real time. Automatic triggers and real-time data make it possible for emergency personnel, including those in the government, to act quickly and efficiently. There are a number of firms that are mixing artificial intelligence (AI) with traditional modeling approaches to simulate the impact of extreme weather events on infrastructure and other systems in order to advise on disaster risk management strategies, such as IBM and Palantir. Our research has identified a number of environmental applications for these technologies, including storm prediction, farming, livestock monitoring, and more. The authors feel that IoT devices, AI, and ML-based technologies can be game changers for environmental maintenance and enhancement due to their effectiveness in achieving diverse goals. Economic progress is inextricably linked to environmental progress, and we will soon be able to accomplish all of humanity's future economic and societal goals thanks to all of these technologies.

References

1. Analytics Vidhya. https://www.analyticsvidhya.com/blog/2020/11/artificial-intelligence-in-agriculture-using-modern-day-ai-to-solve-traditional-farming-problems/. Access time: 14:03, Access Date: January 07 2022.
2. Windspeed and Direction Sensor and Rain Gauge Temperature and Humidity Sensor. https://www.google.com/imgres?imgurl=https://www.mdpi.com/sensors/sensors-20-03113/article_deploy/html/images/sensors-20-03113-g001-550.jpg&imgrefurl=https://www.mdpi.com/1424-8220/20/11/3113/htm&h=340&w=550&tbnid=1LEH5LVGw_el5M&tbnh=176&tbnw=286&osm=1&hcb=1&source=lens-native&usg=AI4_-kS7d7iOB5mfQ1y6vg641R4K3WrvNQ&docid=6-FRYqS-nDqCBM. Access time: 21:45 Access Date: January 06 2022.
3. Fgdc.gov. https://www.fgdc.gov/metadata/csdgm/. Access time: 16:45 Access Date: January 05 2022.
4. Psiborg. https://psiborg.in/livestock-monitoring-using-iot/. Access time: 14:10, Access Date: January 08 2022.
5. Air quality monitoring solution architecture for smart city. https://www.google.com/imgres?imgurl=https://www.mdpi.com/sensors/sensors-20-03113/article_deploy/html/images/sensors-20-03113-g001-550.jpg&imgrefurl=https://www.mdpi.com/1424-8220/20/11/3113/htm&h=340&w=550&tbnid=1LEH5LVGw_el5M&tbnh=176&tbnw=286&osm=1&hcb=1&source=lens-native&usg=AI4_-kS7d7iOB5mfQ1y6vg641R4K3WrvNQ&docid=6-FRYqS-nDqCBM. Access time: 22:45 Access Date: January 06 2022.
6. Parmentola, A. et al. (2021). Is blockchain able to enhance environmental sustainability? A systematic review and research agenda from the perspective of sustainable development goals (Sdgs). *Business Strategy and the Environment*, *31*(1), 194–217.
7. Frontiers. https://doi.org/10.3389/frobt.2019.00032/full. Access time: 15:10, Access Date: January 08 2022.

8. Wark, T., et al. (2007). Transforming agriculture through pervasive wireless sensor networks. *IEEE Pervasive Computing, 6*(2), 50–57.
9. Hewlett packard enterprise. https://www.hpe.com/us/en/insights/articles/why-ai-is-an-increa singly-important-tool-in-weather-prediction-2007.html. Access time: 14:14, Access Date: January 06 2022.
10. Rew, R., Hartnett, E., & Caron, J. (2006). NetCDF-4: Software implementing an enhanced data model for the geosciences.
11. V7. https://www.v7labs.com/blog/computer-vision-applications#agriculture%20dicast%20i mage:%20https://ral.ucar.edu/solutions/products/dynamic-integrated-forecast-dicast-system. Access time: 15:26, Access Date: January 06 2022.
12. Parris, T. M., & Kates, R. W. (2003). Characterizing and measuring sustainable development. *Annual Review of Environment and Resources, 28*, 559–586.
13. Frew, J. E., & Dozier, J. (2012). Environmental informatics. *Annual Review of Environment and Resources, 37*(1), 449–472.
14. Logics Power. https://poweramr.in/weather-station?gclid=CjwKCAiA5t-OBhByEiwAhR-hm4IyC0yXG1tx59f1OrfJdOEGum64YInr_O5aRrQoweTqtXTJdfLLJxoCS7AQAvD_BwE. Access time: 13:44, Access Date: January 08 2022.
15. Portele, C. (Ed.). (2007). OpenGIS® Geography Markup Language (GML) Encoding Standard, Version 3.2.1. Wayland, MA, Open Geospatial Consortium, (p. 437) (OGC 07–036).
16. Kempler, S., Lynnes, C., Vollmer, B., Alcott, G., & Berrick, S. (2009). Evolution of information management at the GSFC earth sciences (GES) data and information services center (DISC): 20062007. *IEEE Transactions on Geoscience and Remote Sensing, 47*, 21–28.
17. Duerr, R. E., et al. (2009). Ensuring long-term access to remotely sensed data with layout maps. *IEEE Transactions on Geoscience and Remote Sensing, 47*(1), 123–129.
18. INHABITAT. https://inhabitat.com/odd-bot-the-weed-pulling-robot-that-could-eliminate-her bicides/. Access time: 16:07, Access Date: January 06 2022.
19. AI-Enabled water quality monitoring system. https://www.google.com/imgres?imgurl= https://www.mdpi.com/sensors/sensors-20-03113/article_deploy/html/images/sensors-20-03113-g001-550.jpg&imgrefurl=https://www.mdpi.com/1424-8220/20/11/3113/htm&h=340&w=550&tbnid=1LEH5LVGw_el5M&tbnh=176&tbnw=286&osm=1&hcb=1&sou rce=lens-native&usg=AI4_-kS7d7iOB5mfQ1y6vg641R4K3WrvNQ&docid=6-FRYqS-nDq CBM. Access time: 19:40 Access Date: January 06 2022.
20. AI-Enabled Air quality monitoring system. https://www.google.com/imgres?imgurl=https:// www.mdpi.com/sensors/sensors-20-03113/article_deploy/html/images/sensors-20-03113-g001-550.jpg&imgrefurl=https://www.mdpi.com/1424-8220/20/11/3113/htm&h=340&w=550&tbnid=1LEH5LVGw_el5M&tbnh=176&tbnw=286&osm=1&hcb=1&source=lens-nat ive&usg=AI4_-kS7d7iOB5mfQ1y6vg641R4K3WrvNQ&docid=6-FRYqS-nDqCBM. Access time: 19:45 Access Date: January 06 2022.
21. National Research Council. (2010). *Advancing the science of climate change* (p. 528). National Academics Press.
22. ML Techniques exploited for IoT issues. Machine learning for smart environments in B5G networks: connectivity and QoS https://pubmed.ncbi.nlm.nih.gov/34589123/. Access time: 09:40 Access Date: January 05 2022.
23. Date, C. J. (2000). An Introduction to Database Systems, (p. 938) 7th edn. Addison-Wesley.
24. National Research Council. (2010). *Monitoring climate change impacts: Metrics at the intersection of the human and earth systems* (p. 110). National Academics Press.
25. ML based identification and connectivity of IoT devices. Machine learning for smart environments in B5G networks: connectivity and QoS https://pubmed.ncbi.nlm.nih.gov/34589123/. Access time: 09:40 Access Date: January 05 2022.
26. ML for enhancing for IoT QoS. https://pubmed.ncbi.nlm.nih.gov/34589123/. Machine learning for smart environments in B5G networks: connectivity and QoS Access time: 09:40 Access Date: January 05 2022.
27. Herweijer, C., Combes, B., & Gillham, J. (2018). *How AI can enable a sustainable future.* Microsoft and PWC.

28. Michener, W., et al. (2011). DataONE: data observation network for earth—Preserving data and enabling innovation in the biological and environmental sciences. *D-Lib Magazine, 17*(1/2).
29. Use of ML in IoT. https://pubmed.ncbi.nlm.nih.gov/34589123/. Machine learning for smart environments in B5G networks: connectivity and QoS Access time: 09:40 Access Date: January 05 2022.
30. Gray, J. (2000). Rules of thumb in data engineering. Rep. MS-T R-99-100, Microsoft Res.
31. Turyshev, S. G., & Viktor, T. T. (2010). The pioneer anomaly. *Living Reviews in Relativity, 13*(1).
32. Becker, C., et al. (2009). Systematic planning for digital preservation: Evaluating potential strategies and building preservation plans. *International Journal on Digital Libraries, 10*(4), 133–157.
33. Kulmukhametov, A., et al. (2021). Improving data quality in large-scale repositories through conflict resolution. *International Journal on Digital Libraries, 22*(4), 365–383.
34. Keeling, C. D. (1960). The concentration and isotopic abundances of carbon dioxide in the atmosphere. *Tellus, 12*(2), 200–203.
35. Koyamparambath, A., Adibi, N., Adibi, S., & Sonnemann, G. (2021). Implementing artificial intelligence techniques to predict environmental impacts: case of construction products.
36. Keeling, C. D. (1998). Rewards and penalties of monitoring the earth. *Annual Review of Energy and the Environment, 23*(1), 25–82.
37. Keeling, R. F. (2008). Recording earth's vital signs. *Science, 319*(5871), 1771–1772.
38. Molina, M. J., & Rowland, F. S. (1974). Stratospheric sink for chlorofluoromethanes: chlorine atom-catalysed destruction of ozone. *Nature, 249*(5460), 810-812+.
39. Rowland, F. S. (1996). Stratospheric ozone depletion by chlorofluorocarbons (Nobel lecture). *Angewandte Chemie, 35*, 1786–1798.
40. Stolarski, R. S., & Cicerone, R. J. (1974). Stratospheric chlorine: A possible sink for ozone. *Canadian Journal of Chemistry, 52*, 1610–1615.
41. Farman, J. C., Gardiner, B. G., & Shanklin, J. D. (1985). Large losses of total ozone in Antarctica reveal seasonal ClOx/NOx interaction. *Nature, 315*, 207–210.
42. Solomon, S. (1999). Stratospheric ozone depletion: A review of concepts and history. *Reviews of Geophysics, 37*, 275–316.
43. Fire Map. https://www.springer.com/gp/authors-editors/book-authors-editors/your-public ation-journey/manuscript-preparation. Access time: 16:30, Access Date: January 04 2022.
44. National Research Council. (1996). *The ozone depletion phenomenon* (p. 8). National Academic Press.
45. Allflex Livestock Intelligence. https://www.allflex.global/in/product_cat/dairy-cow-monito ring/. Access time: 12:27, Access Date: January 07 2022.
46. Priority action areas for environmental challenge. https://earth.org/data_visualization/ai-can-it-help-achieve-environmental-sustainable/. Access time: 16:10 Access Date: January 05 2022.
47. Impact of AI on the achievement of each target from the SDG. https://earth.org/data_visu alization/ai-can-it-help-achieve-environmental-sustainable/. Access time: 15:45 Access Date: January 05 2022.
48. Gille, J. C. (2008). *How satellites saved the ozone layer*. Presented at American association advanced science, annual meeting, Boston.
49. P4 INFRASTRUCTURE. https://www.p4infrastructure.com/products/rain-mx/. Access time: 12:56, Access Date: January 07 2022.
50. Stommel, H. (1972). The Gulf stream: A physical and dynamical description, (p. 248) 2nd edn. University California Press.
51. Using AI for environment applications in four key sector. https://msit.powerbi.com/view? r=eyJrIjoiZDgxYjVkODQtNDg1MS00MWQ0LTkyYjktYTcwYzAyZmZmZjA5IiwidCI 6IjcyZjk4OGJmLTg2ZjEtNDFhZi05MWFiLTJkN2NkMDExZGI0NyIsImMiOjV9. Access time: 19:20 Access Date: January 03 2022.
52. Lee, T., & Cornillon, P. (1995). Temporal variation of meandering intensity and domain-wide lateral oscillations of the Gulf stream. *Journal of Geophysical Research, 100*, 13603–13613.

53. Joyce, T., Backus, R., Baker, K., Blackwelder, P., Brown, O., et al. (1984). Rapid evolution of a Gulf stream warm-core ring. *Nature, 308*, 837–840.
54. Yoder, J. A., Doney, S. C., Siegel, D. A., & Wilson, C. (2010). Study of marine ecosystems and biogeochemistry now and in the future: Examples of the unique contributions from space. *Oceanography, 23*, 104–117.

Chapter 6
AI in Waste Management: The Savage of Environment

Sharda Bharti⑩**, Shourat Fatma, and Vinay Kumar**⑩

Abstract Disposal of waste has become a major challenge throughout the world due to uncontrolled disposal of domestic as well as industrial wastes in open spaces. Exposure to a variety of wastes may eventually lead to the spread of various diseases and, thus, may pose serious health hazards to the public and adversely affect the environment. To minimize such issues, integrated waste management system could be a sustainable solution. It is well known that sorting of the waste at the source can be the first and the most important step to start with for efficient management of waste. This can simply be done by putting a number of labeled bins specified for each kind of wastes at the point of generation itself. However, this is the most tedious step among all the steps involved in effective waste management. To fasten the process and efficiently manage the waste collection, artificial intelligence (AI) may play a critical role in waste management which starts with the use of smart garbage bins. These bins are often combined with an app that helps the users know the availability of nearest location of the waste bins, thus preventing the bins from overflowing. AI can also play an incredible role in sorting of the wastes, as sorting is another major issue for most of the waste management facilities. AI-based sensors can discriminate items composed of different materials and distinguish the items of the same material whether an item has been chemically contaminated, ensuring purity of the waste stream. A number of waste management companies have been using such techniques and are taking the advantage of Internet of Things (IoT) sensors to monitor the fullness of trash receptacles throughout the city. The advantage of using such smart bins have effectively optimized the routes, timing and frequencies of waste collection, and reducing the load of municipalities. Such automated process would provide the best use of technology for effective waste management to prevent the human health risks as well as to protect the environment. This review article includes

S. Bharti (✉)
Department of Biotechnology, NIT Raipur, Raipur 492010, India
e-mail: sbharti.bt@nitrr.ac.in

S. Fatma
Department of Life Sciences, NIT Rourkela, Rourkela 769008, India

V. Kumar
Department of Computer Science and Engineering, NIT Jamshedpur, Jamshedpur 831014, India

© The Author(s), under exclusive license to Springer Nature Singapore Pte Ltd. 2022
P. K. Paul et al. (eds.), *Environmental Informatics*,
https://doi.org/10.1007/978-981-19-2083-7_6

details on various techniques based on machine learning and the use of artificial intelligence for efficient waste management than could significantly minimize the risks associated with human health and environment.

Keywords Artificial intelligence · Internet of Things · Sensors · Smart waste management

6.1 Introduction

Solid waste generation is one of emerging issues nowadays. Population growth, industrialization, rapid urbanization, and lack of financial resources have enormously increased the generation of solid waste worldwide. The expanding population across the globe are becoming a main reason for accumulation of solid wastes and should be answerable for a clean, healthy, and safe climate [1–3]. The globe generates around 2.01 billion tons of municipal waste annually, with about 33% of the waste produced being dumped into unmonitored landfills and unchecked waste dumps (The World Bank) [4]. The more the amount of waste produced, the more resources they will invest in finding solutions. Global waste is expected to increase by 3.4 billion tons by 2050 (World Bank Group, 2022) [5]. As everyone is getting vaccinated for the COVID-19 virus, there has been a spike in the clinical waste exposure to various wastes that may eventually lead to the spread of various diseases, such as, tuberculosis, pneumonia, and diarrhea. Moreover, it also adversely affects the environment leading to soil contamination, land pollution, thereby causing the loss of aquatic, and terrestrial lives. The lack in handling waste materials and keeping the lanes clean leads to breeding mosquitoes, which is the sole reason for diseases such as dengue and malaria [3, 6]. Hence, there is an urgent need for the implementation of proper solid waste management. Insufficient operation and inadequate planning are the reasons behind poor solid waste management. Everything must be appropriately managed starting from the initial steps, i.e., waste disposal, collection, and preventing overflow of bins to proper disposal of the waste following the waste hierarchy [7]. In evolved countries, several smart waste management strategies are being invented, implemented, and adopted and enormous benefits are achieved. However, the waste management appears to be a challenge for developed and developing countries.

The waste hierarchy indicates that prevention (reduction) of waste material, reusing them, recycling, recovery, and adequate disposal can decrease the amount of solid waste generated. The overall waste management system comprises various parameters and are connected by complex processes affected by multiple socioeconomic factors. Sorting the waste at the source can be the first and the most crucial step to start with for the efficient management of waste management. Exposure of open municipal solid waste causes several diseases and adversely affects the environment [6, 7]. Hence, the industrial waste, biomedical waste, the radioactive waste, and non-radioactive waste should be segregated properly and handled carefully as the radioactive wastes may emit radiation leading to lethal skin diseases and increase the risk of skin diseases, abnormalities in birth and child maturity, and cancer [10]. In addition, direct disposal and poor waste management practices may also contaminate the soil and water, thereby causing land/soil pollution and thus deteriorating the land and water quality.

Therefore, classifying the solid wastes into domestic, industrial, and biomedical/hazardous waste is the primary step toward waste management. Moreover, the solid waste can be reused and recycled efficiently. For instance, urban solid waste (USW) can be converted into a different form of energy via biochemical, thermochemical, and mechanical ways.

6.2 Waste Management

Each step of waste management is crucial; however, reuse and recycle of waste have provided an additional advantage of economic gain to solid waste management [11]. Among these, waste-to-energy technology is identified as an excellent opportunity for sustainable and economical solid waste management. In this approach, the waste is converted into energy primarily via biochemical, mechanical, and thermochemical ways (as demonstrated in the Fig. 6.1). Incineration, pyrolysis, and gasification for conversion of organic matter contain less biodegradable substances, converted via

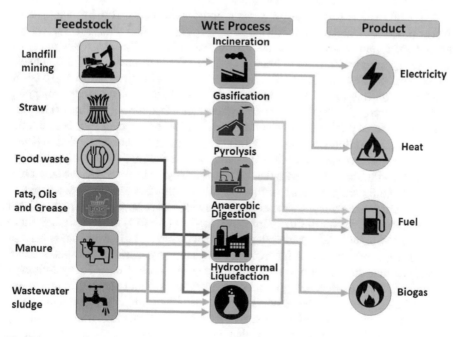

Fig. 6.1 Waste to energy conversion [12]

thermal conversion; the anaerobic digestion is done for waste with more extensive moisture content and biodegradable substance for methane gas production [2, 6, 9].

The waste-to-energy conversion system includes several key steps such as collection of waste input, bio-conversion approaches, conversion of waste to energy, and the energy carriers [13]. The waste input includes the carbonaceous wastes which are collected from sources such as municipal solid waste, agricultural residues, sewage sludge, forest residues, and wastes from food industries [2, 6]. Subsequently, the waste materials based on their characteristics are segregated into three groups for the conversion process. These processes involve the biochemical processes, thermochemical processes, and mechanical processes. The energy carriers are produced based on abovementioned process. Ultimately, a controlled disposal of waste in sanitary landfills and preventing overflow should be done.

With the advancement in research toward revolutionizing urban waste management, artificial intelligence (AI) and machine learning (ML) are being widely explored for a sustainable waste management [14–17]. Different methods based on the Internet of Things (IoT) and AI are being developed to increase waste management efficiency [17]. All the smart bins are connected digitally through the Internet to display the level of waste in the bins and their respective locations [18]. The IoT-integrated smart bins send the volume of the bins to the Internet over the servers. With efficient optimization techniques and associated algorithms, different methods are being proposed.

Recent machine learning techniques including neural networks is being explored in temporal models to predict the generation of solid waste. The Artificial Neural Network (ANN), Genetic Algorithm, fuzzy logic (FL), and other AI models can solve human traits such as problem-solving, reasoning, and understanding [12, 16, 17]. As reported by Zade and Noori [21], a feed-forward artificial neural network (ANN) was employed for prediction of waste generation pattern on weekly basis in Mashhad city, Iran [21]. Expert systems, such as, FL can solve complex mapping systems and provide results wherever systems like Genetic Algorithms (GA) use the Darwin theory of natural selection to select the set of data that best fits the procedure for handling certain conditions [22]. The sensors are connected to detect different types of waste materials. Computer vision annotation and intelligent algorithms allow the sensors to sense the different garbage to be placed in the smart bin. Artificial intelligence has been employed to solve several issues on large scale such as air pollution, soil erosion, wastewater management, and several environment-related problems. Adaptive Neuro-Fuzzy Inference System (ANFIS) models are helpful to forecast and optimize wastewater plant treatment processes [19]. Multi-layer perception (MLP) algorithm is used for weather forecasting, measuring the levels of atmospheric carbon dioxide and nitrous oxides [23]. Adaptive Neuro-Fuzzy Inference System (ANFIS) models help predict particulate matters and check different waste management processes. There are still several techniques reported in literature, and many are in their way of being prepared. ANFIS is widely used to remove turbidity in chemical industries and check methane gas production and other solids during anaerobic digestion for biochemical conversions of waste to energy [12, 20]. Thus, AI-based models offer an effective alternative approach with stand-alone and hybrid models to optimize urban waste management (UWM) models [11, 21]. This article has discussed waste segregation, waste hierarchy, and recent trends in artificial intelligence and machine learning for USW (Urban Solid Waste) management, recycling waste, and fewer case studies. To explore the prospective application of AI models in solid waste management, the emerging application of AI and machine learning techniques is crucial to employ for an efficient solid waste management. Various models based on artificial intelligence and machine learning algorithms have been explored to improve existing SWM schemes for each of the stages, starting from waste collection to final disposal. Hybrid AI-based models and various comparative studies employing AI/non-AI models have been included in this article for better understanding of the waste management.

6.3 Waste Management

6.3.1 Classification of Waste Management

If we go by the definition of waste, there would be several definitions for waste. According to the waste framework directive of the European Union, "Waste" means any substance or object which the holder discards or intends or is required to discard (European Union, 2011). Defining waste can be a case-to-case decision as well. Removal of waste improves the quality of life. Efforts are being made globally at their best to reorient the face of solid waste management (SWM) toward sustainability. There are different ways of managing the solid waste produced in the developed countries, such as USA, South Korea, and Japan, and in the developing countries like India and China. In most cases, a significant amount of waste generated is taken care by the respective municipal bodies, as per the Government norms. The facility for the recyclable materials (papers, drink cans, and plastics) is better in the developed economies, whereas compostable organic matter is minor in countries with lower GDP (INTOSAI, 2020). In developed countries, recycling occurs almost at every stage of product usage, whereas this system is lacking in developed countries, leading us to understand a more solid waste approach. The classification of solid waste can be done on different criteria such as source of waste generation, composition of wastes, hazardous properties of the waste, and who manages the waste. Broadly, the solid waste is categorized into Hazardous and Non-Hazardous waste (Depicted in Figure 6.2). Further classifications occur under these two divisions. Hazardous waste constitutes radioactive waste, e-waste, and biomedical waste. The radioactive waste emits radiation that can either degrade the DNA of the cells or mutate them. Exposure to such radiation may cause acute respiratory syndrome (ARS) or even cancer (CDC, 2021). Hence, the separate bins have been used to collect these wastes separately. Finally, they should be disposed of properly. Incineration, high temperature, and chemical treatments are a few ways to treat them.

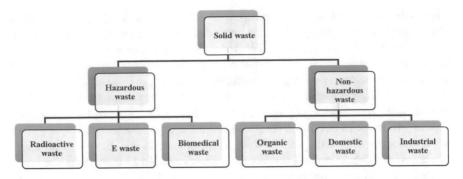

Fig. 6.2 Classification of waste

6.3.1.1   Types of Hazardous Waste

E-waste: These waste materials are produced from unused or broken electrical and electronic appliances after the end of their useful life. It contains toxic elements like radium, barium, mercury, lead, cadmium, arsenic, and certain carcinogens leaching into the environment and may cause cancer in humans and animals, and particularly, lead can trigger neurological damage [5, 8, 23]. When these electronic parts are mishandled during disposal, they mix with soil, water, and air and adversely affect the human, animal health, and environment.

Biomedical waste: It arises from medicals, hospitals, pharmaceuticals, bandages, and body fluids. They can be infectious and may contain toxic and radioactive microorganisms [27]. Exposure to these chemical compounds can interfere with the immune system and cause diseases [25, 26]. Around 16 billion injections are dispensed globally every year, however, most of these needles and syringes are not discarded properly after their use. Sometimes, these are burnt directly through open burning or incineration of such biomedical wastes which may result in the emission of dioxins, furans, and particulate matter in some cases (WHO, 2018).

Radioactive waste: These wastes are produced from nuclear activities including earth mining, nuclear research, fuel processing plants, and nuclear power generation. They need special treatment for handling and disposal processes. Storage of used fuel is usually done underwater for five years and then in dry storage. Deep geological is the widely used method for their disposal [30].

6.3.1.2   Non-Hazardous Waste

Non-hazardous wastes include waste that can be recycled and reused but may lead to profound environmental and health impacts when left untreated and uncontrolled. They are broadly classified into municipal waste and industrial waste. Municipal waste can be classified into organic, packaging, and industrial wastes. They are disposed of in different ways, like taking them to a disposable site, recycling the waste, and working with a disposable company [31].

6.3.2   The Waste Hierarchy

With increasing population and urbanization, the solid waste generation rate has been increasing tremendously. Currently, as several countries have chosen to follow social distancing and declare a lockdown as a protection measure from COVID-19, waste production has increased again. The pandemic effect is forcing retailers to use low-grade plastic materials for packaging. The recycling of waste products has slowed down due to COVID virus transmission and disturbance in the supply chain. The International Finance Corporation (IFC, 2021) report also notes an uptick in single-use plastic production, mainly prompted by the increased use of plastic-based personal protective equipment (PPE) and packaging materials (ISM Waste and Recycling, 2021). The waste hierarchy is a ranking system used for different environmental waste management options at the individual and organizational levels. Prevention is the most preferred, followed by reuse, recycling, recovery, and disposal. All these five priorities are often illustrated as the five-tier pyramid, as depicted in Figure 6.3 (ISM Waste and Recycling, 2021). Thus, the maximum benefits can be extracted from the products we use while minimizing the waste output produced when waste management hierarchy is being followed. These includes:

1.   Reduce: We can prevent extra packaging materials, reuse materials, less dispos-
     able, and less filled landfill sites. Avoiding waste is the essential and most
     preferred option in the waste industry. Wherever possible, reducing the usage

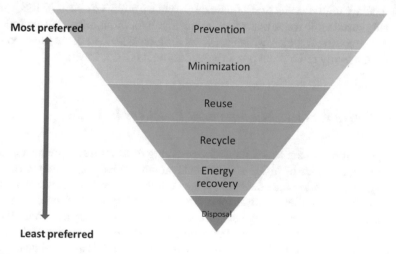

Fig. 6.3 Waste management hierarchy

of materials before it becomes waste is preferred. Cleaning, repairing, and refurbishing items can increase reusability.

2. Reuse: Reusing materials is the second-best waste management approach. It also allows us to avoid spending money on new goods and materials.

3. Recycling: Recycling turns into new items that can be used for different purposes, thus reducing the industrial raw materials used. Plastics, paper, cardboard, and metal products can be recycled.

4. Recover: Wherever we cannot use the 3R's (reduce, reuse, and recycle), we should recover the waste material in the form of waste to energy. Waste to energy helps reduce fossil fuel emissions and carbon footprints. Domestic wastes from the kitchen can be converted to compost and manures by composting.

5. Disposal: The least sustainable option in the waste hierarchy is disposal which is the most expensive method. This is the most unsustainable method done by incineration and filling landfills. For example, one ton of landfilled food waste can produce 450 kg of carbon emissions.

6.3.3 Conventional Waste Management Scenario

The waste management and its handling rules in India are governed by the Ministry of Environment and Forests (MOEF). Waste is a potential resource; the primary goal is to extract waste and effective waste management. In small towns, per capita MSW generation in India is approximately 0.17 kg per person per day whereas, in cities, it is about 0.62 kg per person per day (CJES, 2009). Solid waste management (SWM) is a major challenge for urban places in India due to the rising population, industrialization, and economic growth. Currently, the solid waste produced in India is approximately 42 million tons on annual basis, which fluctuates from 200 to 600 kg/capita/day, with a collection efficiency varying from 50 to 90% (CJES, 2009). Achieving sustainable waste management is difficult for India, with a high population density. The informal sector, which accounts for almost 90% of the waste produced, dumps it randomly rather than properly landfilling it (CJES, 2009).

6.3.4 Current Waste Management Practices in India

Municipal authorities are accountable for enforcing the laws issued by management and handling the rules of MOEF. The municipal authorities formulate the rules for executing the regulations and develop the methods and techniques for waste management including the collection, transportation, segregation, storage, processing, and final disposal. However, rag pickers are usually seen collecting the domestic and industrial waste manually and sell the collected waste to earn money and thus are dependent on waste for their social and economic benefits despite the health risks associated with it. Some of them collect from home, some work in recycling industries and waste management associations. Sometimes, it is the only source of livelihood for a significant population.

Biodegradable waste and inert waste are often dumped openly at several places. Municipality and other local bodies are involved in their management who put an expenses of around Rs. 500–1000 per tons on SWM, and out of which, 70% of the total amount is spent on collection and 20% on transport [2]. Nowadays, engineered landfills provide an alternative way of solid waste disposal that minimizes the exposure of the waste with the environment. Properly managed landfills in India will allow the safe removal and protect groundwater from contamination, avoid odors, fire hazards, and air emissions, and reduce the emission of greenhouse gases [2]. Properly managed landfills will slowly replace waste dumping areas in India.

6.3.5 Barrier and Challenges for Waste Management in India

India is facing challenges in waste management due to lack of awareness among people, lack of proper knowledge, and training for the workers which is required for an efficient waste management. Ever-increasing urbanization is driving additional force on landfill sites situated in urban areas. Waste management becomes difficult when the waste segregation is not accomplished and different kinds of wastes including the recyclables, biodegradable waste, and industrial and toxic wastes all are dumped together [32]. In general, people directly throw the household/domestic solid waste in a common bin which is not a good option. It is because domestic waste also contains several hazardous/toxic materials which should not be discarded directly with other non-toxic/non-hazardous household wastes. Dumping them together makes all the wastes hazardous in nature. This reduces the possibility of recycling of the wastes or conversion of wastes into other usable forms. Hence, household wastes must be segregated at the household level itself and should be collected separately into wet, dry and domestic/household hazardous waste categories. As per the Solid Waste Management Rules, 2016, domestic hazardous waste includes the discarded paint drums, pesticide cans, compact fluorescent lightbulbs, tube lights, expired medicines, broken mercury thermometers, used batteries, used needles and syringes and so on generated in houses. Hence, the most crucial barriers in rural part of India are recognized as household hazardous waste, inadequate assets for research on SWM, lack of local architecture, a shortage of staff capability, and a lack of a standard operating process for data collection and analysis. There is insufficient budget allocated for managing the urban waste produced. Limited qualified waste management professionals, lack of environmental awareness, and less motivation among people have hindered the adoption of new technologies to solve waste management in India. In such cases, AI, along with machine learning techniques, could give a unique shape to waste management in India. When coupled with proper management skills from the people and waste management association, an effective and sustainable waste management may be achieved in India.

6.4 Waste-to-Energy Technologies: Transformation Through Biochemical, Thermochemical, and Mechanical Pathways

At a time, it was being predicted that the developing countries like India and other such countries in the world may also increase the waste generation rate and may reach to value comparable to the MSW generation rate of developed countries [8, 9]. The rate of solid waste generation is projected to achieve 2.2 billion tons per year by 2025 and 4.2 billion tons per year by 2050 [33]. In other words, the solid waste produced is directly proportional to the country's Gross Domestic Product (GDP). Hence, this might have become a huge problem for developing countries like India. Subsequently, people around the globe realize the power of waste to energy (WTE) as the energy supply in the current situation is less as compared to the real energy expected for consumption. This paves the way for WTE from Urban Solid waste (USW). WTE is sustainable, ecofriendly, and economically attainable for developing countries like India. They can be applied to different kinds of waste: from solid (thickened sludge from treatment plants) to liquid (sewage discharge) and gaseous (refinery gases) waste. Various methods including thermochemical, biochemical, and mechanical conversion approaches can be employed toward energy generation from the waste (as summarized in Table 6.1), which are discussed in the following sub-section.

6.4.1 Thermochemical Conversion

Thermochemical conversion involves thermal organic matter treatment into heat energy, fuel, and gases. They are mainly used for dry waste with a high concentration of non-biodegradable waste. It involves three treatment processes that differ among

Table 6.1 Methods for waste-to-energy conversion

Methods for waste-to-energy conversion	Thermal conversion	Biochemical conversion	Mechanical conversion
Temperature	Incineration:850–1200 °C Pyrolysis: 400–800 °C Gasification: 800–1600 °C	150–450 °C	900–1200 °C
Type of waste	Dry waste	Organic waste	Organic and dry waste
Methodologies used	Incineration Pyrolysis Gasification	Decomposition Anaerobic sludge digestion	Aerobic degradation Fermentation Acetogenesis Methanogenesis Oxidation

the temperature and oxygen content used. Incineration is the complete oxidative combustion at 850–1200 °C of any kinds of solid combustible wastes (i.e., solid, liquid, or gaseous) predominantly to carbon dioxide, water vapor, other gases, and a relatively small, non-combustible residue known as ashes. The ashes are disposed further in the landfills in an ecofriendly manner. The incineration process includes two key processes: primary and secondary processes. Primary processes include a number of stages, which comprises drying, volatilization, combustion of fixed carbon, and burnout of char of the solids, whereas a secondary process includes the complete combustion of the products generated during the primary process, i.e., vapor, gases, and particulates driven off.

Pyrolysis is the degradation of organic matter without oxygen at 400–800 °C, which can be used for any kinds of solid waste and are easy to be adapted to any changes in their composition. Gasification is the partial oxidation at 800–1600 °C. Gasification can be described as the thermo chemical conversion of a solid or liquid carbon-based waste material (feedstock) into a combustible gaseous product (combustible gas) in the presence of suitable gasification agent. It converts solid wastes into combustible gases, integrated into other technology sources (Paul and Helmet, 2015).

In thermochemical conversion, all types of waste materials, i.e., of the organic matter, biodegradable as well as non-biodegradable, produces the energy output. However, the amount of energy recovered is dependent on the efficiency of the selected process in SW management schemes; In other words, the efficiency of energy recovery is dependent on the rate of conversion of heat energy contained in fuel into usable energy. The two key factors influencing process efficiency are as follows: (a) electrical efficiency of the power generation technology and (b) amount of heat recovery [28]. The choice of technology such as incineration vs. gasification is an important determining factor for determining the process efficacy, and the degree of productive utilization of generated heat and electricity.

6.4.2 Biochemical Conversion

This is the decomposition of the organic waste of USW by microbial decomposition, mainly used when the waste is filled with biodegradable organic materials and moisture content. Anaerobic digestion degrades organic biowaste without oxygen that produces biogas and stabilizes the sludge. The sludge can be used as manure in agricultural fields. As reported in literature, the anaerobic digestion is more efficient as it can generate 2–4 times of the methane per tons of solid waste in just 3 weeks as compared to that produced through landfill approach in 6–7 years [35].

6.4.3 Mechanical Conversion

Sanitary landfilling is the regulated disposal of the waste on landfills for decreasing environmental impacts through the leachate method and biogas recovery. The degradation of organic matter in landfills produces landfill gas (LFG) by five different methods: aerobic degradation, fermentation, acetogenesis, methanogenesis, and oxidation [1, 6].

Although waste-to-energy technologies for solid waste management are developing nowadays, the inconsistent composition of solid waste, the complexities of the designing of the system, and specific social, economic, and environmental issues may limit the applicability of the waste-to-energy technologies [12]. To improve the efficacy of such technologies, there is a critical requirements of the analysis of composition of the solid wastes, and accordingly, a suitable preprocessing step should be included along with minimal impacts on environment. Overall, the overall process should be done in such as a way that the technologies should provide a solution to the solid waste management in a cost-effective and environmentally sustainable manner. Energy technologies may arise afterward. All these factors should be considered for the development of this technology.

6.5 Opportunities of Digitalization in Waste Sorting

Increasing population is increasing the human need in day-to-day life. Hence, there is advancement in science and technology to improve the lifestyle and fulfill the human need in a more efficient manner and in less time. As we all know that the twenty-first century is becoming a digital world, so many processes are being improved due to digitalization. This is not only improving human life in their daily life, but also has created revolution in industrial sector including the environmental sector. As a result of digitization, there is significant revolutionary changes in the waste management sector [35]. This is because the digitalization will empower any economy in recovering the economic gain and useful substances through efficient conversion of the waste materials. In addition, this will provide additional advantages that it minimized the amount of waste to be handled and raw materials, thereby reducing the adverse impacts on the environment and climate. The consequences have been felt in all developed economies.

Waste management processes are a complex managerial task that involves significant involvement of manpower which increases the expenses tremendously, thereby putting an economical pressure [36]. This created an urgent need of alternative technologies that may reduce the requirement of manpower. Digitalization may play a crucial role in minimizing such economic stress due to high manpower requirements. Moreover, it also increases the job opportunities in high-value sectors of the supply chain. One of such opportunities is waste sorting, which is associated with increased possibility of reuse and recycling of the waste [32, 33].

Several major manufacturers of commodities, such as electronics, are already using artificial intelligence-based image processing techniques aided by robotic sorters [35]. Alternative possibilities include using watermarks, quick-response codes, or other digitally readable identifiers on product labels. The advantage of using such technique assist the automated sorters by sending the required data on composition of waste material and the product setup, allowing high-value materials to be recovered more easily [2, 25]. Robotic sorters could also generate data about the materials they have sorted, allowing them to improve artificial intelligence or optimize following procedures. For example, these data streams may be used to identify trends in incoming garbage loads and to acquire the information about efficacy of the waste sorting in order to forecast how sorting lines should be set up. When these information are connected to other data, such as market prices of secondary raw material, the procedures to be employed may be modified as and when required. As illustrated in Fig. 6.4, various digital technologies which includes robotics, the IoTs, artificial intelligence, cloud computing, and data analytics can be employed to predict the influence on the efficiency of the industries in waste management in future (European Environmental Agency, 2021).

Recently, the digital technology has been widely explored in various stages of waste management including garbage collection [36]. An advancement in digital technologies has improved various stages of garbage collection, predominantly the logistics which involves the process of organizing, creating scheduled collection,

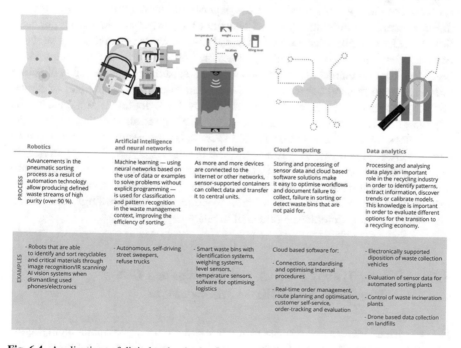

Robotics	Artificial intelligence and neural networks	Internet of things	Cloud computing	Data analytics
PROCESS Advancements in the pneumatic sorting process as a result of automation technology allow producing defined waste streams of high purity (over 90 %).	Machine learning — using neural networks based on the use of data or examples to solve problems without explicit programming — is used for classification and pattern recognition in the waste management context, improving the efficiency of sorting.	As more and more devices are connected to the internet or other networks, sensor-supported containers can collect data and transfer it to central units.	Storing and processing of sensor data and cloud based software solutions make it easy to optimise workflows collect, failure in sorting or detect waste bins that are not paid for.	Processing and analysing data plays an important role in the recycling industry in order to identify patterns, extract information, discover trends or calibrate models. This knowledge is important in order to evaluate different options for the transition to a recycling economy.
EXAMPLES - Robots that are able to identify and sort recyclables and critical materials through image recognition/IR scanning/ AI vision systems when dismantling used phones/electronics	- Autonomous, self-driving street sweepers, refuse trucks	- Smart waste bins with identification systems, weighing systems, level sensors, temperature sensors, sofware for optimising logistics	Cloud based software for: - Connection, standardising and optimising internal procedures - Real-time order management, route planning and optimisation, customer self-service, order-tracking and evaluation	- Electronically supported diposition of waste collection vehicles - Evaluation of sensor data for automated sorting plants - Control of waste incineration plants - Drone based data collection on landfills

Fig. 6.4 Applications of digital technologies for waste sorting and management [38]

and demonstrating the tasks, persons, and vehicles for garbage collection. Using the digital technologies has enhanced the overall efficiency by storing the data, processing, analyzing, and optimizing the required information. The data can be monitored in real time for the garbage collection process, including the progress of the task or any incidents. The overall process starts becoming complicated in nature as more and more data is collected over time. The use of optimization algorithms may be helpful in defining and selecting the most suitable options for allocating various resources, such as manpower or vehicles in such cases. Hence, application of telematics plays a crucial role, which involves vehicle routing systems, navigation and use of vehicle tracking software, enterprise resource planning systems, and other associated digital technologies. The outcome of employing such technologies can be seen in terms of enhanced efficiency of the overall process.

The application of IoTs in improving the efficacy of waste management is another suitable example, which incorporates various applications, such as use of smart bins for waste collection at the site of waste generation and use of robotics for semi-autonomous trash collection vehicles [15, 35]. However, there is still a lot of scope of improvement to further enhance the efficiency of the garbage collection and linking them in the future, as per the demands of a circular economy. Hence, it must be flexible in adapting the new and emerging technologies rapidly to everchanging pattern of waste generation and waste management purposes, including the installation of required system and services to make the customized services better. The conventional garbage collection process involves paperwork, communication, and billing processes as their part. A transition from paper-based management systems toward digital systems on continuous basis will enhance the process efficiency and information flow. Use of digital identity tags for trash bins and waste containers, digital mode of order processing, invoicing, and payment can improve the efficacy. The digital user interfaces for communication with the customer and linking the garbage collection corporations to the appropriate governmental databases are also the part of digital technologies that can improve the overall efficacy. Such digital technologies can be exploited toward collection of the data related to waste generation/collection from the public working in documentation-related sectors. Subsequently, the collected data can be converted into valuable information by the data analytics. This might help in endorsing a circular economy by offering an "improved knowledge of the geographical and temporal patterns of trash creation" [39]. Moreover, the local governments may be provided with further information by collecting multiple single-data points than only providing cumulative values of the data.

6.5.1 Recent Trends of Artificial Intelligence Usage in Municipal Solid Waste Management

Due to various interrelated processes at work and the highly variable demographic and socioeconomic aspects affecting the complete waste management systems, the waste management processes have complex processes and non-linear features. Moreover, it is a challenging task to achieve good performance in solid waste management systems without threatening other health and environmental issues. Therefore, artificial intelligence approaches are being explored to determine their suitability in the solid waste management system sector [11–13, 21]. AI is concerned with the use of computer systems and programs to imitate human characteristics to solve the problem, gain knowledge, perceive, understand, find reasons, and awareness of their environment. Therefore, application of various AI models, such as the artificial neural network, expert system, genetic algorithm, and fuzzy logic, can solve the complex issues, create complicated maps, and anticipate consequences [19].

Recent trends suggest that there are six main AI application sectors in municipal solid waste management. Detection of levels in the waste bin, prediction of the waste characteristics, forecasting the process parameters, process output, vehicle routing, and approaches used for solid waste management are the key sectors where AI can be applicable. Detection of levels in the waste bin is linked with the monitoring the filling of waste bins, whereas prediction of the waste characteristic involves categorization of the waste, waste compression ratio, waste creation, patterns, or trends. The heating value and co-melting temperature of the waste comes under the projected process parameters. Similarly, simulation and optimization of biogas generation in the landfill and leachate creation over time comes under the process output forecast. The garbage collection route and frequency optimization problem is part of the vehicle routing problem. Finally, waste management planning includes the placement of waste facilities, the build-up of garbage, and unlawful dumping locations, as well as the financial and environmental implications of collection, transportation, treatment, and disposal. Evidently, there has been a recent surge in enthusiasm for AI research in solid waste management [23].

6.5.2 Machine Learning for Forecasting the Generation of Municipal Solid Wastes

Most of the applications of artificial intelligence toward solid waste management have focused on forecasting the characteristics of municipal solid waste. Predicting municipal solid waste generation is an application that has received the greatest attention in these investigations. In such applications, artificial neural networks are commonly used, followed by support vector machines. Spectral analysis, correlation analysis, response surface modeling, generalized linear modeling, gene expression programming, partial least squares, hybridized wavelet de-noising, Gaussian mixture models, hidden Markov models, Viterbi algorithms, and principal component analysis are all used in conjunction with the AI models [36–38]. For waste generation, several short-term and long-term forecasting periods are used. The rarity of research mimicking everyday waste generation is almost certainly a result of the unavailability of such data.

Various research studies have examined a broad variety of input variables impacting waste creation. They have also examined the categorization of waste materials for automated sorting systems that reduce the need for manual waste segregation. The bulk of these investigations employed artificial neural networks to classify various waste components. One such study employed hyperspectral imaging and multi-layer artificial neural networks to identify different varieties of plastics in e-waste [41]. The suggested technique identified these materials with an accuracy of 99%. Another group of researchers used deep convoluted neural networks to attempt to automate the waste sorting procedure [16]. In comparison with human sorting, the automated procedure significantly reduced the time required for garbage sorting and categorization. Similar to the previous example, deep convoluted neural networks were utilized to differentiate multiple kinds of paper and cardboard [36]. The model's mean accuracy varied between 61.9 and 77.5%; these low results were linked to the training database's small size comprising of only 24 images. Chu et al. employed convoluted neural networks to extract features and MLP to classify garbage into recyclable and non-recyclable components [44]. The hybrid technique achieved a maximum accuracy of 98.2%, which was almost 10% greater than the accuracy achieved using simply convoluted neural networks. Additionally, a few researchers evaluated the usefulness of other machine learning algorithms for garbage categorization [25, 45]. Singh et al. demonstrated that RF, Nu-, and C-LibSVM were all capable of classifying with an accuracy of better than 90%. On the other side, Naïve Bayes and closest neighbor algorithms scored badly, with accuracy rates of 44.8 and 84.8%, respectively. Only a few studies have been conducted to determine the influence of various characteristics on waste creation. Márquez et al. used data mining algorithms like cluster analysis and decision tree classifier to associate sociodemographic and behavioral characteristics with garbage creation [46]. The tree classifier performed admirably, with an error rate of as low as 3.6%. Another research employed data mining techniques to ascertain garbage generation patterns by home type and seasonal fluctuations [47]. Furthermore, decision tree has been used in conjunction with Quinlan's M5 method to anticipate the MSW compression ratio, a valuable tool for assessing settlement of the waste during municipal landfill design [48]. The model has been trained and validated using a variety of solid waste elements and properties, including dry density, moisture content, and biodegradable proportion. The model performed satisfactorily throughout the testing phase, with a correlation value of 0.92.

6.5.3 Smart Waste Management Using Artificial Intelligence

Smart waste management is the practice of proposing solutions to the current waste management problem using Internet, Smart Sensors, and Mobile Applications [10, 35, 38, 44]. Waste management, in general, can be broken down into several problems, such as waste segregation where we have to separate solid waste from the wet one, a task which when left to every individual sees minimal success and when given to an organization seems very difficult to overcome due to the overwhelming amount of waste that is created on a daily basis which will only increase with the increase in population. Some other problems are the collection of waste, when to collect them, the amount of manpower required, and the next obvious problem after collection is disposing of waste. Smart waste management tries to solve some of these problems using sensors, Internet, and continuous monitoring but there is only so much that can be done by this because there comes a point where human intelligence needs to intervene like sending a truck when a Smart Bin notifies that is has been filled or deciding the route for the truck, the manpower required, etc.

Artificial intelligence can play a big role in Smart Waste Management [13, 45]. Deep learning models for Image Recognition and Object Identification can be used to help segregate the waste inside a Smart Bin, or Predictive models that can predict changes in environment to see how carbon dioxide emissions will change and schedule a pickup [46, 47]. Fuzzy Logic Algorithm that can markup the route for the pickup truck reducing manpower and fuel consumption can also decide the destination of the dumping ground using sensors information provided by the ones fixed there and calculating which would be the most optimum for waste dumping at that given point of time [19, 20].

6.5.3.1 Smart Bins

The most logical problem to conquer would be waste segregation. And the point where we can tackle is right at the start of the waste management cycle that would be the bin where people dump their garbage. Many researches have been done to create a smart bin let us talk about one that particularly picked my interest, where the researchers have developed smart bin using Internet-based smart system [52], where they have created a Smart Bin (Figs. 6.5 and 6.6). In the smart bin, the camera module is associated with Raspberry Pi to catch the waste picture with the end goal of item location and recognizable proof. After the waste is distinguished, servo engines constrained by the Raspberry Pi will activate the opening and shutting of the top of the waste compartment. The kickoff of the cover permits waste to tumble from the waste location compartment into its particular waste compartment.

A radio-frequency identification (RFID) module is associated with the Raspberry Pi to distinguish approved staff having access cards. When approved faculty are distinguished, RFID module will activate Arduino Uno to open the electronic compartment. Correspondence of the RFID module comprises two sections, a RFID module that has a receiving wire liable for sending and getting a transmission through radio waves, and a detached RFID label that contains a receiving wire and incorporated circuit that stores the ID code and other data. Since the motivation behind the RFID module is to just permit approved staff to get to the receptacle using RFID labels, a rundown of distinguishing proof codes that accompany the RFID labels are encoded into the framework so the framework will possibly react when it experiences enlisted RFID labels. The framework reacts by opening the electronic compartment. Since the RFID module depends on a backscattered framework, the power sent

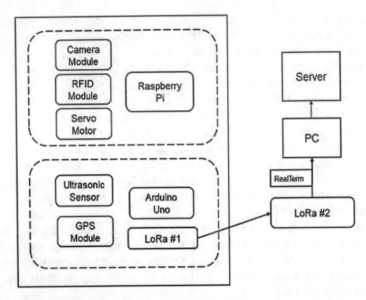

Fig. 6.5 Block diagram representing the overall system [52]

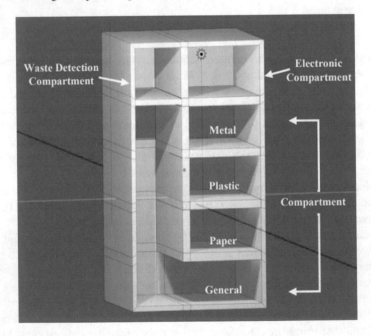

Fig. 6.6 3D model of smart bin [52]

between the RFID module and tag may fluctuate with its position, which eventually influences the exhibition of the RFID module. To settle this issue, we have arranged the RFID so it is effectively reachable and has no hindrances over the outer layer of the RFID module. The last option guarantees great transmission of force between both the RFID module and tag [15, 35, 47, 48].

The ultrasonic sensor is associated with Arduino Uno to screen the filling level of every one of the receptacle's waste compartments. It involves a plastic, metal, paper, and general waste compartment. The ultrasonic sensor utilizes sonar to quantify the time taken for the sign to go from the transmitter end to the recipient end, and the time contrast is utilized to compute the filling level of waste inside the container. A GPS module gives data on the area (scope, longitude) just as the constant of the canister from the satellite. The filling level, area, and constant container are gathered and moved through a LoRa module from the canister to the Waspmote passage, which is associated with the PC [15, 47].

Waste ID is performed utilizing the TensorFlow article discovery API running on the Raspberry Pi. This object identification API runs on a pre-prepared item location model, SSD MobileNetV2, which is lightweight and appropriate to run on low registering power gadgets like Raspberry Pi [54]. The engineering of MobileNetV2 depends on straight bottlenecks profundity distinguishable convolution with upset residuals, and it is an improvement over the past variant, MobileNetV1 [55]. Profundity distinct convolution requires less calculation by parting convolution into two separate layers, depth wise convolution and point wise convolution (Fig. 6.7).

Here, the ArticleNote button was provided to capture any note to the content that will be placed in the first page according to the author preference.

6.5.4 Vehicle Routing

The most logical problem to conquer would be waste segregation. And the point where we can tackle is right at the start of the waste management cycle that would be the bin where people dump their garbage. Vehicle Routing is another area that can be focused on using AI models and algorithms let us have a look at Assessment of waste characteristics and their impact on GIS vehicle collection route optimization using ANN waste forecasts [56].

Information of week-by-week gathered trash, fortnightly gathered single stream recyclables, waste structure, and number of families were gathered from Austin's Open Data Portal. Thickness of waste was determined in light of waste organization and explicit load of every material from a USEPA study (2016b). The following stage is an ANN time-series examination to figure out future recyclables and trash age paces of each sub-assortment region in the year 2023 (Fig. 6.8).

Various situations are considered with various recyclables and trash creations. The anticipated recyclables and trash age rates were utilized to process anticipated volume of waste in trucks. The anticipated waste volumes from the objective regions are inputs into the GIS–Network Analysis apparatus (ArcGIS—adaptation 10.5.1) to

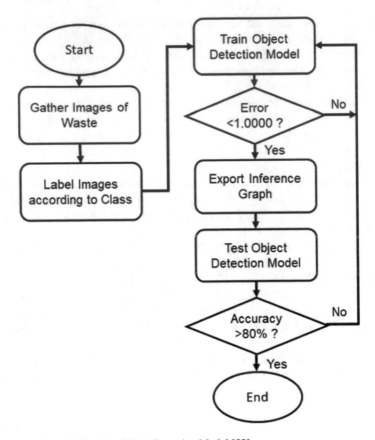

Fig. 6.7 Flowchart of obtaining Object Detection Model [52]

foster ideal truck courses. It is expected that during the 5-year gauging period there will be no significant changes in the road and street network setup and the number and area of assortment focuses (family) are comparative. A sum of 36 situations are created to analyze the impacts of changing waste attributes on ideal truck courses with negligible travel distance [20, 51].

ANN time series and GIS–Network Analysis–VRP models were joined to look at what waste creation and the mass of waste means for truck courses just as air outflows from the trucks. The ANN model showed better outcomes when waste info information had less outrageous qualities for both recyclables and trash. The coordination of ANN expectation model with GIS streamlining detailed in this study uncovers the interrelationships between waste organization and GIS enhanced courses and permits WMS chiefs to better reaction to the progressions in waste synthesis [20, 51].

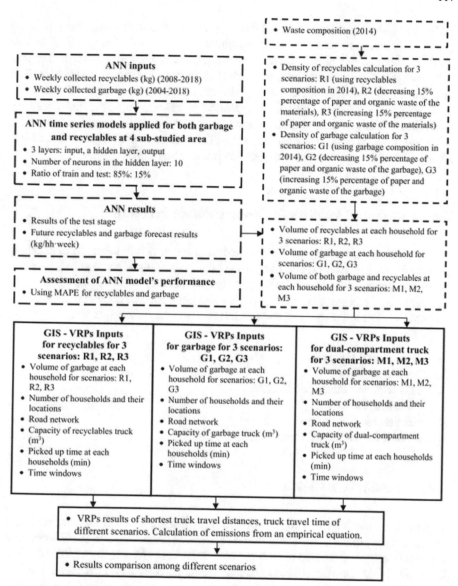

Fig. 6.8 Methodology flowchart for vehicle routing [56]

6.5.5   By-Product Utilization

Being able to predict useful by-products of garbage and the harmful ones and the quantity in which they are used is big part of waste management and AI can help us with this too. Combining fuzzy logic (FL) and artificial neural networks (ANN) in modeling landfill gas production is a great paper that has worked on this exact field

[23]. The work utilizes a cross breed ANN-FL model that utilizes information-based FL to depict the interaction subjectively and carries out the learning calculation of ANN to enhance model boundaries. The model was created to recreate and anticipate the landfill gas creation at a given time in view of functional boundaries. The exploratory information utilized were ordered from laboratory-scale try that elaborate different working situations. The created model was approved and genuinely examined utilizing F-test, straight relapse among real and anticipated information, and mean squared error measures. Generally, the reenacted landfill gas creation rates showed sensible concurrence with genuine information [23].

6.6 Influential Factors for Smart Waste Prediction

Waste prediction is a very important field of study which helps us to predict and prepare the required steps for management of upcoming bulk of garbage being created at an unprecedented rate but to predict this we need to pinpoint the factors that contribute to water generation. There are some common factors that determine the amount of waste generated and disposed. Below are the factors that influence waste generation:

1. Institutional Factors
2. Social conditions
3. Financial and Economic Factors
4. Technical Factors
5. Geographic Conditions
6. Environmental Conditions.

6.6.1 Institutional Factors

There are certain rules and regulations for proper solid waste management. These laws and policy come under institutional factors that give consent to the Government for effective implementation of an Integrated Solid Waste Management. The possible steps to be taken for effective waste management involve launching a national and/or local policy and permit laws on SWM standards and practices, identifying the roles and responsibilities for each level of government and ensuring that the authority and resources for the implementation of an ISWM plan are provided with the local governments.

6.6.2   Social Conditions

Social conditions dictate how people manage their wastes, when their culture for example has festivals how much waste is created during them and what kind of waste is being created. Social conditions also tell us to what extents rule governing waste management are followed (people littering on road, people segregating their waste).

6.6.3   Financial and Economic Factors

The next factor that influences the waste predictions is the funds that need to be used to dispose the waste. More fund equals more waste disposal equals less waste generation. Economic factors affecting solid waste management system should be differentiated from the financial factors. It is because the economic factors include the financial turnout of the integrated solid waste management (IWSM) plans, for instance, the creation of jobs creation, improvement of public trade and tourism, political gain, and so on. The local government must determine the requirements of the initial capital investment and operating, and maintenance costs associated with each activity conducted for waste management in long term. Furthermore, the ability of people and their willingness to pay for the services and to determine the efficacy of job creation for the activities based on handling waste are the additional factors that needs to be taken into consideration.

6.6.4   Technical Factors

The technical factors include finding the requirements of equipment and the required facilities for an effective execution of the ISWM plan and determining the locations where these equipment and facilities will be kept; however, it will depend on geological factors, distances used for transportation, and forecast of waste generation.

6.6.5   Environmental Factors

Each stage of the ISWM plan significantly affects the natural resources, human health, and the environment. One must consider the environmental cost of these activities, for instance, landfilling or combustion of waste materials and attempt should be made to reduce their impact on public health and environment. Therefore, there should be an established practices to validate the groundwater and drinking water protection and the local authority should examine the compliance with the national standards assuring the minimum impacts on the human health.

6.6.6 Geographic Conditions

The area of the land, the population, and the location play a great role in predicting how much waste will be generated. The climate of the land will also determine how the waste needs to be disposed of and thus in turn affecting the financial resources required.

6.7 Conclusion

Generation of waste is increasing day by day with increasing population and urbanization. These wastes can be categorized into different categories, hazardous and non-hazardous waste. Hazardous wastes are particularly toxic and may pose severe adverse impacts on human, animal, and environmental health. Waste such as e-waste, Plastic waste, and Metal waste can cause a significant risk to the ecosystem if they are not managed properly. The most logical problem to conquer such issues would be waste segregation where we can tackle right at the start of the waste management cycle that would be the bin where people dump their garbage. Artificial intelligence and machine learning can be employed for smart waste management. Deep learning models for Image Recognition and Object Identification can be used to help segregate the waste inside a Smart Bin, or Predictive models that can predict changes in environment to see how carbon dioxide emissions will change and schedule a pickup. Fuzzy Logic Algorithm that can markup the route for the pickup truck reducing manpower and fuel consumption can also decide the destination of the dumping ground using sensors information provided by the ones fixed there and calculating which would be the most optimum for waste dumping at that given point of time. Such automated segregation and monitoring system implementation in the bin significantly decrease the operating cost and improve the overall waste management system. Furthermore, an automated routing system can be created to find and determine the shortest path to the bin for the purpose of maintenance. Thus, the convention solid waste management system can be enhanced and risk to the society due to waste exposure, thereby providing a greener and healthier life.

References

1. Kenny, C., & Priyadarshini, A. (2021). Review of current healthcare waste management methods and their effect on global health. *Healthcare, 9*(3).

2. Kumar, S., et al. (2017) Challenges and opportunities associated with waste management in India. *Royal Society Open Science, 4*(3).
3. Ferronato, N., & Torretta, V. (2019). Waste mismanagement in developing countries: A review of global issues. *International Journal of Environmental Research and Public Health,* 16(6)
4. Kaza, S., Yao, L., Bhada-Tata, P., & Van Woerden, F. (2018). *What a waste 2.0.* International Bank for Reconstruction and Development/The World Bank.
5. The World Bank. (2019). Solid waste management.
6. Shekdar, A. V. (2009). Sustainable solid waste management: An integrated approach for Asian countries. *Waste Management, 29*(4), 1438–1448.
7. EPA. (2021). Sustainable materials management: Non-hazardous materials and waste management hierarchy.
8. Abdel-Shafy, H. I., & Mansour, M. S. M. (2018). Solid waste issue: Sources, composition, disposal, recycling, and valorization. *Egyptian Journal of Petroleum, 27*(4), 1275–1290.
9. Vinti, G., et al. (2021). Municipal solid waste management and adverse health outcomes: A systematic review. *International Journal of Environmental Research and Public Health, 18*(8), 1–26.
10. Dutta, S., Upadhyay, V. P., & Sridharan, U. (2006). Environmental management of industrial hazardous wastes in India. *Journal of Environmental Science and Engineering, 48*(2), 143–150.
11. Kinnaman, T. C. (2009). The economics of municipal solid waste management. *Waste Management, 29*(10), 2615–2617.
12. Foster, W., et al. (2021). Waste-to-energy conversion technologies in the UK: Processes and barriers—A review. *Renewable and Sustainable Energy Reviews, 135*(X), 1–3.
13. Jude, A. B., et al. (2021). An artificial intelligence based predictive approach for smart waste management. *Wireless Personal Communication,* 0123456789.
14. Bijos, J. C. B. F., Queiroz, L. M., Zanta, V. M., & Oliveira-Esquerre, K. P. (2021). Towards artificial intelligence in Urban waste management: An early prospect for Latin America. *IOP Conference Series Materials Science Engineering, 1196*(1), 012030.
15. Abdallah, M., Abu Talib, M., Feroz, S., Nasir, Q., Abdalla, H., & Mahfood, B. (2020). Artificial intelligence applications in solid waste management: A systematic research review. *Waste Management, 109*, 231–246.
16. Sudha, S., Vidhyalakshmi, M., Pavithra, K., Sangeetha, K., & Swaathi, V. (2016). An automatic classification method for environment: Friendly waste segregation using deep learning. *Processing—2016 IEEE international conference technology innovation ICT agriculture rural development,* (pp 65–70). *TIAR 2016,* no. Tiar.
17. Wilts, H., Garcia, B. R., Garlito, R. G., Gómez, L. S., & Prieto, E. G. (2021). Artificial intelligence in the sorting of municipalwaste as an enabler of the circular economy. *Resources, 10*(4), 1–9.
18. White, G., Cabrera, C., Palade, A., Li, F., & Clarke, S. (2020). WasteNet: Waste classification at the edge for smart bins.
19. Yetilmezsoy, K., Ozkaya, B., & Cakmakci, M. (2011). Artificial intelligence-based prediction models. *Neural Network World, 3*(11), 193–218.
20. Ahmed, A. A. A., & Asadullah, A. (2020). Artificial intelligence and machine learning in waste management and recycling. *Engineering International, 8*(1), 43–52.
21. Jalili Ghazi Zade, M., & Noori, R. (2008). Prediction of municipal solid waste generation by use of artificial neural network: A case study of Mashhad. *International Journal Environment Research, 2*(1), 13–22.
22. Kalogirou, S. A. (2003). Artificial intelligence for the modeling and control of combustion processes: A review. *Progress in Energy and Combustion Science, 29*(6), 515–566.
23. Abdallah, M., Warith, M., Narbaitz, R., Petriu, E., & Kennedy, K. (2011). Combining fuzzy logic and neural networks in modeling landfill gas production. *World Academic Science Engineering Technology, 78*(6), 559–565.
24. Vitorino de Souza Melaré, A., Montenegro González, S., Faceli, K., & Casadei, V. (2017). Technologies and decision support systems to aid solid-waste management: a systematic review. *Waste Management, 59*, 567–584.

25. Singh, S., et al. (2017). Identifying uncollected garbage in urban areas using crowdsourcing and machine learning. In *IEEE international symposium on technologies for smart cities*, (pp 3–7).
26. Fazzo, L., et al. (2017). Hazardous waste and health impact: A systematic review of the scientific literature. *Environment Health A Global Access Science Source, 16*(1), 1–2.
27. WHO. (2018). Health-care waste. *WHO*, no. February, pp 1–6.
28. Begum, S., Rasul, M. G., & Akbar, D. (2012). An investigation on thermo chemical conversions of solidwaste for energy recovery. *World Academic Science Engineering Technology International Journal Environment Ecology Engineering, 62*(2), 624–630.
29. Manzoor, J., & Sharma, M. (2019). Impact of biomedical waste on environment and human health. *Environment Claims Journal, 31*(4), 311–334.
30. International Atomic Energy Agency. (2011). Radioactive waste management objectives. *IAEA Nuclear Energy Series, 18*(5), 32.
31. Bulucea, C. A., Mastorakis, N. E., Bulucea, C. A., Boteanu, N., & Stinga, A. (2010). Systemic approach of hazardous and non-hazardous waste management. *4th WSEAS international conference energy planning, energy saving, environment education EPESE'10, 4th WSEAS international conference renewable energy sources, RES '10*, no. November, (pp. 181–189).
32. Asnani, P. U. (2008). India infrastructure report: Ch. 8 solid waste management. *India Infrastructure Report, 3*(2005), 160–189.
33. Hoorweg, D., & Bhada-Tata, P. (2012). What a waste: A global review of solid waste management. *World Bank*, 29–43.
34. Rahman, M. W., Islam, R., Hasan, A., Bithi, N. I., Hasan, M. M., & Rahman, M. M. (2020). Intelligent waste management system using deep learning with IoT. *Journal of King Saud University—Computer and Information Sciences*.
35. Wilts, H., Garcia, B. R., Garlito, R. G., Gómez, L. S., & Prieto, E. G. (2021). Artificial intelligence in the sorting of municipal waste as an enabler of the circular economy. *Resources, 10*(4).
36. Vrancken, C., Longhurst, P., & Wagland, S. (2019). Deep learning in material recovery: Development of method to create training database. *Expert Systems with Applications, 125*, 268–280.
37. Singh, S. et al. (2017). Identifying uncollected garbage in urban areas using crowdsourcing and machine learning. *TENSYMP 2017—IEEE international symposium technology smart cities*, (pp. 3–7).
38. European Environmental Agency. (2021). Digital technologies will deliver more efficient waste management in Europe. *Briefing No 26/2020*, (pp. 1–6).
39. Law, H. J., & Ross, D. E. (2019). International solid waste association's 'closing dumpsites' initiative: Status of progress. *Waste Management and Research, 37*(6), 565–568.
40. Countries, G. C., Mt, M., & Mt, M. (2021). Smart waste management using AI and ML. *Telecom Review*, (4), 1–4.
41. Tehrani, A. & Karbasi, H. (2018). A novel integration of hyper-spectral imaging and neural networks to process waste electrical and electronic plastics. *2017 IEEE conference technology sustainability sustech 2017*, (vol. 2018-Janua, pp. 1–5).
42. Gupta, P. K., Shree, V., Hiremath, L., & Rajendran, S. (2019). The use of modern technology in smart waste management and recycling: Artificial intelligence and machine learning. *Studies Computer Intelligence, 823*(4), 173–188.
43. Albadr, M. A., Tiun, S., Ayob, M., & Al-Dhief, F. (2020). Genetic algorithm based on natural selection theory for optimization problems. *Symmetry (Basel), 12*(11), 1–31.
44. Chu, Y., Huang, C., Xie, X., Tan, B., Kamal, S., & Xiong, X. (2018). Multilayer hybrid deep-learning method for waste classification and recycling. *Computational Intelligence and Neuroscience*, 2018.
45. Kuritcyn, P., Anding, K., Linß, E., & Latyev, S. M. (2015). Increasing the safety in recycling of construction and demolition waste by using supervised machine learning. *Journal of Physics: Conference Series, 588*(1).

46. Márquez, M. Y., Ojeda, S., & Hidalgo, H. (2008). Identification of behavior patterns in household solid waste generation in Mexicali's city: Study case. *Resources, Conservation and Recycling, 52*(11), 1299–1306.
47. Korhonen, P., & Kaila, J. (2015). Waste container weighing data processing to create reliable information of household waste generation. *Waste Management, 39*, 15–25.
48. Heshmati, R. A. A., Mokhtari, M., & Shakiba Rad, S. (2014). Prediction of the compression ratio for municipal solid waste using decision tree. *Waste Management and Research, 32*(1), 64–69.
49. Kodali, R. K., & Gorantla, V. S. K. (2018). Smart solid waste management. *Processing 2017 3rd international conference applications theornational computer communication technology iCATccT 2017*, pp. 200–204.
50. Jenny, H., Wang, Y., Alonso, E. G., & Minguez, R. (2020). Using artificial intelligence for smart water management systems. 5(143).
51. Gupta, T. et al. (2021). A deep learning approach based hardware solution to categorise garbage in environment. *Complex and Intelligent Systems*.
52. Sheng, T. J., et al. (2020). An internet of things based smart waste management system using LoRa and tensorflow deep learning model. *IEEE Access, 8*, 148793–148811.
53. Gupta, P. K., Shree, V., Hiremath, L., & Rajendran, S. (2019). *The use of modern technology in smart waste management and recycling: Artificial intelligence and machine learning*, (vol. 823). Springer International Publishing.
54. Sandler, M., Howard, A., Zhu, M., Zhmoginov, A., & Chen, L. C. (2018). MobileNetV2: Inverted residuals and linear bottlenecks. *Processing IEEE computer society conference computer visual pattern recognition*, (pp. 4510–4520).
55. Howard, A. G. et al. (2017). MobileNets: Efficient convolutional neural networks for mobile vision applications.
56. Vu, H. L., Bolingbroke, D., Ng, K. T. W., & Fallah, B. (2019). Assessment of waste characteristics and their impact on GIS vehicle collection route optimization using ANN waste forecasts. *Waste Management, 88*(2019), 118–130.

Chapter 7
Smart Energy Conservation in Irrigation Management for Greenhouse Agriculture

T. George Princess, E. Poovammal, and G. Kothai

Abstract Greenhouse farming is a closed structure for protecting the plants from extreme weather conditions such as wind, hailstorms sects, and pest attacks. The smart greenhouse improves the current agricultural practices with smart technologies for better yield. Modern technologies incur several distinct sensors to reduce the cost of inputs and improve soil productivity, conserving soil and water resources like energy. Smart irrigation in farming reduces the intake of fertilization, optimizes the growth of the crop, increases productivity, and protects the greenhouse environment. The sensors in the field acquire the measurement of soil moisture and fix the threshold level. According to the level of threshold, the water is streamed to the plants. Planning of cultivation of distinct crops in the farm can be done based on water and other nutrient requirements. Optimized planning can be achieved based on any artificial intelligence techniques. Greenhouse farm faces many challenges in irrigation such as controlling the intelligent system for dripping, sprinkling, handling the drainage problem, and waterlogging. The congestion between dripping, sprinkling and water logging can be avoided along with reduction of water wastage. Internet of Things (IoT) in smart irrigation systems with intelligent sensors will overcome the above issues by maintaining the moisture in the soil, preserving the temperature of plants. The existing sensor control irrigation system exposes the plants with controlled water deficit. Fertigation is an emerging technology in a smart irrigation system for integrating the environmental control system to optimize the level of nutrient and water management. This paper aims to deliver an optimum level of water along with nutrients to the root of the crop for reducing the wastage and evaporation of water.

Keywords Nutrient · Evaporation · Fertigation · Crop · Control · Water

T. George Princess (✉) · E. Poovammal · G. Kothai
Department of Computing Technologies, SRM Institute of Science and Technology,
Kattankulathur 603203, India
e-mail: gt4254@srmist.edu.in

E. Poovammal
e-mail: poovamme@srmist.edu.in

G. Kothai
e-mail: kg2247@srmist.edu.in

© The Author(s), under exclusive license to Springer Nature Singapore Pte Ltd. 2022
P. K. Paul et al. (eds.), *Environmental Informatics*,
https://doi.org/10.1007/978-981-19-2083-7_7

7.1 Introduction

Agriculture is Asia's primary source of revenue, with 70% of farmers and common people relying on it [1]. In the agriculture sector, the development of organic food and the breeding of animals will build a sustainable lifestyle. Production of food and feeding the lives are the ancient procedures that are followed for years in the field. The use of information and communication technology (ICT) has proven for extending the potentiality and promoting agriculture in various aspects from several areas of developing countries. Technology has resolved obstacles by utilizing wireless technology, networking, mobile, and other methods to reduce the use of energy, electricity, and cost-intensive equipment that will be beneficial in developing agriculture. The current issues lie in the usage of resources such as labor and water, which are lacking in many parts of the country. To monitor the agriculture system on the regular basis without labor resources, the developed automated structure is preferred [2].

A wireless sensor network (WSN) can be used to measure environmental conditions. The application of WSNs in precision agriculture helps the farmers to make a better-informed decision with statistical support. Establishing a new integrated information system (IIS) based on IoT helps in regional monitoring and managing the environment, to obtain higher efficiency in difficult operations. The IIS combines several distinct technologies such as cloud technology, IoT, and e-Science for monitoring and managing the contexture. It also incurs test cases on climate change in a specific region and its biological reactions where it is further considered one of the most often hotly debated topics in science. The consequences of the IIS system demonstrate the significant benefits not merely for data analysis assisted by IoTs as well as in online services along with the applications which are based on e-Science platforms and cloud computing [3].

The process in IIS incurs three parts: a control box, a Web-based and mobile application. The electronic device called the control box is present in a watertight container, which can be located anywhere on the farm and all the sensors will be connected to it. Node microcontroller unit (MCU) gets the agriculture data from the Web-based application. It gets connected to the Internet via Wi-Fi connection. This application can be used for managing the agricultural plots, watering the crops, and also analyzing the suitable level for watering the crop. The last one is the mobile application which acts as an interface between the farmers for switching the ON–OFF control. This mobile application helps the farmer for monitoring the field and controlling them with an ON and OFF switch. Monitoring water, setting up the details of the crop from every plot, and receiving notifications through the LINE application are the key features of the program [4–6]. Figure 7.1 demonstrates the process of collecting data and transmission of it to the users.

Smart farming is one among IoT which is presented as one of many paradigms that can be explored to manage crops in real time. The IoT is a network of physical products that are integrated with electronics, software, sensors, and network connectivity for intelligence gathering and dissemination. The need for food is strongly

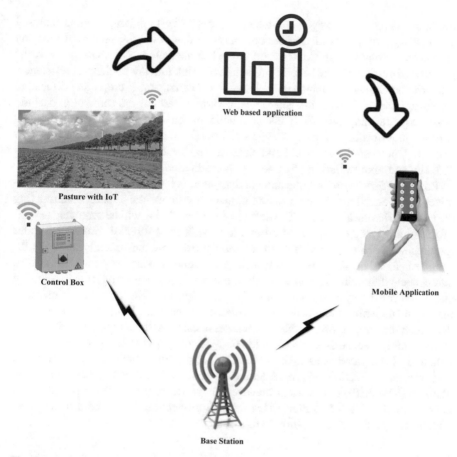

Fig. 7.1 Process of collection and transmission of field data

connected to the population, therefore finds it is challenging for growers to meet the requirement. This necessitates the use of artificial intelligence to combine environmental and crop-growing data. The moisture content of soil and its temperature is to be monitored for better yield in agriculture which can be achieved by relevant sensors and IoT. The usage of fertilizers, water, and pesticides can be controlled in smart farming. Greenhouse farming is an agricultural method that is capable of providing socio-ecological sustainability in the future and also for reducing manual labor [7].

Greenhouse farming is a distinct agricultural method that involves cultivating crops in shielded structures that are partially or entirely transparent. The fundamental function of greenhouses is to furnish favorable growing conditions which also helps in protecting crops from inclement weather and various pests. The most common greenhouse crops are tomatoes, lettuce, spinach, peppers, cucumbers, herbs, and so on. Every agricultural pasture has distinct properties which might be evaluated as crop volume and crop quality. The moisture content level of the soil is properly maintained for the growth of several vegetables which reduces the costs and improves agricultural productivity. Numerous distinct factors helps to determine the potentiality of any unique crop along with the kind of soil through some factors such as the pasture, the presence of nutrients in the soil, the procedure of irrigation flow, and insect resistance. Site-specific analysis demands the cultivation of a comparable crop that will lead to the emergence of diverse characteristics in various parts of the same crop pasture. Greenhouse farming can be expensive to build, expensive to heat, and also requires constant monitoring with care. The garden's aesthetic appeal may suffer as a result and could increase the water bill and electricity so these are the important problems faced by greenhouse farming that need to be addressed [8].

Precision agriculture is a mechanism that controls the productivity of land, maximizes the revenue from the field, and minimizes environmental impact by automating the entire agricultural process. The data obtained from WSN will be evaluated to optimize and adapt the environment around it, as well as to forecast future crop water requirements. A timely prediction of soil moisture content may quickly show significantly enhanced irrigation optimization by taking into consideration all of the critical components that influence soil water flow. It is used for precise irrigation monitoring and a controlling system consisting of distinct sensors. The farmland consists of many sensors which incur sensors to estimate the moisture level of soil, sensors to determine the temperature of the soil, sensors to evaluate the environmental temperature, humidity sensor, carbon-di-oxide (CO_2) sensor, and sunlight sensor [3]. The structure of the paper is as follows. Section 7.2 describes the smart irrigation and distinct sensors. Section 7.3 gives a detailed overview of the smart water management platform (SWAMP). Smart water management platform (SWAMP) is postulated in Sect. 7.4, and Sect. 7.5 demonstrates fertigation. Section 7.6 concludes the paper with the integration of fertigation with IoT.

7.2 Smart Irrigation and Sensors

Irrigation is an agricultural practice that is followed for applying the regulated amounts of water toward land to aid in crop production as well as the growth of landscape plants and lawns, which is termed watering. The agriculture work is independent of IoT for enabling the WSN framework. This framework consists of distinct several sensors to acquire the farm information in real time through multi-point measurement. The precise irrigation is used for controlling and monitoring the mechanism by developing a low-cost and high-accuracy intelligent system. This system helps the farmers in the agricultural environment for monitoring the soil and water requirements through IoT-enabled wireless sensor network, and it also develops an optimization technique for forecasting the level of moisture content on time. It generates the control commands on site-specific irrigation toward the regulation of valves in irrigation systems by the use of structural_similarity_index measure (SSIM) to collaborate the interpolated data. It implements the distinct conditions of weather during the control of irrigation by the weather model of fuzzy logic. The two major factors of WSN nodes are battery capacity and backup. The farmland consists of sensors such as soil MC sensor, soil temperature sensor, environment temperature sensor, environmental humidity sensor, environmental co_2 sensor, and the sunlight sensor. The microcontroller of the analog–digital converter (ADC) connects all the respective sensors for collecting the farm-related data. During the daytime, the solar panel supplies the requisite current to the battery unit through charging. In real time, the energy supply is monitored by an ADC controller; whereas, the Zigbee is eventually consumed a low current of $0.02\mu A$ when it is in deep sleep mode. The monitoring circuit sends a 5 V supply to the microcontroller and a 3.3 V supply to Zigbee S2 wireless transceiver. The device firmware is updated by a mesh network that is configured by ZigBee transceivers. The sensor parameters are used for analyzing the real-time gateway process that sends the data over the Internet server. The whole unit is bounded in a panel box, and a solar panel has maximum power efficiency for the panel box [3, 9].

Uferah et al. proposed the multi-model mechanism for assessing agricultural production by employing low-altitude remote sensing technologies by incorporating machine learning, drone technology, and IoT methods to monitor crop health. The drone furnishes multispectral data for establishing the indices of vegetation and the normalized difference vegetation index (NDVI) for analyzing crop health. The IoT sensor furnishes the real-time status of environmental parameters for detecting crop health. The variable-length are used for obtaining the multispectral data images and are converted to a fixed size for mapping the crop health. A technology-based agricultural solution is capable of extracting and complementing the information from multi-modal datasets and reduces the crop ground survey work. This work is especially advantageous in the agriculture field which is huge [10].

The main components of the system are IoT agri-nodes, communication channels for sending data, drones with multispectral cameras, local servers for archiving data, and web portals for visualizing data. By using LoRa technology, the data is transmitted from the slave node to the master node. The master node is connected directly to the local server for using GSM technology through IoT sensors and finally, it processes the data to the end user. IoT agri-node consists of soil moisture sensor, soil temperature sensor, air temperature, and humidity sensor. All these sensor nodes are connected to slave nodes. To overcome this limitation to some extent, we placed the SD card on the master node and saved the data for 24 h in case of a communication failure on the master node side [11, 12].

To provide insights, data cards from IoT sensors have been developed, and it also helps in identifying factors that affect plant health. A digital sensor called DHT11 are be used to record the temperature and humidity of the air. A digital sensor known as the DS18B20 is used to measure the temperature of the soil. The capacitive soil moisture sensor is used to detect the amount of moisture in the soil. The data is collected independently from the cultivation by IoT sensors and drones, and it is preprocessed to remove outliers and perform radiometric adjustments. Multi-model data integration collects and integrates those preprocessed data in dataset preparation. Validated maps can reveal the state of the crop's health, and NVDI maps will provide information on the crop's health [11, 13].

Rekha et al. estimate the quality characteristics of water such as turbidity, BOD, pH, TDS, and temperature for determining the variation. When an irregularity in the cultivation attains, it can assist by sending out an alarm message to the users. The GSM is a traditional module that incurs pH for moisture estimation, temperature for determining the environmental temperature, TDS for collecting the farm-related data, and turbidity sensors for estimating the scattered light. Estimating the mobility of hydrogen molecules in the instance water will be used for determining the acidity of the sample. Temperature and turbidity sensors confirm and detect the quality of water and temperature depending on the assessment in water. TDS sensors detect the number of milligrams and disintegrate the dissolvable solids in water.

The Arduino incurs several distinct sensors. These sensors will determine the amount of water and produce distinct findings by calculating the techniques. The acquired data can be transferred to the cloud for the controller for further analysis. The messages are transmitted from the cloud to the customers who have registered on it. Whenever any anomalies are discovered during cultivation, a warning or alert message is sent to government officials and users. The warning or alert message will assist users in taking corrective action and overcoming crop damage and also help the cultivators for increasing their business and sales process [14].

NFC sensors, meta-meters, beamforming, and LoRaWAN are technologies employed for supporting the use of resources. Trading between charging and NFC, changing environmental conditions, beamforming, and route planning are the drawbacks in the harsh atmosphere. The authors Hongzhi Guo and Albert A. Ofori explore two applications for establishing a network of mobile readers beneath. The NFC sensors help in monitoring the roadway infrastructure and the quality of water. The shift of power-hungry devices on the reader side of the smartphone minimizes the power consumption of sensors. This system also enables ubiquitous sensing and computation that are discussed in an extreme atmosphere [15]. Table 7.1 describes the several distinct sensors in smart agriculture.

Muhammad Shoaib Farooq et al. [20] evaluate around 170 publications on IoT for agriculture technologies and illustrate the components in smart farming. In a smart farming system, the major critical technologies employed are massive data storage and analytics along with cloud computing.

In an IoT cultivation network, the architecture incurs the layers such as application, network, transport, adaptation, MAC, and physical layer. Around 20 publications are carried out on cultivator experience, predictive and intercultural analysis, monitoring, sensing, and physical implementations in big data analytics. The applications of primary cultivation are precision monitoring, greenhouse monitoring, and livestock monitoring.

Monitoring the greenhouse aids in the management of water resources, the observation of plants, and various climatic conditions. The temperature of animals is monitored with the stress level of heat, physical gesture recognition, ruminating, detecting the heart rate, and GPS tracking. In this IoT platform, crop conductivity will be improved [20, 21].

Table 7.1 Distinct sensors in Smart agriculture

Author and Year	Sensor	Objective	Advantage
Uferah Shafi et al. [11, 16] 2020	• DHT11 • DS18B20 • Capacitive soil moisture • Volumetric sensor • Tensiometric sensor	• To monitor the health of crops in the aquaculture	• Real health status of the crop can be determined
Rekha P et al. [3] 2020	• pH sensor • Turbidity sensor • Temperature sensor • TDS sensor	• To estimate the quality of water • To generate an alert message for the abnormalities that are identified in the aquaculture	• The remedial measures can be taken by alerting • Business and sales are increased for the farmers
Shan he et al. [17] 2020	• Flexible sensor	• To detect the sulfate ions from the sample water	• The concentration ranges can be determined
Carlos andrés gonzález-amarillo et al. [18] 2018	• Soil humidity sensor • DTH11 sensor • Temperature sensor • LDR sensor	• To track and maintain the record of growth stages about seedling and various aquaculture products	• Consumption of energy and water resources are reduced in the whole aquaculture process
Hongzhi Guo and Albert A. Ofori [19] 2021	• NFC sensor • Passive sensor	• To monitor and exploit the status and resources	• The power consumption of sensor get reduced
Noramalina Abdullah et al. [14] 2021	• Humidity sensor • Capacitive soil moisture sensor • Ultrasonic sensor	• To control the switching time of pumps	• To reduce the consumption of water
Muhammad Shoaib Farooq et al. [15] 2019	• pH sensor • Gas sensor • Ultra violet sensor • Motion detector sensor • Passive infrared sensor • Soil moisture sensor • Temperature sensor • Humidity sensor • Barometric pressure sensor	• To present the IoT technologies and components for smart aquaculture	• Enhances the crop conductivity in the IoT platform

As a result, numerous sensors and their applications are explored for detecting the health state of the crop in the pasture, maximizing business for the cultivators, and improving crop production.

7.2.1   Fuzzification

The technique that converts a crisp quantity into a fuzzy quantity is termed fuzzification. The inverse technique of fuzzification mapping is done for converting a fuzzy quantity to a crisp quantity is called defuzzification. Fuzzy logic model forecasts the weather-dependent soil MC level. The rule-based model of fuzzy logic is given as inputs and outputs are bounded with rule sets. Human employs real-time sensor values as model inputs, and valve control commands are created based on the ruleset to establish the rules in the fuzzy logic model. The weather data, as well as the expected soil moisture content (MC), are exploited to develop irrigation valve control ON/OFF commands. By employing an optimization algorithm, the valve control is done successfully and contemplate the soil MC prediction method [3]. Figure 7.2 depicts the process of fuzzification and fuzzy inferences.

7.3   Smart Water Management Platform (SWAMP)

The SWAMP is a developing IoT-based intelligent water management platform for precision irrigation in agriculture with a practical approach based on four pilot

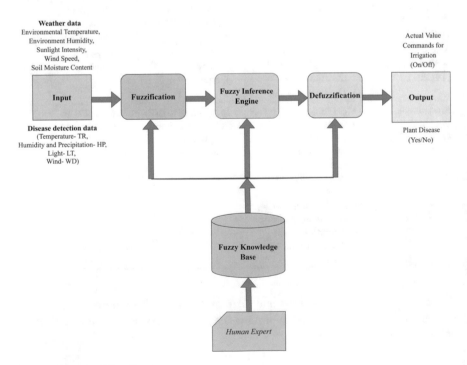

Fig. 7.2 Process of fuzzification

projects in Brazil and Europe. This work presents a SWAMP architecture, platform, and system deployment that focuses on platform reproducibility. Scalability is a major concern for IoT applications and it includes performance analysis of the FIWARE components that are used on the platform techniques. It also employs fewer computational resources to provide high scalability. IoT system for precision irrigation is either too generic or too specific and do not address the easy deployment system for replicability and streamlining the deployment of new systems. The SWAMP water management of agriculture is of three phases:

- Water reserve,
- Water distribution, and
- Water consumption.

The SWAMP architecture is divided into five layers:

- Layer 1: Device and communication
- Layer 2: Acquisition security management
- Layer 3: Data management
- Layer 4: Water irrigation data models
- Layer 5: Water application services.

The first three layers of architecture are interconnected within them and it uses fully replicated services in different settings, the fourth layer as fully customizable services that provides services that may vary for different models and techniques of water irrigation and distribution, and finally, the fifth layer is application-specific services where the pilot as to be addressed.

The device and communication layer incurs a variety of sensor and actuator robotics to acquire soil, plant, weather as well as LPWAN robotics. The acquisition security management layer's key characteristics are protocol and software components of data acquisition for adding the security and device management functions. The third layer does data management. It is distributed infrastructure of cloud servers and fog nodes that works together for processing large amounts of data and making it available at higher levels. The water irrigation and distribution model layer contains traditional agricultural models for estimating plant water demand by measuring soil moisture using drone images (crop-based approach) and soil sensors (soil-based approach). Optimized models and techniques for water distribution based on plant water requirements are always essential when the collective networks replace individual water sources.

The final layer provides water application services. These irrigation services are more useful to farmers. The water utilities are analyzed through the user interface for further decision-making [10, 22].

7.4 Smart Water Management Platform (SWAMP)

In precision agriculture, a self-designed moisture wireless sensor is utilized based on the investigation and applications. The WSN architecture is built, and the smart irrigation control system is developed using the network. Wireless sensors, cluster heads, repeaters, base stations, data center servers, and data backup services are constituting the moisture wireless sensor network. The network is structured by employing a chain topology, which is simple and effective in huge moisture monitoring in the field. The moisture data from sensors are calibrated by the data center server. The database application system matches the expert data with real-time moisture data for identifying whether plants require irrigation and the amount of irrigation to be applied per square meter.

A decision support system is essential for transmitting the watering instructions to the base station and also helps in selecting the total volume of water and watering location. Irrigation instructions are transferred from the base station to the irrigation control system. The irrigation control system activates the electronic hydro valve in the desired irrigation position and then closes when it reaches the threshold. Local irrigation is stopped and the above process is repeated. It mainly demonstrates the calibration of sensors and compares the real-time humidity data. The watering decision incurs the demand for water for the plant. The WSN-based intelligent irrigation control system implements irrigation decisions through real-time moisture and expert data. The technology has been demonstrated to save water and be furnish suitable commands for precision agriculture [18, 23].

The traditional methodology increases crop yield and sets higher quality requirements for crops with high economic value. The indoor environment system of a greenhouse is monitored and managed, and users will obtain crop seedlings and harvesting information. In the cultivation process, energy and water usage are minimized. The communication procedure with the actors improves the agriculture value chain [24, 25].

Noramalina Abdullah et al. [24] presents a system based on advanced fuzzy logic for managing pump switching time based on user-defined factors. The flow rate is determined and helps the cultivators to regulate volume, flow rate, and velocity. Determining the link between air temperature, humidity, soil moisture, and crop watering time is possible. An IoT-based application is created with the data that provides the users to attain a better knowledge of the technology. The subsurface and underwater resources are determined using a DHT11 sensor for evaluating the temperature of air and a capacitive sensor for evaluating the soil moisture. The humidity sensor and an ultrasonic sensor are employed for further evaluation in the field. Smart water management can be carried out and it dramatically saves water usage and watering time.

7.5 Fertigation

The term fertigation is derived from fertilization + irrigation. It is an application method of fertilizer in which fertilizer is introduced into irrigation water via the drip system. The fertilizer solution is spread uniformly in irrigation with the drip technique. As the nutrients are easily accessible, it is efficiency increases. Farmers continue to utilize conventional agricultural management practices, resulting in input waste and low productivity that owes to inaccuracies in several types. The amounts of fertilizers contributed to the field are dependent on the analysis of soil and plant yield. However, it consumes higher time to determine the type of fertilizer from the results of soil analysis. This work helps to design an automated fertigation system based on IoT and efficient use of water. It is controlled and achieved by a moisture sensor to pre-arrange the data according to plant needs.

The system is divided into four components; they are three sensor nodes and one master node. The three sensor nodes are the soil moisture sensor, electrical conductivity (EC) sensor, and pH sensor. Sensor networks are installed in the input field to read soil characteristics into the microcontroller (Arduino UNO). The intelligent system performs based on the characteristics and controls the process of irrigation and fertilization. All the characteristics of sensors are transmitted over the master network through XBee that monitors the website, stores in a database, and carried for further analysis. Every XBee scans its channels and connects to the master node. All sensors started to measure the values.

All these values are stored in Arduino memory. Depending upon these stored values, the Arduino takes immediate action to perform the task. Some are listed below.

If the soil moisture reaches the limit of 70% in irrigation, it automatically switches OFF the water pump; whereas, if it is less than 30% of it automatically switches ON the pump.

According to the preceding phases, the fertilizing procedure must preserve the soil in an optimal condition.

Each sensor node sends the frame wirelessly through XBee to the master node, then the master node gets the ID and measured value of each sensor node and sends them to the server by Ethernet. Finally, all the frames are stored in a database and displayed on a website. This technique saves money and time, conserves water, saves electricity, and prevents over-fertilization. Furthermore, the system is easily adaptable to any climate change. All components are easily accessible and reasonably priced in the local market [2, 17].

Mohanraj et al. [14] postulate the WSNs for monitoring environmental conditions with defined knowledge. It helps in estimating the growth of plants and yield characteristics. The system automates WSN irrigation and fertilization that resulting in relatively higher yields than other traditional methods. The drip irrigation and fertigation process are combined using WSN to obtain the optimum amount of water for crops by the irrigation system without affecting its growth and soil erosion. The system adds the fields of rich crops such as sugarcane and maize. The WSN comprises several distinct sensor network nodes and each node in the network incurs a processing unit, sensing unit, power, and communication unit. The demand-based algorithm is implemented, and the processor will save its machine cycles to increase the lifetime of WSN. The knowledge of faulty dripper lines will improve the irrigation dependability. The threat of water overflow will avoid crop spoilage during flooding.

Sureshkumar et al. [14] depicts the fertilization that incurs the moist soil volume that results in determining the root volume, and the optimization. Precision farming is characterized based on the efficient use of applied inputs, particularly water and fertilizers. Henceforth, the fertilizer is dissolved in water at the appropriate concentration and applied to a micro-irrigation system that supplies through the irrigation water. This process is termed fertigation. Both nutrients and water are supplied in the precise quantity at the right time to the root zone to ensure optimal absorption of supplying nutrients and water that result in producing (crop) per drop of water. The cost of solid fertilizer is usually low, and the selected fertilizers for fertigation are mainly dependent on four main factors.

Fertilizers with good quality and solubility will have lower salt levels. A suitable pH should be chosen to fit cost-effectiveness into the farm management system.

7.5.1 Essentials and Needs of Fertigation

Inadequate and unbalanced typical application fertilizer results in an increase in fertilizer use and reliance on fertilizer imports. This results in unequal growth due to the consumption of fertilizer between states and crops. Adding to it, the reduction in the response of plants is due to imbalanced fertilization as it weakens the relationship between the use of fertilizer and potential yield. It reduces the need for water-soluble and balanced fertilization as there is a gradual need for culture in active root areas. It achieves maximum efficiency of water and use of fertilizer.

7.5.2 Principles of Fertigation

Nutrient requirements are evaluated based on the amount and ratio vary by each crop, variety, and growth stage. This results in a variation of fertigation by the volume and combination of the usage of fertilizers. Precision farming systems help in regulating fertilizer and water delivery during each development stage. It is carried out based on the needs of the plant and allows the plant to maximize all physiological functions that result in the maximum potential yield. The consequence of the process leads to the determination of nutritional prescriptions that depends on the desired yields [26].

7.5.3 Relationship in Irrigation

Water with a higher level of calcium, magnesium carbonates, and bicarbonates are found in tube well water that causes precipitation in fertilizer tanks. Particularly, water with phosphates is identified along with the blockage in the system owing to an elevation in pH. In contrast, water with lower pH, higher iron and aluminum concentrations are identified in tropical climates. It produces toxicity owing to these metals along with phosphorus precipitation. This result suggests carrying out the fertigation by using water that has a pH close to neutral.

7.5.4 Advantages of Fertigation

1. Fertigation ensures whether the nutrients are applied uniformly to the rhizo-sphere, as the active roots need to be concentrated.
2. The soil space that accounts for more than 80% of root activity and is referred to as a plant's effective foraging space (EFS). As a result, nutrients are supplied in the EFS for ensuring nearly perfect absorption by crop requirement during the planting period.
3. It also assures a greater and higher quality output, as well as time and labor savings, making it economically viable.
4. It furnishes not only the optimal employment of two valuable inputs such as water resources and level of nutrients, utilization of the synergy that is created by their simultaneous availability to plants.

The main limiting factor of fertigation is expensiveness in investment for fertilizer injection systems and safety devices.

7.5.5 Preventive Measures

1. The time taken to apply the fertilizer should be less than the time it takes to provide enough water to the field; otherwise, salinity may develop.
2. It prevents overwatering.
3. The consumption of water and nutrient efficiency depends on the function of yield during fertigation.
4. Effects of soil root volume.
5. Reducing the dangers of salinity [23].

7.5.6 Applications of Fertilizers

For promoting the growth of plants, fertilizer is used and it is usually applied in the soil. Fertilizers are buried around the roots of a tree when it is planted, deposited in boreholes near tree roots, distributed on the ground, or sprayed by the person.

Fertilizers are of two types

- Organic fertilizer
- Inorganic fertilizer.

Organic fertilizers are more complex chemical compounds that take longer to break down into plant-useable forms. Some of the instances are cottonseed meal, blood meal, manure, fish emulsion, and sewage sludge.

Inorganic fertilizers dissolve quickly in water and are accessible to plants for absorption almost immediately. The majority of our man-made, commercial fertilizers are inorganic fertilizers, which are obtained from non-living sources.

It has four methods in practice

Broadcasting—the whole crop has a uniform distribution.

Foliar application—with a small or large amount of atomizer, fertilizer is sprayed over the plant.

Placement—used for ribbons and pockets near plants and rows of plants.

Fertigation—fertilizer is the injection of fertilizer through an irrigation system.

7.6 Conclusion

This article illustrates whether emerging IoT trends are facilitating smart greenhouse cultivation. It includes the different greenhouse crop production with the associated innovative technology. Several diverse sensors are audited for smart greenhouse cultivation to enhance crop productivity, restrict water consumption, evaluate crop health status, enhance business, and prevent power usage. To assist growers, numerous and diverse decision-making applications are suggested in sustainable farming. It delivers potential assistance to economic researchers, growers, and engineers in terms of increasing crop productivity. Furthermore, smart greenhouse farming has great challenges, such as the need for low-cost accurate sensors for monitoring and securely broadcasting ambient climatic conditions and distinct soil conditions.

References

1. Rao, R. N., & Sridhar, B. (2018). IoT based smart crop-field monitoring and automation irrigation system. In *2018 2nd International Conference on Inventive Systems and Control (ICISC)* (pp. 478–483).
2. Mohanraj, I., Ashokumar, K., & Naren, J. (2016). Field monitoring and automation using IOT in agriculture domain. *Procedia Computer Science, 93*, 931–939
3. Keswani, B., Mohapatra, A., Mohanty, A., Khanna, A., Rodrigues, J., Gupta, D., & De Albuquerque, V. H. C. (2019). Adapting weather conditions based IoT enabled smart irrigation technique in precision agriculture mechanisms. In *Neural computing and applications.*
4. Muangprathub, J., Boonnam, N., Kajornkasirat, S., Lekbangpong, N., Wanichsombat, A., & Nillaor, P. (2019). IoT and agriculture data analysis for smart farm. *Computers and Electronics in Agriculture, 156*, 467–474.
5. Deshpande, S., Lokare, S. R. (2019). Fertigation—nutrient dispensary management using internet on things, a novel approach. *JETIR, 6*(2)
6. Mudholkar, P. (2020). Smart air fertigation (SAF) without using fertilizers through air irrigation for sustainable agriculture. *International Journal of Recent Technology and Engineering (IJRTE)*
7. Rayhana, R., Xiao, G., & Liu, Z. (2020). Internet of Things Empowered Smart Greenhouse Farming. *IEEE Journal of Radio Frequency Identification, 4*(3), 195–211.
8. da Silva, J. G. (2017). *The future of food and agriculture–trends and challenges.* Food Agr. Org., Rome, Italy.
9. He, S. (2020). IoT-based laser-inscribed sensors for detection of sulfate in water bodies. *IEEE Access, 8*, 228879–228890.
10. Kamienski, C. (2018). SWAMP: an IoT-based smart water management platform for precision irrigation in agriculture. In *Global internet of things summit (GIoTS)* (pp. 1–6).
11. Shafi, U. (2020). A multi-modal approach for crop health mapping using low altitude remote sensing, internet of things (IoT) and machine learning. *IEEE Access, 8*, 112708–112724.
12. Kashyap, P. K., Kumar, S., Jaiswal, A., Prasad, M., & Gandomi, A. H. (2021). Towards precision agriculture: IoT-enabled intelligent irrigation systems using deep learning neural network. *IEEE Sensors Journal, 21*(16), 17479–17491.
13. Huong, T. T., Huu Thanh, N., Van, N. T., Tien Dat, N., Long, N. V., & Marshall, A. (2018). Water and energy-efficient irrigation based on Markov Decision model for precision agriculture. In *2018 IEEE Seventh International Conference on Communications and Electronics (ICCE)* (pp. 51–56).
14. Rekha, P., Sumathi, K., Samyuktha, S., Saranya, A., Tharunya, G., & Prabha, R. (2020). Sensor based waste water monitoring for agriculture using IoT. In *2020 6th International Conference on Advanced Computing and Communication Systems (ICACCS)* (pp. 436–439).
15. Guo, H., & Ofori, A. A. (2021). The internet of things in extreme environments using low-power long-range near field communication. *IEEE Internet of Things Mag, 4*(1), 34–38.
16. Ghosh, S., & Sayyed, S. (2016). Smart irrigation: A smart drip irrigation system using cloud, android and data mining. In *2016 IEEE International Conference on Advances in Electronics, Communication and Computer Technology (ICAECCT).*
17. Elhassan Ahmed, O. M., Osman, A. A., & Awadalkarim, S. D. (2018). A design of an automated fertigation system using IoT. In *2018 International Conference on Computer, Control, Electrical, and Electronics Engineering (ICCCEEE)* (pp. 1–5).
18. Kehui, X., Deqin, X., & Xiwen, L. (2010). Smart water-saving irrigation system in precision agriculture based on wireless sensor network. *Transactions of the CSAE, 26*(11), 170–175.
19. Sureshkumar, P., Geetha, P., Narayanan Kutty, M. C., Narayanan Kutty, C., & Pradeepkumar, T. (2016). Fertigation—the key component of precision farming. *Journal of Tropical Agriculture, 54*(2), 103–114.
20. Farooq, M. S., Riaz, S., Abid, A., Abid, K., & Naeem, M. A. (2019). A survey on the role of IoT in agriculture for the implementation of smart farming. *IEEE Access, 7*, 156237–156271.

21. Nachankar, P. J., Somani, M. G., Singh, D. M., & Katkar, S. N. (2018). IOT in agriculture. *International Research Journal of Engineering and Technology (IRJET), 5*(4).

22. Jayalakshmi,·M., & Gomathi, V. (2020). Sensor cloud based precision agriculture approach for intelligent water management. *International Journal of Plant Production, 14*,177–186.

23. Mohanraj, I., Gokul, V., Ezhilarasie, R., & Umamakeswari, A. (2017). Intelligent drip irrigation and fertigation using wireless sensor networks. In *2017 IEEE Technological Innovations in ICT for Agriculture and Rural Development (TIAR)* (pp. 36–41).

24. Abdullah, N. (2021). Towards smart agriculture monitoring using fuzzy systems. *IEEE Access, 9*, 4097–4111.

25. González-Amarillo, C. A. (2018). An IoT-based traceability system for greenhouse seedling crops. *IEEE Access, 6*, 67528–67535.

26. Kumar, A., Kamal, K., Arshad, M. O., Mathavan, S., & Vadamala, T. (2014). Smart irrigation using low-cost moisture sensors and XBee-based communication. In *IEEE Global Humanitarian Technology Conference (GHTC 2014)* (pp. 333–337).

Chapter 8
Role of Artificial Intelligence and IoT in Environmental Monitoring—A Survey

S. Karkuzhali and S. Senthilkumar

Abstract According to a World Health Organization (WHO) survey, 97 percent of cities in low- and middle-income countries with over 100,000 inhabitants do not comply with WHO recommendations on air quality. Air pollution can lead to chronic and respiratory problems, such as asthma, and fatal diseases, such as lung cancer. Internet of Things (IoT) and artificial intelligence (AI) will provide tools for real-time monitoring of air pollution. The tools can identify sources of air pollution quickly and accurately. Consequently, nations around the world have raised their directions in titles of monitoring as well as regulating and handling consequent infection like COVID-19. Over the last several years, people have developed dramatically with the rush of the manufacturing revolt in which a new production of wireless communication facilitates ubiquitous connectivity among things. The recent advancement of the IoT and AI plays the major role in various enterprises. IoT is regularly perceived as practical things, broadly distributed, with low repository ability and processing potential, with the purpose of enhancing authenticity, enforcement, and preservation of the smart living and its foundations. The paper directs to a significant criterion change of how to monitor, sense, and track dynamic phenomena in real time in the environment. From this review, it is visible that there are remaining a few exciting opportunities and challenges on improvement and deployment of IoT and AI algorithms for smart and dynamic environmental monitoring.

Keywords Artificial intelligence (AI) · Internet of Things (IoT) · Environmental monitoring · Pollution · Environmental protection

S. Karkuzhali (✉)
Department of Computer Science and Engineering, Mepco Schlenk Engineering College, Sivakasi, Tamil Nadu, India
e-mail: karkuzhali@mepcoeng.ac.in; vijikarkuzhali@gmail.com

S. Senthilkumar
Department of Chemistry, Ayya Nadar Janaki Ammal College, Sivakasi, Tamil Nadu, India
e-mail: senthilkumar_ts208@anjaconline.org

8.1   Introduction

Traffic-related air contamination has been a genuine worry among strategy creators and general society because of its physiological and natural effects. An early admonition framework dependent on precise estimating apparatuses must along these lines be actualized to bypass the unfavorable impacts of introduction to significant air contaminations. The multi layer percptron network is used to monitor the Air pollution and meteorological information over a two-year period from an observing site in Marylebone Road, Central London, to foresee side of the road focus estimations of NO_2 24 h ahead [1]. Ongoing years have seen an expanding enthusiasm for Demand Response (DR) as a way to give adaptability and henceforth improve the unwavering quality of vitality frameworks in a savvy way. The artificial intelligence (AI) and machine learning (ML) plays a vital role in providing solutions for DR. Computer-based intelligence techniques can be utilized to handle different challenges, running from choosing the ideal arrangement of buyers to react, learning their characteristics and inclinations, dynamic evaluating, planning and control of gadgets, and figuring out how to boost members in the DR plans and how to compensate them in a reasonable and financially productive manner [2].

8.2   Literature Review

The advancing period of globalization and industrialization has brought upon its result in mask of its different associative harmful components which have maladjusted the harmony of the earth. The ill-advised treatment of such components has come about into the pollution of essential components of biological system which serve the premise to continue human life. This is one of the most urgent issues relating to endurance of agreeable conjunction among human and nature. The chief advance to figure the arrangement is deciding the degree of harm and dependent on it taking preventive and destruction measures. The developing innovation of Internet of Things (IoT) gives an extent of part possible models for the above expressed purposes [3]. The stage breaks the current latent strategy and gathers information progressively through IoT sensor based on 5G remote system. So as to forestall phony and altering of gathered information, we use blockchain innovation to encode what's more, transmit to cloud and give continuous air contamination list estimation stage. You can separate innovation escalated information through edge and distributed computing [4]. IoT gives a broad incorporated system to human services to battle with COVID-19 pandemic. Every clinical gadget is associated to the Web, and during any basic circumstance, it naturally passes on a message to the clinical staff [5].

Savvy producing is basic in improving the nature of the procedure business. In keen assembling, there is a pattern to consolidate various types of new-age data advancements into process safety investigation, well-being, and ecological guidelines. Rising data innovations, for example, man-made consciousness (AI), are very encouraging as a methods for defeating these challenges [6, 7].

Liu et al. (2020) used to review various algorithms in AI, machine learning, and deep learning in the lack of irrigation and water management in agriculture [8].

Nourani et al. (2019) discussed that vehicular traffic clamor is the fundamental wellspring of commotion contamination in significant urban areas around the world [9].

Shafiq et al. (2020) proposed another system model and a crossover calculation to settle this issue. Right off the bat BoT-IoT recognizable proof informational collection is applied, and its 44 viable highlights are chosen from various highlights for the AI calculation. To discover which ML calculation is powerful and ought to be utilized to choose for IoT oddity and interruption traffic recognizable proof, a bijective delicate set approach and the calculation are applied. At that point, we applied the proposed calculation dependent on bijective delicate set methodology. The author uses 5 different machine learning approaches and calculated the performance measures for all the five approches out of them using the best one [10].

Nishant et al. (2020) reviewed that AI offers a freeing chance to fabricate frameworks of insight that will create information required for saving life. Be that as it may, for AI to convey even a small amount of the possible advantages for manageability, information system (IS) research must investigate approaches to go around restrictions and conceptualize novel approaches to use AI. Artificial intelligence cannot be a panacea for all our complex natural maintainability issues. Generally, new advancements have vowed to tackle quick issues, yet in time demonstrated impractical. In this manner, we must perceive the confinements of this development, investigate approaches to go around a few imperatives, and conceptualize novel approaches to use AI [11].

Goralski et al. (2020) examination consolidates the viewpoints of business system and open approach to investigate the effects of AI on practical advancement with a particular concentrate on the progression of the SDGs. It likewise draws a few exercises on administrative learning and authority improvement for worldwide maintainability. Simulated intelligence presents a wide exhibit of uses that can fill in as distinct advantages for the quest for economical turn of events, which will include numerous entertainers from various nations, societies, and divisions. Through the UN Global Compact, organizations around the globe have been called to assume a job in accomplishing the Sustainable Development Goals (SDGs). As should be obvious in the three contextual analyses featured above, AI can be an amazing empowering agent of the worldwide exertion to advance monetary turn of events and simultaneously economically address the effect of our creation what's more, utilization on our social orders, administration frameworks, and nature. The advances made by the trendsetters, activists, and worldwide victors of advancement utilizing AI-empowered applications put them at the wilderness of the reasonable improvement work. Their advancements have improved effectiveness of ventures and areas, assisted with moderating valuable, non-sustainable assets, diffuse information and skill, connect the worldwide holes in assets and innovation, and assisted with fashioning compelling multi-area associations (governments, private part, respectful society, and residents) that add to worldwide supportability.

As we enter the Age of Sustainable Development, in which the 17 Sustainable Development Goals (SDGs) are defining the development agenda for the nations of the world, AI is also rapidly opening up a new frontier in the fields of business, corporate practices, and governmental policy. The intelligence of machines and robotics with deep learning capabilities is already solving cognitive problems commonly associated with human intelligence. They run from aloofness, latency, and the obliviousness of individuals and the absence of assets also, political will of governments, to the quest for momentary benefit by enterprises and the limited focal point of restricted national interests by the country states to the disregard of the worldwide basic great. The battle for worldwide maintainability and the eventual fate of humankind on the planet will require a dedication from a scope of open and private segment associations, national governments and common society, and all the assets they can gather. The appearance of SDGs comprises an extremely noteworthy business open door for the early AI industry. Man-made intelligence can create information for additional canny focusing of mediation (as on account of Plant Village), diminish waste and misfortunes underway and utilization (for example, in Smart Water Management), make new applications that will change whole enterprises and callings, and give the vital upgrades in availability and cost decreases (Clean Water AI) that brings the advantages of the quick pace of mechanical improvement to numerous individuals around the world. These SDG-propelling advancements and activities, in any case, may must be embraced and systematized at an expense and bear a few likely dangers. Computer-based intelligence is a twofold edged blade. It can accompany multifaceted entanglements and complex issues that must be thoroughly examined and figured out how to contain its negative and unintended outcomes. Its invigorating and supportability advancing applications may likewise be utilized for evil, in exercises that will intensify the most exceedingly awful effects of an unnatural weather change, contamination, unbridled utilization, what's more, untrustworthy creation strategies to take care of the way of life of ceaseless development endemic to the entrepreneur worldwide financial request of today. As we find in the three cases above, even probably the most straightforward and minimal effort developments would require motivating forces and organizations between governments, partnerships, networks, laborers, businesses, and the scholarly community to embrace, oversee also, and continue these transformative applications. [12].

Mohanta et al. (2020) portrayed an outline of the IoT innovation and the zone of its application. The essential security issue CIA (Confidentiality, Integrity and Availability) and layer-wise issues are recognized [13].

Moazamnia et al. (2020) developed AI-based displaying procedure that is a two-level learning process; at Levels 1 and 2, it give apparatuses to plan the SWI weakness of the examination region. Model exhibitions in the paper are considered utilizing RMSE and R2 measurements, where the models at Level 1 are seen as fit-for-reason and the SVM at Level 2 is improved especially as for the decreased size of dissipates in the outcomes. Assessing the outcome and groundwater tests by Piper outline affirms the correspondence of SWI status with powerlessness list [14].

Díaz-Alcaide et al. (2019) discussed about fecal contamination an inexorably significant danger to groundwater supplies in some low-pay districts of the world. Thirteen machine inclining classifiers, including diverse measurable calculations, case-based students, and tree-based models, were utilized to decide the spatial dissemination of fecal contamination according to five informative factors (restroom thickness, separation to the nearest restroom, borehole yield, water table profundity, and populace thickness). The best performing classifiers, chosen on test scores, were then used to create prescient guides. Irregular woodland and strategic relapse rendered forecast scores for fecal contamination in abundance of 0.90. Multilayer perceptrons, bolster vector machines, and quadratic discriminant investigations additionally demonstrated proficient at guaging fecal contamination. Outfit planning shows that 30–50 m cradles around household wells might be adequate to forestall defilement of local supplies in most cases. This shows AI may give an adaptable methodological option in contrast to customary Darcian approaches. Then again, the down to earth troubles associated with looking after wellhead assurance zones recommend the need to actualize funneled water supplies [15].

Tung et al. (2020) review covers the model structure, input fluctuation, execution measurements, local speculation examination, and far reaching appraisals of AI model progress in stream quality exploration. Henceforth, this is profoundly accentuating the inclusion of AI model improvement which can manage missing information, ready to incorporate the highlights of a discovery model [16].

Shu et al. (2020) developed image recognition technology dependent on implanted innovation that has the attributes of high separation, high adequacy, and solid touch sense. It is of extraordinary criticalness for natural observing to utilize image recognition technology. Image recognition technology is to catch a similar situation by computerized camera and afterward store the got picture captured by PC [17].

Senthilkumar et al. (2020) proposed an inserted framework, where sensors gather the air quality data inside period time and send it over the mist hubs [18].

Zhao et al. (2020) presented an orderly survey of four parts of the use of man-made reasoning to wastewater treatment: innovation, economy, the executives, and wastewater reuse. At long last, we give points of view on the likely future bearings of new exploration wildernesses in the use of man-made brainpower in wastewater treatment plants that at the same time address toxin expulsion, cost decrease, water reuse, and the board challenges in complex reasonable applications [19].

Zahmatkesh et al. (2020) discussed about fog computing that can act as a link between IoT smart devices and cloud data centers in order to provide services that have better delay performance [20].

Sharma et al. (2020) presented a WSN and IoT-based stage for the discovery of flames at a beginning phase. Remote sensor arrange has been effectively actualized utilizing sense nut equipment stage. The sensor hubs are conveyed in an outside situation for gathering and examination of constant information. Information gathered through the sensor hubs is put away on the cloud for investigation. A few plots have been planned utilizing the ThingSpeak cloud application for smoke, temperature, mugginess, and light power. The framework is prepared to do detecting different ecological boundaries and effective in the identification of an occasion, by breaking down constant information. A fire discovery framework is planned dependent on the cloud stage and IoT gadgets [21].

Abdallah et al. (2020) orderly survey concentrated on evaluating the different AI models utilized in SWM applications, extricated from 85 examinations distributed somewhere in the range of 2004 and 2019. The presentation of AI algorithms was looked at, and the qualities and constraints of AI applications in squander the executives were examined [22].

Cabaneros et al. (2019) assessed air pollution by multilayer perceptron. By far, most of the distinguished works used meteorological and source discharge indicators only. Besides, impromptu methodologies are seen as dominatingly utilized for deciding ideal model indicators, proper information subsets, and the ideal model structure. Multilayer perceptron and gathering type models are overwhelmingly actualized. In general, the discoveries feature the need for creating deliberate conventions for growing incredible ANN models [23].

Elkiran et al. (2019) used three single Artificial Intelligence (AI) based models i.e., Back Propagation Neural Network(BPNN), Adaptive Neuro Fuzzy Inference System (ANFIS), Support Vector Machine (SVM) and a linear AutoRegressive Integrated Moving Average (ARIMA) model as well as three different ensemble techniques i.e., Simple average ensemble (SAE), Weighted Average Ensemble (WAE) and Neural Network Ensemble (NNE) are applied for single and multi-step ahead modeling of dissolve oxygen (DO) in the Yamuna River, India [24].

Fan et al. (2018) performed survey that depicts the essentials, preferences, and constraints of AI instruments. Counterfeit neural systems (ANNs) are the AI apparatuses every now and again embraced to foresee the contamination evacuation forms as a result of their capacities of self-learning and self-adjusting, while hereditary calculation (GA) and molecule swarm improvement (PSO) are likewise helpful AI procedures in effective quest for the worldwide optima [25].

Kaab et al. (2019) used AI for determining energy and environmental impacts. The vitality utilization and yield vitality age for sugarcane creation in planted homesteads are 172,856.14 MJ ha^{-1} and 120,000 MJ ha^{-1}, respectively, while, in ratoon ranches, the relating esteems are 122,801.15 MJ ha^{-1} and 98,850 MJ ha^{-1}, individually [26].

Kalia et al. (2020) recommended compact gadget which is skilled to check and measure the air quality and particulate matter concentration in understanding ecological boundary like encompassing temperature, dampness, environmental weight, and dew point [27].

Wang et al. (2019) investigation, a ANN called the incorporated long momentary memory arrange (LSTM), utilizing cross-connection and affiliation rules (a priori), was used to distinguish the qualities of water poisons and follow modern point wellsprings of contaminations. Water quality observing information from Shandong Province, China, was utilized to check the relevance of the man-made reasoning framework utilizing a cross-connection technique to build up a water quality cross-relationship map [28]. Wang et al. (2020) give a scientific classification examination of the current solar power forecasting determining models dependent on AI calculations. Scientific classification is a procedure of efficiently partitioning sun oriented vitality expectation techniques, analyzers, and forecast systems into a few classifications dependent on their disparities and likenesses [29].

Zhao et al. (2019) built up a wind energy decision system in light of multitude insight improvement, which incorporates two modules: wind vitality appraisal and wind speed anticipating. In the breeze vitality evaluation module, the boundaries of the Weibull dispersion were enhanced by utilizing a numerous multitude knowledge advancement calculation, and the ideal Weibull dispersion was gotten [30].

Kishorebabu et al. (2020) secure distinctive ecological data utilizing the various sensors and push the information to the site page or as SMS. The experiments were conducted for 15 days in June 2018 and changes in climatic conditions of Hyderabad region were studied and information was handed over to local farmers in terms of indica-tors which can be used as an assist to decide which crop (cotton, jowar, redgram) has to be cultivated/reaped. The result of the tests was summarized, and data about the earth is given to ranchers of Hyderabad district who utilized the information as a help to develop/procure provincial yields, for example, Cotton, Jowar, Corn, and Red Gram. The curiosity of the strategy is that it presented two checks (downpour and wind) with information approved utilizing meteorologist. The framework is an information assortment model, gathers information, and deciphers a marker to the ranchers that causes them in their yield. The accuracy of the framework is consistently a little concerned, and a significant level information examination must be performed to connect information between rancher's harvest and sensor information [31]. Tastan et al. (2019) developed an IoT device for monitoring air quality checking [32].

Karar et al. (2020) introduced another versatile application, to be specific GASDUINO to permit the clients to quantify and characterize the degree of the air contamination utilizing the Internet of Things (IoT) effectively [33].

Parmar et al. (2017) utilize minimal effort air quality observing hubs which include minimal effort semiconductor gas sensor with Wi-Fi modules. This frame-work measures centralizations of gases which utilize semiconductor sensors. The sensors will accumulate the information of different natural boundaries and give it to Raspberry Pi which go about as a base station. Acknowledgment of information accumulated by sensors is shown on Raspberry Pi 3-based Webserver. A MEAN stack is created to show information over site. The basic part of proposed work is to give minimal effort framework to empower the information assortment and scattering to all partners [34].

Geetha et al. (2016) presented a nitty gritty diagram of ongoing works completed in the field of shrewd water quality checking [35].

Singh et al. (2018) comprised utilizing different sensors and IoT devices air quality monitoring [36].

Ejaz et al. (2019) describe the idea of the keen city that was acquainted as the possible arrangement with the challenges made by urbanization with perplexing and exorbitant activities. The imagined objective of keen city is to be savvy, wise, and independent with convenience giving better personal satisfaction. Most definitions for brilliant city include the utilization of data and correspondence advances (ICTs) to improve the nature of urban existence with diminished expense and asset utilization. As of late, ICT union with the IoT has been successfully abused to furnish numerous novel highlights with least human intercession in brilliant urban communities. This book portrays various parts of IoT for shrewd urban areas including sensor advances, correspondence innovations, large information examination, and security. The book is sorted out into five sections that are depicted underneath. IoT offers savvy answers for urban areas as far as administration, financial development, natural manageability, personal satisfaction, transportation, force, and water utilization [37].

Blessy Evangelin et al. (2019) made the created air quality checking and perception framework precisely estimated the grouping of poisons carbon monoxide, carbon dioxide, smoke and residue in climate. The sensor has been composed with IoT structure which has capably been used to evaluate and screen the pollutions ceaselessly. This framework beats the issue of contamination observing, wellbeing checking, vocation estimation, supportability evaluations and estimation related fields. The data's are normally put away in the database; this information can be used by the specialists to take brief exercises. It also causes the run of the mill people to consider the proportion of defilements in their overall region and to take control measures. This is an incredible structure which is useful in organizations due to the growing pollution on account of addition in adventures. This structure is straightforward, and cost of the thing is sensible. This structure is checking only five boundaries and from now on can be stretched out by considering more boundaries that reason the tainting especially by the endeavors [38].

Lai et al. (2019) build up a minimal effort air quality checking and expectation framework by means of Raspberry Pi, which is an edge gadget to run the Kalman filter calculation [39].

Malche et al. (2019) developed IoT-based environmental monitoring system. Consequently, so as to keep away from well-being dangers because of the dirtied condition, it is basic to screen its state. In the present situation, monitoring of data on the state of the environment is not a well-researched field. Along these lines, it is required to build up a framework which can proficiently gather and dissect information on the earth so as to maintain a strategic distance from any likely dangers. The Internet is one of the vital and significant apparatuses which can be utilized to build up a framework equipped for observing and sharing data on natural contamination. This examination proposes IoT-based condition observing and ready framework. The proposed framework screens the district explicit condition for air quality, what's more, stable contamination, while likewise encouraging secure information transmission over the system which explains the security issues in IoT framework [40].

Mokrani et al. (2019) reviewed a testing of examination papers about the utilization of IOT frameworks for contamination and air quality observing, introduced the significant toxins and the related sicknesses, and set up a few groupings. In addition, a basic part of an examination business was introduced related to the effect of poisons on well-being. Additionally, we talked about the surveyed papers what's more, featured a few disadvantages. Be that as it may, this audit is not thorough: Some difficult perspectives, for example, security and protection, flexibility, and different classifications about vitality utilization recipes, or client suggestion are not talked about here and are arranged in an all-inclusive rendition of this paper as well as some measurable investigations. At the light of the overviewed papers, a few difficulties are brought up in the last segments. They speak to promising issues of existing frame works that should be tended to and issues for expected logical coordinated efforts and exploration headings [41].

Kumar et al. (2019) introduced IoT device for air quality monitoring. IoT use smart devices and internet to provide innovative solutions to various chal-lenges and issues related to various business, governmental and public/private industries across the world [42].

Kim et al. (2018) sum up the most recent exploration on sensor-based home IoT and concentrate on indoor air quality, what's more, specialized examinations on arbitrary information age. Moreover, they build up an air quality improvement model that can be promptly applied to the market by obtaining introductory inves-tigative information and building foundations utilizing range/thickness investigation and the regular cubic spline strategy. Likewise, they produce related information dependent on client social qualities. They incorporate the rationale into the current home IoT framework to empower clients to effectively get to the framework through the Web or portable applications. They expect that the current presentation of a useful promoting application strategy will contribute to upgrade the development of the home IoT market. Moreover, they build up an air quality improvement model that can be promptly applied to the market by obtaining beginning logical informa-tion and building frameworks utilizing range/thickness investigation and the normal cubic spline strategy. The author generate related data based on user behavioral values. They integrate the logic into the existing home IoT system to enable users to easily access the system through the Web or mobile applications. They incorporate the rationale into the current home IoT framework to empower clients to effectively get to the framework through the Web or versatile applications. It is expected that the current presentation of a reasonable promoting application strategy will contribute to upgrade the extension of the home IoT showcase [43].

Fang et al. (2014) present a novel IIS that joins Internet of Things (IoT), distributed computing, geoinformatics [remote detecting (RS), geological data framework (GIS), and worldwide situating framework (GPS)], and e-Science for ecological observing and the board, with a contextual investigation on territorial environmental change and its biological impacts [44].

Elmustafa et al. (2019) discussed that savvy condition sensors incorporated with IoT innovation can give another idea in following, detecting, and observing objects of condition. This can give potential advantages paving the way to the chance of accomplishing a green world and a manageable way of life. IoT permits ecological sensors to associate with different frameworks such advanced mobile phones through Bluetooth or Wi-Fi to send tremendous measures of information to the system and can permit us to have a superior comprehension of our environmental factors and find appropriate answers for now natural issues. In this audit article, we will give a short origination of condition regions of study dependent on IoT innovation and examine the defense behind utilizing IoT in the field of ecological examinations. Additionally, we will examine many proposed utilizations of ecological exploration dependent on IoT [45].

Wu et al. (2019) presented a low-power wearable sensor hub for natural Internet of Things (IoT) applications, framing remote sensor organize (WSN) in light of XBee. Natural information is checked by the wearable sensor hub and afterward transmitted to a distant cloud worker by means of WSN. The information is shown to approved clients through an online application situated in cloud worker. The trial results demonstrate that the introduced wearable sensor organize framework can screen natural conditions reliably. The creator presents a wearable IoT remote sensor organize for condition checking. Utilizing remote sensor arrange for wearable ecological observing is talked about. Low force procedures, programming, and detecting ecological boundaries are portrayed. The proposed work gives a powerful and dependable answer for long haul observing that presents numerous open doors in security-related checking applications [46].

Zhao et al. (2018) concentrated on the issue of urban air contamination which is exceptionally connected with the exorbitant specific issue noticeable all around; the work builds up an ongoing remote sensor framework mimicking the situation of a green rooftop that screens the PM 2.5 and other related sensor boundaries, for example, relative mugginess, temperature, and wind speed which depend on the IEEE 802.11b/g/n principles. The principle objective of the methodology is to legitimize the security, extensibility, what's more, information exactness of the model which was structured what's more, tried. Considering the way that genuine utilizations of green rooftops in the city require different simultaneous gadgets on the Web, the exhibition in the observing framework was extravagantly assessed and upgraded for simultaneous associations. What's more, we did various examinations to inspect and assess the framework as far as key focuses by and large execution, idleness, and exactness of information. Additionally, by cautiously choosing the equipment parts just as programming arrangement, the arrangement is profoundly powerful and has a wide possible application [47] (Figs. 8.1, 8.2, 8.3, 8.4 and 8.5).

8.3   Discussion

This chapter surveys the work of many researchers to get a brief overview about the current implementation of automation in agriculture, diagnosis of COVID-19, pollution monitoring using IoT devices, AI, and ML. IoT devices are dynamic and have constrained capacity and handling abilities. In any case, these regular brought together mists have a few troubles, similar to high idleness and system disappointment. So as to illuminate explicit challenges, haze processing has been advanced as an expansion of the cloud, in spite of nearer to the IoT devices in which all information handling will be done at haze hubs, through lessening dormancy, especially for time-delicate applications.

The proposed framework has no limits on the establishment place. IoT cloud was utilized to look at air quality information and to deliver the information perceptions for the end user. When breaking down the air quality pattern and practically identical elective information, some interesting realities will unfurl it. Likely, long haul and

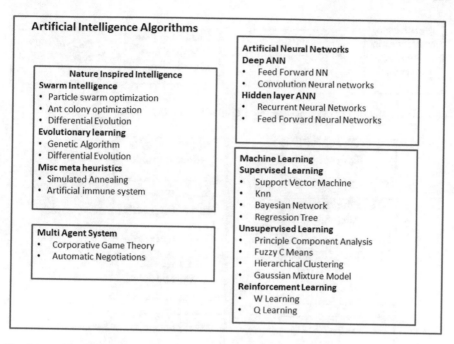

Fig. 8.1   Artificial intelligence algorithms used in diabetic retinopathy

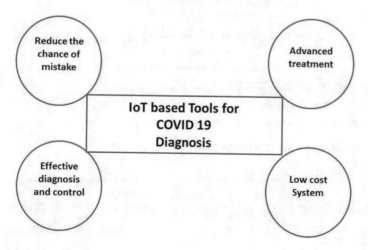

Fig. 8.2   Advantages of using IoT for COVID-19

huge scope air observing will altogether bolster us know air contamination and comprehend the procedure to take care of the issue of air contamination at any rate incompletely.

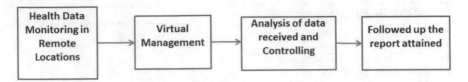

Fig. 8.3 Step-by-step process for detecting COVID-19 using IoT

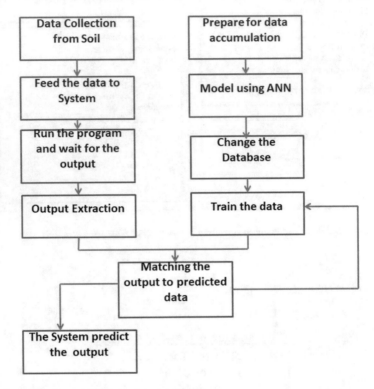

Fig. 8.4 AI-based crop predictor using smartphone

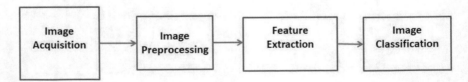

Fig. 8.5 Architecture of image recognition tool

8.4 Conclusion

Research activity in the field of environmental monitoring forecasted using IoT, AI, artificial neural networks (ANNs), and deep learning has increased dramatically in recent years. The discoveries of this audit likewise propose that more endeavors are coordinated toward the utilization of worldwide streamlining techniques in deciding the ideal model structure, as most of the distinguished papers done specially appointed methodologies and principally experimentation technique. This pattern may not be valuable to future modelers in the field of environmental pollution, estimating as specially appointed techniques offer restricted repeatability due to their case-explicit nature. Thus, there is a requirement for a more set up and precise convention for recognizing ideal model structures that takes into account a wide scope of model indicator yield elements.

References

1. Cabaneros, S. M. S., Calautit, J. K. S., & Hughes, B. R. (2017). Hybrid artificial neural network models for effective prediction and mitigation of urban roadside NO_2 pollution. *Energy Procedia, 142*, 3524–3530.
2. Antonopoulos, I., Robu, V., Couraud, B., Kirli, D., Norbu, S., Kiprakis, A., Flynn, D., Elizondo-Gonzalez, S., & Wattam, S. (2020). Artificial intelligence and machine learning approaches to energy demand-side response: A systematic review. *Renewable and Sustainable Energy Reviews, 130*, 109899.
3. Arora, J., Pandya, U., Shah, S., & Doshi, N. (2019). Survey-pollution monitoring using IoT. *Procedia Computer Science, 155*, 710–715.
4. Han, Y., Park, B., & Jeong, J. (2019). A novel architecture of air pollution measurement platform using 5G and blockchain for industrial IoT applications. *Procedia Computer Science, 155*, 728–733.
5. Singh, R. P., Javaid, M., Haleem, A., & Suman, R. (2020). Internet of things (IoT) applications to fight against COVID-19 pandemic. *Diabetes and Metabolic Syndrome: Clinical Research and Reviews, 14*(4), 521–524.
6. Mao, S., Wang, B., Tang, Y., & Qian, F. (2019). Opportunities and challenges of artificial intelligence for green manufacturing in the process industry. *Engineering, 5*(6), 995–1002.
7. Liu, P., Jiang, W., Wang, X., Li, H., & Sun, H. (2020). Research and application of artificial intelligence service platform for the power field. *Global Energy Interconnection, 3*(2), 175–185.
8. Jha, K., Doshi, A., Patel, P., & Shah, M. (2019). A comprehensive review on automation in agriculture using artificial intelligence. *Artificial Intelligence in Agriculture, 2*, 1–12.
9. Nourani, V., Gökçekuş, H., & Umar, I. K. (2020). Artificial intelligence based ensemble model for prediction of vehicular traffic noise. *Environmental Research, 180*, 108852.
10. Shafiq, M., Tian, Z., Sun, Y., Du, X., & Guizani, M. (2020). Selection of effective machine learning algorithm and Bot-IoT attacks traffic identification for internet of things in smart city. *Future Generation Computer Systems, 107*, 433–442.
11. Nishant, R., Kennedy, M., & Corbett, J. (2020). Artificial intelligence for sustainability: Challenges, opportunities, and a research agenda. *International Journal of Information Management, 53*, 102104.
12. Goralski, M. A., & Tan, T. K. (2020). Artificial intelligence and sustainable development. *The International Journal of Management Education, 18*(1), 100330.

13. Mohanta, B. K., Jena, D., Satapathy, U., & Patnaik, S. (2020). Survey on IoT security: Challenges and, artificial solution using machine learning intelligence and blockchain technology. *Internet of Things*, 100227.
14. Moazamnia, M., Hassanzadeh, Y., Nadiri, A. A., & Sadeghfam, S. (2020). Vulnerability indexing to saltwater intrusion from models at two levels using artificial intelligence multiple model (AIMM). *Journal of Environmental Management, 255*, 109871.
15. Díaz-Alcaide, S., & Martínez-Santos, P. (2019). Mapping fecal pollution in rural groundwater supplies by means of artificial intelligence classifiers. *Journal of Hydrology, 577*, 124006.
16. Tung, T. M., & Yaseen, Z. M. (2020). A survey on river water quality modelling using artificial intelligence models: 2000–2020. *Journal of Hydrology, 585*, 124670.
17. Shu, Y., Chen, Y., & Xiong, C. (2020). Application of image recognition technology based on embedded Technology in environmental pollution detection. *Microprocessors and Microsystems*, 103061.
18. Senthilkumar, R. (2020). Intelligent based novel embedded system based IoT enabled air pollution monitoring system. *Microprocessors and Microsystems*, 103172.
19. Zhao, L., Dai, T., Qiao, Z., Sun, P., Hao, J., & Yang, Y. (2020). Application of artificial intelligence to wastewater treatment: A bibliometric analysis and systematic review of technology, economy, management, and wastewater reuse. *Process Safety and Environmental Protection, 133*, 169–182.
20. Zahmatkesh, H., & Al-Turjman, F. (2020). Fog computing for sustainable smart cities in the IoT era: Caching techniques and enabling technologies-an overview. *Sustainable Cities and Society*, 102139.
21. Sharma, A., Singh, P. K., & Kumar, Y. (2020). An integrated fire detection system using IoT and image processing technique for smart cities. *Sustainable Cities and Society*, 102332.
22. Abdallah, M., Talib, M. A., Feroz, S., Nasir, Q., Abdalla, H., & Mahfood, B. (2020). Artificial intelligence applications in solid waste management: A systematic research review. *Waste Management, 109*, 231–246.
23. Cabaneros, S. M., Calautit, J. K., & Hughes, B. R. (2019). A review of artificial neural network models for ambient air pollution prediction. *Environmental Modelling & Software, 119*, 285–304.
24. Elkiran, G., Nourani, V., & Abba, S. I. (2019). Multi-step ahead modelling of river water quality parameters using ensemble artificial intelligence-based approach. *Journal of Hydrology, 577*, 123962.
25. Fan, M., Hu, J., Cao, R., Ruan, W., & Wei, X. (2018). A review on experimental design for pollutants removal in water treatment with the aid of artificial intelligence. *Chemosphere, 200*, 330–343.
26. Kaab, A., Sharifi, M., Mobli, H., Nabavi-Pelesaraei, A., & Chau, K. W. (2019). Combined life cycle assessment and artificial intelligence for prediction of output energy and environmental impacts of sugarcane production. *Science of the Total Environment, 664*, 1005–1019.
27. Kalia, P., & Ansari, M. A. (2020). IOT based air quality and particulate matter concentration monitoring system. *Materials Today: Proceedings, 32*, 468–475.
28. Wang, P., Yao, J., Wang, G., Hao, F., Shrestha, S., Xue, B., & Peng, Y. (2019). Exploring the application of artificial intelligence technology for identification of water pollution characteristics and tracing the source of water quality pollutants. *Science of the Total Environment, 693*, 133440.
29. Wang, H., Liu, Y., Zhou, B., Li, C., Cao, G., Voropai, N., & Barakhtenko, E. (2020). Taxonomy research of artificial intelligence for deterministic solar power forecasting. *Energy Conversion and Management, 214*, 112909.
30. Zhao, X., Wang, C., Su, J., & Wang, J. (2019). Research and application based on the swarm intelligence algorithm and artificial intelligence for wind farm decision system. *Renewable Energy, 134*, 681–697.
31. Kishorebabu, V., & Sravanthi, R. (2020). Real time monitoring of environmental parameters using IOT. *Wireless Personal Communications, 112*(2), 785–808.

32. Taştan, M. (2018). An IoT based air quality measurement and warning system for ambient assisted living. *Avrupa Bilim ve Teknoloji Dergisi, 16*, 960–968.
33. Karar, M. E., Al-Masaad, A. M., & Reyad, O. (2020). *GASDUINO-Wireless Air Quality Monitoring System Using Internet of Things.* arXiv preprint arXiv, 2005,04126
34. Parmar, G., Lakhani, S., & Chattopadhyay, M. K. (2017). An IoT based low cost air pollution monitoring system. In *International Conference on Recent Innovations in Signal processing and Embedded Systems (RISE)* (pp. 524–528). IEEE.
35. Geetha, S., & Gouthami, S. (2016). Internet of things enabled real time water quality monitoring system. *Smart Water, 2*(1), 1.
36. Singh, N. K., Singh, A., & Singh, R., & Gehlot, A. (2018). Design and development of air quality management devices with sensors and web of things. *International Journal of Engineering and Technology (UAE), 7*(2).
37. Ejaz, W., & Anpalagan, A. (2019). *Internet of things for smart cities: technologies, big data and security* (pp. 1–15). Springer International Publishing.
38. Blessy Evangelin, K., & Pandian, M. T. (2019). *IoT based air pollution monitoring system to create a smart environment.*
39. Lai, X., Yang, T., Wang, Z., & Chen, P. (2019). IoT implementation of kalman filter to improve accuracy of air quality monitoring and prediction. *Applied Sciences, 9*(9), 1831.
40. Malche, T., Maheshwary, P., & Kumar, R. (2019). Environmental monitoring system for smart city based on secure internet of things (IoT) architecture. *Wireless Personal Communications, 107*(4), 2143–2172.
41. Mokrani, H., Lounas, R., Bennai, M. T., Salhi, D. E., & Djerbi, R. (2019). Air quality monitoring using IoT: A survey. In *IEEE International Conference on Smart Internet of Things (Smart IoT)* (pp. 127–134) IEEE.
42. Kumar, S., Tiwari, P., & Zymbler, M. (2019). Internet of Things is a revolutionary approach for future technology enhancement: A review. *Journal of Big Data, 6*(1), 111.
43. Kim, J., & Hwangbo, H. (2018). Sensor-based optimization model for air quality improvement in home IoT. *Sensors, 18*(4), 959.
44. Fang, S., Da Xu, L., Zhu, Y., Ahati, J., Pei, H., Yan, J., & Liu, Z. (2014). An integrated system for regional environmental monitoring and management based on internet of things. *IEEE Transactions on Industrial Informatics, 10*(2), 1596–1605.
45. Elmustafa, S. A. A., & Mujtaba, E. Y. (2019). Internet of things in smart environment: Concept, applications, challenges, and future directions. *World Scientific News, 134*(1), 1–51.
46. Wu, F., Rüdiger, C., Redouté, J. M., & Yuce, M. R. (2019). A wearable multi-sensor IoT network system for environmental monitoring. In: Advances in body area networks, pp 29–38
47. Zhao, Z., Wang, J., Fu, C., Liu, Z., Liu, D., & Li, B. (2018). Design of a smart sensor network system for real-time air quality monitoring on green roof. *Journal of Sensors.*

Chapter 9
E-Waste Management in Digital Healthcare System and Sustainability Paradigm

Abhijit Bandyopadhyay, Ritam Chatterjee, and Nilanjan Das

Abstract In modern time, the e-waste generation has become one of the most serious environmental issues and difficulties to attaining long-term development. There is a pressing need for study in this area, given the potential for negative ecotoxicological consequences and a wide range of health implications. This study aims to determine the requirement for appropriate methodology, implementing basic models and approaches. In this study, an effort was made to comprehend the issues of e-waste, its production, source and destination of e-wastes, the harmful toxic contaminants present in e-wastes and their hazardous effects on human health and the environment, and the lack of collective consciousness that exacerbates the problem. A thorough investigation has been conducted into how policies may be implemented at the national, regional, and global levels to address e-waste issues. In vulnerable nations, such as developing countries in South Asia in general and India in particular, formal sequential inventory activities have been determined to be necessary. Incorporating policy implementation with an emphasis on knowledge and awareness creation has proved to be beneficial with sustainable management solutions. Furthermore, economic, environmental, and technical cooperation between high-income nations that create and supply e-waste and those that are affected by it, particularly low-income ones, might be improved. The conclusion of this study chapter is that comprehensive global e-waste management and legislation may be the best method for attaining sustainable development and mitigating the risks of e-waste.

Keywords E-waste · Sustainable practices · Hazardous waste · Waste disposal · Occupational hazards · Biomagnifications · Toxicity · Reverse logistics

A. Bandyopadhyay (✉)
Raniganj Institute of Computer and Information Science (RICIS), Raniganj, West Bengal 713358, India
e-mail: abhijit.ricis@gmail.com

R. Chatterjee
Raiganj University, Raiganj, West Bengal 733134, India

N. Das
Siliguri Institute of Technology, Siliguri, West Bengal 734009, India

9.1 Introduction

The age of digital revolution has brought the menace of e-waste. Industries are trying to adopt the zero-waste initiatives while healthcare professionals face pressure to implement digital innovation in consonance with increasing awareness about environmental sustainability. Much of this is contributed by health care. It is very difficult to apply digital healthcare system without proper strategic plan to manage e-waste. It is required to know about the characteristics of e-waste in health care to ensure the zero e-waste in the future and eliminate harmful effects of digital technology [1]. The formidable amount of e-waste comes from the built-in obsolescence of many devices. India is at a transitional stage in the adoption and implementation of universal digital healthcare system. It is very tough work to dumping e-waste materials in India due to huge amount of e-waste generated daily. It is major issues for environment and health [2]. In India huge amount of e-waste materials are stored in house. Therefore, e-waste materials are stocked without discarding that create problem. Valuable recyclable metals such as silver and gold generated from mass of e-waste materials that causes to create opportunities in business of recyclables metals in India [3]. Therefore, e-waste materials very harmful to health and environment that indicate to establish the need of strong e-waste management strategies [4].

The life of human is very smooth, easy, and fast due to evolution of the technologies. Various electronics and electrical devices are discovered and developed including home appliances, computer, mobile phones, etc. [5]. The problem lies in their safe handling and responsible disposal. Reselling of e-products may invite many problems, and especially, it can be responsible for the data breaches.

9.2 Objectives

Objectives of this work are as follows-

- To find the problems of e-waste in India
- To explore the causes of e- wastes generation
- To find out the outcome of e-wastes on the life of human
- Examine the risks arises from e-wastes that affects eco-system or environment.
- To suggest ways to minimize e-waste generation
- To recommend adoption of technology in alleviating the e-waste problem
- To conceptualize global partnerships to solve this health menace
- To indicate the future research areas that can be adopted to avoid e-waste peril
- To amalgamate the findings and conclude with a positive note to this global menace.

9.3   Methodology

The first step in applying any research methodology approach is to identify the geographical boundaries of the area of this research. This research area included, here is India specific but e-waste being a global menace so data has been collected from research works done across the geographical boundaries of countries and regional blocks to pragmatically assesses the situation and arrives at a believable conclusion. Hence, previous research works from researchers belonging to different parts of the world have been selected for this secondary research work. Certain statistical techniques like comparative analysis and extrapolation of data have been studied to arrive at meaningful conclusions at the end of the research chapter. This is an evaluative research work which analyses previous works done so far in this field.

9.4   E-Waste

E-waste is electronics or electrical devices and equipments garbage [6]. Various sources of e-waste are equipments of computer devices, mobile phones, parts of mobile phones, tablets, equipments of home appliances including TV [7], refrigerators, washing machine, microwave, batteries, printing devices, and electronics devices are used in health care including X-ray machine parts, MRI machine equipments, ultrasound machine parts, etc.

E-waste created in different sectors such as industry, institutions, households, health care, organizations, etc. The huge amount of e-waste materials is produced due to the use of emerging technologies in every sector. In healthcare sector, various electronics and electrical equipments such as computer, X-ray, MRI, ultrasound, City scan, and others are used to provide operational services. Also in other sectors, the use of electronics or electrical devices are increased gradually [8].

E-waste materials are the sources of obsolete in build equipments of the electronics or electrical devices. E-waste materials generate harmful chemicals or toxic substances [9]. Harmful chemicals of e-waste are lead, cadmium; copper, iron, and other toxic substances [10]. Harmful chemicals have huge impacts in environment and dangerous for human health [11]. Therefore, management of the e-waste materials is very essential to save the environment and life.

9.5   Digital Healthcare System

Healthcare system provides various services related to the health care of people and society. Doctors, health center, and hospitals are associated with this system to deliver the services for the peoples and society. It is very essential to build strong healthcare system for the survival of the society. Also, strong healthcare information system is needed for the research and development purposes [12]. Whole healthcare system is work with the help of the various emerging technologies and the peoples who are associated with healthcare system.

Digital technology is one of the important technologies in the healthcare system that make healthcare system is simpler, easy and very fast. It is required to conduct any operational activities and management related activities [13]. This technology is also help to create proper information system for the health care. Various electronics and electrical devices are used in healthcare system and generate information regarding health issues. Digital technology also helps for research and development in healthcare system. Therefore, digitalization is very essential in healthcare system to provide various health care related services to the society.

Various kinds of electrical and electronics devices like X-rays, MRI, ultrasound, and City scan are used for the operational activities and computer, printer, scanner, and Xerox machine are implemented for management related activities in digital healthcare system.

9.6   E-Waste in Digital Healthcare System

Electronics waste is created due to the utilizations of various devices in healthcare system. Rejected, damaged, and obsolete equipments or parts of the electronics devices generate e-wastes in healthcare system. It is very difficult for the scrapping of e-waste materials in healthcare system.

Hospitals are going paperless [2], yet resulting in an increasing data center footprint that has immense energy demands that inadvertently leads to the unregulated and negative effects of digital healthcare technology. Demands for data storage and computation have exploded over the past decade and energy consumption has doubled in every four years. That significant growth resulted in a carbon footprint larger than the entire aviation industry, largely driven by locations with high-temperature climates, needing more power to maintain centers at optimal temperatures.

The dumping of e-waste spreading dangerous chemicals that has bad effects for the health of the patients and other peoples associated with the healthcare centers and hospitals.

9.7   Sustainable Digital Healthcare E-Waste Management

E-waste management must be required for the sustainability in environment and life. Various initiatives have been taken to manage the e-waste very efficiently. Different types of ways are followed to manage the e-wastes such as reduce, reuse, and recycle. Different organizations, industries, and government take initiative to manage digital e-wastes. Some industries has policies to collect all e-waste materials and redevelop the materials new way for reuse [11]. Industries are also organized campaign to

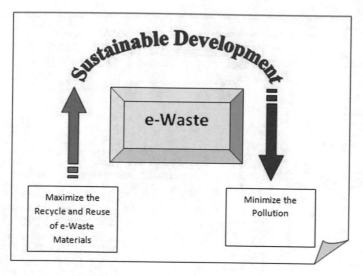

Fig. 9.1 Sustainable e-waste management in digital healthcare system

manage e-wastes and for destroying all the e-wastes or they collect all the e-wastes to dispose. Government organizations also take initiatives to collects e-wastes from home, and other places for proper management of e-wastes [14]. The main aim of the management of e-wastes generated from digital health care is to make sustainable and pollution free environments [15]. Therefore, government organizations, industries, and other organizations take part for sustainable digital e-wastes management. Figure 9.1 shows the sustainable e-waste management in digital healthcare system.

9.8 E-Waste Management in Digital Healthcare System Basic Model

E-waste management in digital healthcare system is very necessary. It needs lots of planning and management policies. Different departments are works simultaneously. It uses lots of modern technologies likes IT, ICT, cloud computing, big data analysis, IoT, wireless IoT, data mining, etc. Figure 9.2 shows the basic model on e-waste management in digital healthcare system.

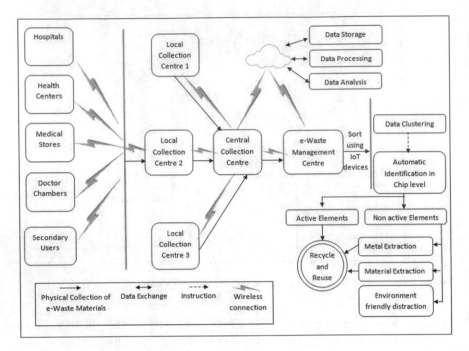

Fig. 9.2 Basic model on e-waste management in digital healthcare system.

9.9 Emerging Technologies Uses in E-Waste Management in Digital Healthcare System

Various modern technologies have used in e-waste management in digital healthcare system. To execute the system properly, it needs to upgrade the existing system by the use of modern technologies. Some of the emerging technologies which will be used in e-waste management in digital healthcare system are as follows:

- **Network technologies**

  – Basic computer network
  – Wireless network
  – IOT and WIOT
  – Cloud computing
  – Cyber security

- **Web technologies**

  – Web-based programming
  – Modern scripting language
  – High level web-based programming language
  – Content management

- **Data base technologies**
  - Data base system
  - Distributed data base
  - Big data analysis

- **Multi-media technologies**
  - Human computer interaction
  - Interaction designing
  - Robotics, etc.

9.10   E-Waste in Digital Healthcare System Stake Holder Model

Stake holders are plays very important role in e-waste management in digital health-care system. Figure 9.3 shows the stake holder model in e-waste management in digital healthcare system.

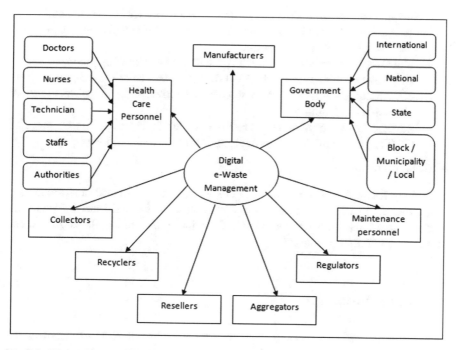

Fig. 9.3 E-waste management in digital healthcare system

9.11 Suggestions and Recommendations

Suggestions and recommendations are as follows

- **Lifecycle Management**

  There are ways to use lifecycle management and move toward a circular economy in which materials that could end up in landfills are reused, remanufactured, or recycled into new, usable products. Some of these methods include—

  - Reducing paper waste through recycling.
  - Repurposing paper.
  - Remarketing devices.
  - Turning to renewable energy sources.
  - Reduce carbon dioxide emissions.
  - Evaluation of device vendor selection by hospitals.
  - Prioritize forward-thinking internet of medical things manufacturers.
  - Accountability for repair, disassembly, and recycling features for devices.

- **Progressing into a circular economy**

  - Hospitals and healthcare organizations that step forward into sustainability and make an effort to offset the negative impact of digital technology can expect to see tangible, financial benefits.
  - Johns Hopkins has been leading the pack in e-waste recycling for over a decade now. They were recognized in 2010 for their practice of recycling a range of medical equipment, and have also earned 13 local and national recognitions related to sustainability.
  - Mayer Alloys Corporation shared that in 2019, the organization recycled 250,000 pounds of single-use devices, furniture, and equipment.

- **Legal measures**

  Moving forward, hospital leaders can expect to see an increase in sustainability initiatives, along with a more mindful approach to digital innovation. Healthcare leaders should take the time to weigh their inclusion among those zero-waste movements within the industry.

9.12 Conclusion

One of the increasingly spreading environmental problems of mother earth is the fatally harmful nature of e-waste. The growing pile of e-waste, compounded by a lack of communal awareness and expertise, is exacerbating the situation. To make a living, a vast number of employees engage in crude dismantling of electrical and electronic gadgets, endangering their health. As a result, there is a critical need to develop a strategy for preventing occupational health dangers associated with e-waste management among these unlucky employees in India and throughout the world. In terms of safe handling and disposal of e-waste, as well as personal health protection measures, these personnel needs to be given accurate and timely information. Many technical solutions may be implemented in the e-waste management system, but the necessary foundations, such as laws, collecting systems, logistics, and labor, must first be in place. This may necessitate operational research and evaluation studies to determine their suitability for long-term e-waste management.

Acknowledgements Special thanks to Dr. P. K. Paul, Department of Computer and Information Science, Raiganj University, West Bengal, India—733134 for his motivation and guidance.

References

1. Healthcare Environmental Resource Center, *Pollution Prevention and Compliance Assistance Information for the healthcare Industry*. Available from: https://www.hercenter.org/wastereduction/paper.php. Last Accessed on January 21, 2022.
2. Dahl, R. (2002). Who pays for e-junk? *Environmental Health Perspectives, 110*, A196–A199.
3. Baud, I., Grafakos, S., Hordjik, M., & Post, J. (2001). Quality of life and alliances in solid waste management. *Cities, 18*, 3–12.
4. Pandve, H. T. (2007). E-waste management in India: An emerging environmental and health issue. *Indian Journal of Occupational and Environmental Medicine, 11*, 116.
5. UN Environment Programme, UN Environment Programme, Available from: https://www.unep.org/news-and-stories/press-release/un-report-time-seize-opportunity-tackle-challenge-e-waste. Last Accessed on January 21, 2022.
6. Balde, C. P., Wang, F., Kuehr, R., & Huisman, J. (2014). *The global e-waste monitor* (p. 80). United Nations University, IAS—SCYCLE, Bonn, Germany.
7. Frazzol, C., Orisakwe, O. E., Dragone, R., & Mantovani, A. (2010). Diagnostic health risk assessment of electronic waste on the general population in developing countries' scenarios. *Environmental Impact Assessment Review, 30*(6), 388–399.
8. Brigden, K., Labunska, I., Santillo, D., & Allsopp, M. (2005). *Recycling of electronic wastes in China and India: workplace and environmental contamination*. Greenpeace.
9. E-waste: A Challenge for Sustainable Development. *Journal of Health and Pollution*. Available from: https://meridian.allenpress.com/jhp/article/5/9/3/67510/E-waste-A-Challenge-for-Sustainable-Development. Last Accessed on January 21, 2022.
10. PMC, US National Library of Medicine, National Institutes of Health, Available from: https://www.ncbi.nlm.nih.gov/pmc/articles/PMC6221494/. Last Accessed on January 21, 2022.
11. Qiu, B., Peng, L., Xu, X., Lin, X., Hong, J., & Huo, X. (2004). Medical investigation of e-waste demanufacturing industry in Guiyu town. In *Proceedings of the International Conference on Electronic Waste and Extended Producer Responsibility* (pp. 79–83). Greenpeace and Chinese Society for Environmental Sciences, April 21–22, 2004.
12. Paul, P. K., Bhuimali, A., Ghose, M., & Poovammal, E. (2016). Health information systems: The issues related to governmental initiatives, political and economics—a theoretical overview. *IRA Academico Research, 4*(2).
13. Paul, P. K., & Aithal, P. S. (2017). Mobile cloud computing vis-à-vis eco friendliness for sustainable development. *International Journal of Engineering Research and Modern Education, 2*(2), 28–32.

14. Paul, P. K., & Aithal, P. S. (2020). Environmental informatics and its features, functions and stakeholders: A comprehensive overview. *Educational Quest: An International Journal of Education and Applied Social Science, 11*(1), 01–05.
15. Paul, P. K., Bhuimali, A., Aithal, P. S., Tiwary, K. S., & Saavedra, R. (2020). Artificial intelligence & cloud computing in environmental systems—towards healthy & sustainable development. *International Journal of Inclusive Development, 6*(1), 69–76.
16. Agarwal, R., Ranjan, R., & Sarkar, P. (2003). *Scrapping the hi-tech myth: Computer waste in India*. New Delhi: Toxics Link.
17. Pinto, V. N. (2008). E-waste hazard: The impending challenge. *Indian Journal of Occupational and Environmental Medicine, 12*(2), 65–70.

Chapter 10
Advances and Applications of Bioremediation: Network of Omics, System Biology, Gene Editing and Nanotechnology

Rohit Chakraborty, Sahita Karmakar, and Waliza Ansar

Abstract Environmental pollution has been on the incline in the recent decades owing to expanded human movements on energy utilizations, perilous agricultural techniques, and surge in industrialization. Heavy metals, pesticides, various nuclear wastes, greenhouse emitting gases, and hydrocarbons are the well-known pollutants that cause environmental and human health problems due to their toxicity. Bioremediation pinpoint the involvement of chemical machinery in environmental decontamination of pollutants by microbial discourse or web through in situ or ex situ outcome. For degrading the pollutants, in situ process required bioaugmentation, biosparging, and bioventing while ex situ bioremediation involves composting, bioreactors, electrodialysis, land farming, and biopiling. Microorganisms utilizing hydrocarbon as the sole resource of carbon and energy have a vital role in the biodegradation of pollutants. Due to the continuous environmental variations, the microorganisms thriving in that environment are well equipped to survive. The actinomycetes, fungus, and thermophylic bacterium like microbes in different biomes have been isolated in biodegradation. With the improvement of scientific technologies, the system biology, omics (proteomics and glycomics) nanotechnology, and gene editing tools are being used in bioremediation of heavy metal pollutants, plastics, petroleum, organic pollutants, or other hydrocarbon, acid leachate, biofilm formation, and xenobiotics. System biology approaches are very promising in decoding the existence of microbial populations under varied environmental setup. Omics such as proteomics and genomics aid in analyzing genetic or protein-level regulation for bioremediation with sequencing, MALDI-TOF, and novel functional genes' involvement in bioremediation pathways of pollutant degradation.

Keyword Bioremediation · Microbial nanotechnology · Gene editing · System biology · Proteomics · Glycomics · Genomics · Xenobiotics · Pollution

R. Chakraborty · S. Karmakar
Department of Environmental Science, Asutosh College, 92, Shyama Prasad, Mukherjee Road, Kolkata 700026, India

W. Ansar (✉)
Department of Zoology, Behala College, 32 Upen Banerjee Road, Parnasree, Behala, Kolkata, West Bengal 700060, India
e-mail: waliza_ansar@yahoo.co.in

© The Author(s), under exclusive license to Springer Nature Singapore Pte Ltd. 2022
P. K. Paul et al. (eds.), *Environmental Informatics*,
https://doi.org/10.1007/978-981-19-2083-7_10

Abbreviations

CMC	Critical micelles concentration
CNTs	Carbon Nanotubes
CRISPR	Clustered Regularly Interspaced Short Palindromic Repeats
DCE	Dichloroethane
DDE	2, 2-Bi's(p-chlorophenyl)-1, 1-dichloroethylene
DDT	Dichlorodiphenyltrichloroethane
HCNTs	Hybrid carbon nanotubes
HDR	Homology directed repair
MB	Microbial bioremediation
MNPs	Magnetic nanoparticles
MWCNTs	Multi-walled carbon nanotubes
NHEJ	Non-homologous end joining
NMs	Nanomaterials
NPs	Nanoparticles
NZVI	Nanoscale zero-valent iron
PAHs	Polycyclic aromatic hydrocarbons
PAM	Protospacer adjacent motif
PAMAM	Polyamidoamine
PBS	Polybutylene succinate
PCE	Perchloroethylene
PCL	Polycaprolactone
PE	Polyethylene
PET	Polyethylene terephthalate
PHA	Polyhydroxyalkanoate
PHB	Polyhydroxybutyrate
PLA	Polylactic acid or polylactide
PP	Polypropylene
PS	Polystyrene
PUR	Polyurethane
PVC	Polyvinyl chloride
SVocs	Semi-Volatile organic compounds
SWCNTs	Single-walled carbon nanotubes
TALENS	Transcription activator-like effector nucleases
TCE	Trichloroethylene
Vocs	Volatile organic compounds
WRF	White rot fungi
ZFNs	Zinc finger nucleases
ZFPs	Zinc finger proteins

10.1 Introduction

Due to extensive urbanization, industrial progress, and recent agricultural processes, an unleashed upsurge of varied contaminants or pollutants are outsourced into the nature. These pollutants can pollute medium in air, water, and soil and cause abhorrent changes in the environment. Pollution can cause biodiversity losses, huge deforestation, soil degradation, and damage to human well-being and wealth. These pollutants are heavy metals (cadmium, zinc, arsenic, nickel, chromium, mercury, lead, and copper), combustion pollutants like carbon monoxide (CO), ammonia, emissions of chlorofluorocarbons (CFCs), hydrocarbons, organic polluting compounds (dioxins, furans, and volatile organic compounds, VOCs), nitrogen and sulfur oxides and dioxides, and various particulate matter. Many of these pollutants may act as potent or suspected mutagens and carcinogens and may modify ecosystem regulations. Thus, a number of eco-friendly cleanup technologies have been advocated using phytoremediation, eradication, bioremediation, bioattenuation, physical and chemical bioremediation. For environmental remediation, age-old unsustainable methods of treatment like pump-and-treat, isolation, and disposal to landfill are gradually turning up to be redundant. Expansion of alternative sustainable treatment techniques provides effective remediation of contamination and also restoring the integrity of natural habitats and niche.

10.2 Bioremediation: The Network of Biochemical Process

Bioremediation is the outcome of different chemical reactions to degrade contaminants by creating a web of metabolic pathway at the contaminated site. Common bioremediation methods include bioleaching, land farming, bioventing, bioreactor, bioslurping, bioaugmentation, composting, natural attenuation, phytoremediation, biostimulation, and rhizofiltration [1]. Bioremediation of an on-site polluted zone mostly interplay in two dissimilar ways. For the first method, optimal conditions (like nutrients, temperature, and presence of oxygen) are utilized to its maximum to stimulate or trigger to the growth of indigenous microorganisms (pollutant-eating microbes) inhabiting the contaminated site. Genetically modified microbes can also be engineered and imposed on the contaminated site for better result. Contaminants can also be carried to a second site, processed according to granularity or contaminant nature, and then, microbes were added to continue the bioremediation process.

Bioremediation (unique eco-friendly method) is the interaction of biochemical processes inclined to decompose and expand microorganisms metabolic network at the polluted site by in situ or ex situ methods [2]. Both aerobic and anaerobic processes of degradation cascade have been shown in the mineralization and stabilization of pollutants. In diverse environmental setup, both aerobic and anaerobic methods may be applicable in single mode or in complex mode. *Pseudomonas, Sphingomonas,*

Alcaligenes, Mycobacterium, and *Rhodococcus* bacterial species degrade hydrocarbons and pesticides under aerobic conditions by utilizing carbon as the main source of energy.

Within the arena of bioremediation, phytoremediation (Fig. 10.1) offers promising benefits with the synergistic employment of plants and microbes. Moreover, various plants exhibit a diversity of additional decontamination methods in comparison with microbes' population alone. The science of system biology and multi-omics provides information about the microbial biology and inter-microbial interactions [3]. All these microbes-related operating systems require optimal conditions to respond and propagate. But in environmental stress (such as extreme temperature, availability of oxygen, and pressure) settings, depending on the type of contaminant and the dose of the inflicting pollutant, microbes respond differentially [4]. Thus, system biology with the aid of genomics, proteomics, transcriptomics, and metabolomics helps in identifying genetic level regulation, target proteins, post-translational modifications, metabolic cascade, and signal transduction pathways analysis for bioremediation. Gene level regulation can be identified by next-generation sequencing and high throughput sequencing to pinpoint the novel functional genes engrossed in bioremediation pathways of assorted relentless contaminants [5].

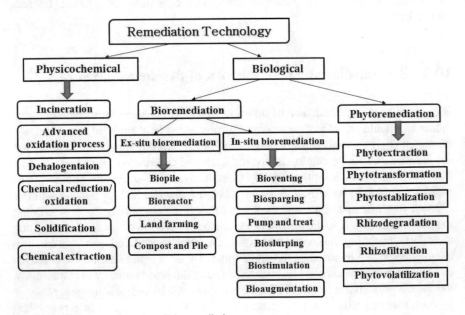

Fig. 10.1 Different techniques of bioremediation

10.3 In Situ and Ex Situ Bioremediation

In situ bioremediation (bioaugmentation, biostimulation, bioslurping, bioventing, biosparging, etc.) cleanup without elimination of soil from the polluted site of contamination and *ex situ* (biopiling, land farming, bioreactors, biofilter, electro-dialysis, and composting) remediate the undigged soil at the polluted place and thereafter transporting them to another location for treatment (Fig. 10.1). Simultaneous multiple bioremediation techniques, along with bacterial dispersal networks in spatial configuration, application of genetically engineered microorganisms as designer biocatalyst, employment of efficient, and novel metabolic, extending the substrate range of accessible pathways may enhance the efficacy of bioremediation even for recalcitrant compounds.

10.3.1 Techniques of Ex Situ Bioremediation

Ex situ process of bioremediation methods are mainly evaluated depending on: kind of physical/chemical pollutant, depth of pollution spread, the expense of detoxification, geographical setting, intensity of pollution, and geological information of the contaminated pollution-loaded site [6].

Land Farming

This technique is the uncomplicated, less equipped treatment process (superficial soil of 10–35 cm) in which polluted sediment, soil, or sludge is unearth and layered on a ready bed and circularly shifted for aerobic degradation and volatilization by autochthonous microorganisms. It depends on pollutant depth, extent of tillage, and irrigation for thorough aeration and nutrient (nitrogen, phosphorous, and potassium) enrichment. It is mainly applicable for hydrocarbon decontamination mainly aromatic hydrocarbons [7–9].

Biopile

Biopile (biomounds, biocells, or bioheaps) implicates up-surface piling (stacking) of blended contaminated soil (with petroleum hydrocarbon mostly), on a treatment area/bed using remediated by forced aeration (by irrigation and tillage), leachate collection, and nutrient amendment (carbon and nitrogen) [10, 11]. It is applicable for different oils (diesel, crude, lubrication) in severe environmental conditions. The process can be enhanced by adding more pollutants degrading microbes, mechanical degradation, ambient environment, aeration, and sieving techniques of contaminated soil before actual procedure of eradication [11, 12].

Bioreactor

Bioreactor (bioremediation modes are batch, fed-batch, continuous, semi-batch and multi-stage batch, sequencing batch), as the name signifies, is a biological reactor where the polluted materials (slurry or dry matter) were incorporated for chemical reactions in optimum growth conditions (temperature, pH, aeration velocity, agitation, substrate, and inoculum application dosage) for natural maintenance and mimicking of cells of indigenous microbes or genetically modified organisms. It is applicable for petroleum, polyaromatic hydrocarbon, linear alkyl benzene sulfonate, total nitrogen, etc [8, 12].

10.3.2 Technology of In Situ Bioremediation

In situ techniques of bioremediation is primarily low-cost technology applied for treatment of contaminations (dyes, heavy metals, hydrocarbons, chlorinated solvents, etc.) without excavation and disturbance in soil texture in suitable environmental conditions (moisture content, pH, temperature, nutrient availability, electron acceptor). But on-site installation of some equipment is required to introduce indigenous microbes for acquaintance [8, 12].

Bioaugmentation

The in situ bioaugmentation techniques increase the biodegradative capabilities of microbes (indigenous or allochthonous or genetically modified form rather than single isolates) of desired catabolic pathways to deteriorate polyaromatic hydrocarbons (effective for high molecular weight and recalcitrant compound also) in such site [13]. Bioaugmentation and biostimulation goes hand in hand for multiple remediation. The thriving of the exogenous species in competition with indigenous microbes causes a risk in this process.

Bioventing

Bioventing is an engineered bioremediation where controlled activation of airflow by providing oxygen to vadose (unsaturated) layer by indigenous aerobic microbes to degrade VOCs and semi-VOCs (SVOCs). It also restores the quality of polluted site, where air flow rate is the prime factor [14, 15].

Bioslurping

Bioslurping is an effective technique (uses a slurp which draws free products and soil gas) for groundwater and soil remediation; it involves soil vapor mining, vacuum-enhanced pumping, and bioventing to activate the biodegradation of contaminant by indirect supply of oxygen. The method is applicable for volatile organic and semi-volatile compounds and also LNAPLs (light non-aqueous phase liquids). This low-cost technique is not proper for degrading soil with excessive soil moisture and low permeability [16].

Biosparging

Biosparging disperses or injects air into subsurface soil (saturated zone) causing upward movement of pollutants to unsaturated zone. Biosparging has been widely applied for toluene, diesel, benzene, ethylbenzene, kerosene, etc., remediation by indigenous microbes. Soil permeability, pollutant bioavailability to microbes and high airflow rate for pollutant volatilization regulate pollutant biodegradability [17].

Biostimulation

Biostimulation involves the supplement of nutrients (nitrogen, potassium, and phosphorus), substrates, electron donors or acceptors, to the contaminated sample to activate metabolic activities of autochthonous microbes. Pollutant concentration directly affects the increased metabolic activity of the microbes, although excessive stimulation of microbes leads to deteriorating bioremediation process [18].

10.4 Pollutant Degradation by Bioremediation

Microorganisms are best applied to the task of bioremediation because they secrete enzymes. Microbes being small, when comes in contact with pollutants, feed on them as their food. Thus, bioremediation with operational technology depends on indigenous microbes growing on the polluted sites, encouraging them to propagate (on site) by providing them with the best-suited nutrients and other relevant chemicals indispensable for their metabolic pathways. Scientists are presently studying ways to implant/introduce contaminated sites with exotic microorganisms including genetically modified/augmented microorganisms mainly designed to degrading the pollutants (even recalcitrant) of concern at pollution site. Microbial degeneration of organic materials mostly occurs because the microbe utilizes the pollutants for their own propagation and growth. Organic contaminants provide a resource of carbon, which is one of the essential nutrient-enriched blocks of cells, and they supply electrons, as source of energy.

Microorganisms uptake energy resources by catalyzing energy-producing oxidation–reduction reaction-like chemical reactions that first break chemical bonding and then transfer electrons away from the pollutants (oxidized). The electron donor is the contaminant, whereas the electron recipient is known as electron acceptor. Microbes release extracellular enzymes which aid in remediation of diverse types of fossil-based and bio plastics [19]. Fungi and bacteria through various enzymatic and metabolic cascades degrade these polymers into aerobic by-products of water and carbon dioxide. The nature and degradation rate of released enzyme concoction diverge based upon the type of microbial isolated species (multiple /single isolate) and even intra-species variations in a proficient and environmentally sustainable way. Thus, degradation of polymer is species-specific. For example, *Bacillus* sp. and *Brevibacillus* sp degrade polymer by proteases while fungi degrade lignin by laccase enzyme. Under stressful intolerable conditions, microbes produce exoenzymes and/or their end products for rapid detoxification of fossil- and bio-based biodegradable polymers through enzymes like proteases, cutinases and lipases [19–22]. Furthermore, enzymes like esterases and lipases, produced by *Achromobacter* sp., *Rhizopus delemar, Candida cylindracea,* and *R. arrhizus,* have been reported to work on complex polymers [19, 23, 24].

Polymers (non-biodegradable and biodegradable) are degraded by bioremediation [25, 26]. The primary mechanistic process involved in plastic biodegradation involves surface colonization and enzymatic hydrolysis of plastics. Plastics polymer are compounds formed of a varied array of synthetic, mixed, semi-synthetic inorganic, and organic compounds [19, 25, 26]. Plastics are extracted mostly from petrochemical materials containing natural gas oil and coal. Various polymer products like poly-caprolactone (PCL), polyethylene (PE), polyhydroxybutyrate (PHB), polyurethane (PUR), etc., are frequently applied for diverse purposes [25, 26]. Most of the bio-fossil-based plastics (like PE and PVC), used at present, are non-biodegradable and accumulate in nature causing impervious damage [22] and causing reduction in soil fertility, human health issues and ecological crisis [27]. Importantly, the surplus amount of plastic polymers hamper plankton growth, disrupting aquatic food chain [21, 28]. So their appropriate waste management, strategic control of garbage, and applicability are needed as they have complex structure and slow mineralization [29].

Degradation products (using oxygen as electron acceptor, under aerobic conditions) of polymers were subsequently low molecular weight monomers, dimers, and oligomers and thus finally to water and carbon dioxide by transformation. Under different anaerobic conditions, anaerobic bacteria utilize sulfate, nitrate, iron, etc., as electron acceptors to degrade polymers under optimum conditions [21, 30]. New microbe released enzyme pathways and strains need to be investigated under optimum state for the bioremediation of non-biodegradable polymers (bio- and fossil-based) for sustainable utilization [31] (Fig. 10.2).

Microorganisms can thrive under different habitats like ice-covered regions, rocks, water bodies, and deserts [32]. Therefore, floating/ immersed plastic trash is colonized by microorganisms and be a part of marine ecosystem and marine food chain [33–37]. Bacterial adherence may start immediately or make take time followed by biofilm production causing change in primary ecosystem [31, 38] (Fig. 10.3; Table 10.1).

In biosorption technology, heavy metal ions were removed from wastewater using mostly non-living algae and inactive biomass. Heavy metal ion accumulation by microbes mostly takes place in two phases. Firstly, the cell metabolism-dependent active sorption technique, which is actually the intracellular heavy metal ion uptake process. The other is cellular metabolism-independent inactive biosorption process which mostly occurs on the cell surface. Both these processes uptake the heavy metal ions inside the cytoplasm of the algae cells for further detoxification [39–44]. However, inadequate biosorption of heavy metal polluted ions through algae cell causes destruction and toxification of live cells [42, 44]. Live algal cell intracellular uptake is dependent of the particular growth phase (mostly growth phase), optimum environmental conditions, and the metal ion absorption potential of the live cell. The process is quite complex. On the contrary, non-living algal cells mostly uptake heavy metal ions extracellularly on its cell surface by forming biomass assemblage and binding of polymers (like cellulose, pectins, glycoproteins, etc.) through adsorption [45, 46]. Both these methods have the prospective of cost-effective effluent treatment.

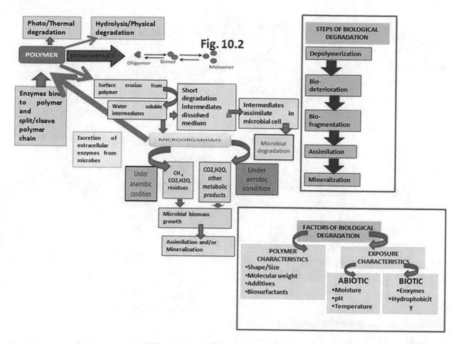

Fig. 10.2 Mechanisms and factors of biodegradation. The first step in plastic biodegradation (apart from photodegradation and physical degradation) is adherence of microbes with polymers consequent by surface microbial colonization. Microbial enzymes attach to the polymer and hydrolyze it. Under aerobic conditions, electron acceptor is oxygen, which is utilized by the bacteria forming water and CO_2 as end products. Under different anaerobic conditions, polymers are crushed down by anaerobic respiration bacteria using manganese, iron, sulfate, nitrate, and carbon dioxide as molecule of electron acceptors. Finally, assimilation and mineralization take place. Factor of abiotic and biotic conditions needs to be optimized [31].

10.5 Computational Biology of Degradation Network

System biology provides all valuable information about the target microbial population (single or mixed) for bioremediation. To characterize the metabolic functions and the elucidation of acclimatization or adjustment for a unique microbe community or unique species in normal habitat, various microorganisms cannot be propagated and the metabolism of those proliferating in single monoculture are most uncommonly to be the same as those microbes propagating in assimilated pattern. Microbial metaproteomics, employing mass spectroscopy-based categorization of amino acid substitutions, somewhat resolved the strain-specific identification of microbes. The performance of microbes selected for bioremediation varies in field as comparison to laboratory setup. Monoculture species perform less in laboratory as its synergistic partners of bioremediation in field was absent during its pure stain-specific isolation.

The main application of different researches is the application of fishing out the peptide sequences and its homology with that particular organism with known

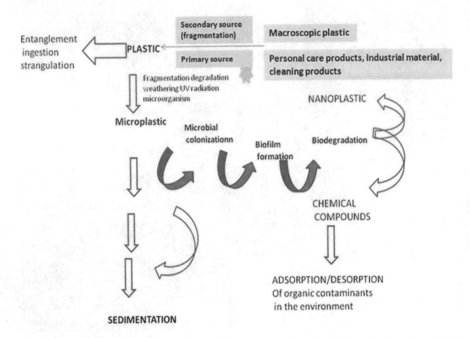

Fig. 10.3 Microplastics and interplay of marine environment. Pollutants from household sources or industry first go into the rivers and finally reach the oceans. Primary microplastics (diameter less than 5 mm) and secondary microplastics (larger diameter) accumulate in the water, loaded with biotic and different abiotic dissolved constituents, and then, biofouling (colonization of the biomass on floating plastic) takes place through bacterial degradation by primary and secondary colonization [31, 34, 35].

sequence when multiple genomic informatics were available from well-known microbes in the same contaminated community are accessible. Acidophilic microorganisms (*Leptospirillium* group II, *Bacillus acidocaldarus,* etc.) can maintain pH homeostasis, regulates proton permeation, and behaves as extremophilic organisms that can acclimatize well in harsh conditions to continue bioremediation. Furthermore, proteomics-analyzed genome typing unwinds an adaption tactic of Leptospirillum group II to intolerant environmental conditions through inter-population recombination [53]. Recently, single-cell sequencing [54] was developed to provide cell-specific genetic data from a single-cell culture of the un-cultured bacteria population, even for the less-abundant microbes. However, both difficulty of metagenomic assembly and the host sequence contamination can be easily avoided by this method [55]. Thus, unique combinations of dual metaproteomics and single-cell sequencing techniques will supply new knowledge into the role via species-specific unique protein identifications as biomarkers among a varied microbial community. Cross-strain identification by community proteomics can be greatly advanced by single-cell sequencing (despite high costs per sample) as compared

Table 10.1 Microorganisms involved in biodegradation of polymer [31]

Plastics and it derivatives	Microbes involved	Application	References
Polyhydroxyalkanoates	*Pseudomonas stutzeri*	Surgery, subsequent wound dressing, drug delivery, and bio-implant patches	[47]
Polyethylene	*Rhodococcus rubber Brevibacillus borstelensis*	Garbage and grocery bags, packaging of different film, insulation for cables and wires	[48]
Terephthalate and polyethylene	*Ideonella sakaiensis*	Packaging foods and beverages,	[47]
Polyethylene succinate	*Pseudomonas* sp.	Processed into films, bags, or boxes,	[49]
Polylactic acid	*Amycolatopsis* sp., *Bacillus brevis, Penicillium Roquefor*	Plastic films, bottles, and biodegradable medical devices	[21]
Polyvinyl alcohol	*Pseudomonas O-3, Pseudomonas putida*	Medicines, coating of ceramic, coating present on adhesives, reprography, and photography	[50]
Nylon	*Pseudomonas* sp. *and Flavobacterium* sp.	clothing, carpets, tire cords, conveyor belts, and brushes	[21]
Polycaprolactone	*Fusarium solani, Clostridium acetobutylicum, C. botulinum*	Long-term usage items, films on agriculture, seedling containers, Fibers and aquatic weeds	[51]
Polyester	*Phanerochaete chyrsosporium, Streptomyces* sp.	Carpets, making air filters, ropes, film making, plastic bottles, preparing fishing nets	[52]
Blends of starch/polyester/citric acid		Present in different fibers and engineering thermoplastics	
Blends of starch/citric acid ternary/poly vinyl alcohol	*Alcaligenes feacalis, Listeria monocytogenes, Escherechia coli*	Agricultural applications, different types of packaging materials	

to metagenomics. In in situ microhabitat, the biofilm formation, the signal transduction cascade, the network of cellular coordination function, and different post-translational modifications (glycosylation, phosphorylation, ubiquitination, acetylation, and glutathionylation) of bacteria can be validated by high throughput of strain-specific proteome data. The alterations in signal transduction may be linked with related environmental factors or stress in modulating important physiological processes [56]. Finally, different omic approaches, like metagenomics, metaproteomics, metatranscriptomics, and metabolomics, are rapidly developing to expand avenues in assembled multi-omic technology in microbial toxicology. With the appreciation of computer-analyzed biology and network, a better revealing of the system biology in a microbial population in a natural habitat may set the future directives to meta-analyze and integrate multiple sets of informatics data.

The execution of microbial bioinformatics and computational tools provide a resource in a progressive technological approach toward the pesticide biodegradation [57–59]. For this, online podium of biodegradative datasets were open-access and analyzed to get back information on bioremediation pathways of xenobiotic pesticides by microbes and detoxifying network created by resistant chemicals [59, 60]. These databases encompass the Biodegradation Network-Molecular Biology database sets (Bionemo; biodegradation genes and their transcription and regulation), by University of Minnesota Biocatalysis/Biodegradation Database (UM-BBD; http://umbbd.msi.umn.edu/predict/ reveal the data for biodegradation pathways and microbial enzyme-catalyzed biocatalytic reactions), Microbial Genome Database (MBGD, comparative investigation of microbes at the genomic level), Pesticide Target interaction database (PTID, annotation of 1347 pesticides along with pesticide target interactions, to propose novel agricultural chemical end products), BioCyc and MetaCyc, Biodegradative Oxygenases Database (OxDBase; biodegradative oxygenase database), and other databases operating in Linux and also compatible with both windows [60, 61]. UM-BBD-Pathway Prediction database is applicable for herbicide, fungicide, algaecide, rodenticide, bactericide, nematicide, etc [61, 62].

In silico process of metabolic engineering of targeted microbes has been applied in various field of microbial toxicology for biodegradation and bioremediation cellular processes employing in silico tools accessible publicly for users for desired data mining sets and dissecting the metabolic cascades of a cellular physiology [62]. Metabolic pathway analysis (MPA), metabolic flux analysis (MFA), and flux balance analysis (FBA) are most commonly used engineering tools for stoichiometric quantitative investigation of metabolic systems of web [63]. Organizing and knowing flux (flow of substance with the side edges carrying a definite value) led to modify the micro-biological cascade dynamics by metabolic engineering [61]. These in silico techniques can also be manipulated to evaluate properties of degrading bacteria [64]. QSAR (quantitative structure–activity relationship) and 3DQSAR chemical atomic models are employed to learn the toxicological level of xenobiotic pesticide molecules at diverse ecological habitats and to trace the level of biomagnification of pesticides in different food web. All these computational tools help to validate different interacting genes, widespread genomic data, and understanding genome scale models [61].

10.6 Gene Editing: Fishing the Functional Gene for Better Bioremediation

Bioremediation is a method that uses microbial population to eradicate, neutralize, or mineralize pollutants from polluted environments. According to many studies, the occurrence of a huge number of unidentified microorganisms aiding in bioremediation in contaminated environments can only be outlined using culture-independent methods [1, 65]. The analysis of 16S rRNA genes has reformed the research of microbial biodiversity in the natural habitat, both by culture-dependent and culture-independent methods [66]. For the study of microbe-related ecology, molecular biology devices have been widely used. Microbial inter-relations within the same communities are also noticed with system biology technology [67]. This method is also productive in analyzing the existence of microbes under extreme pressure and temperature conditions [68]. Omics system biology studies using genomics, transcriptomics, metabolomics, and proteomics support microbial bioremediation network analysis at the genetic level control for bioremediation (Fig. 10.4) [69]. The progressive techniques of culture-independent methods use sequencing and in silico techniques for both sequence and function-driven gene fishing for bioremediation applications [70]. Recent advancements in environmental bioremediation include molecular genetics and knowledge-based study to rationalize protein modification

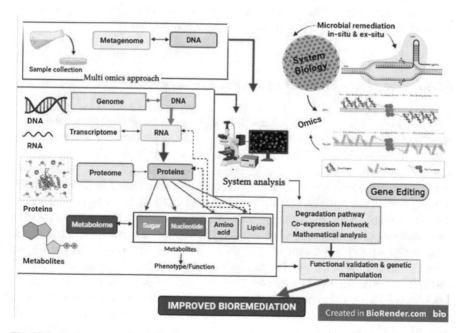

Fig. 10.4 Conjugation of multi-omics system biology and gene editing [69] [created in biorender.com]

to provide a better outcome into the development of designer enzymes/biocatalysts as per requirements [71, 72].

10.6.1   Gene Editing Tools

This is a unique technology that permits the manipulation of DNA sequence through the use of engineered nuclease enzymes employed as molecular scissors. These restriction enzymes have a wide range of functions in animal, plant, and microbial studies [73]. The editing technique encompasses triggered with a self-guided designed sequence which is complementary to the sequence of the novel gene of particular interest, helping a break at an operation site, repairing it with homologous recombination, and manipulating or moderating (deletion or insertion) a desired fragment of sequence [74]. The targeted genome engineering by definite gene editing method led to further usage of microbes in diverse fields like agriculture, food and medicine, clinics, etc [75].

The important gene editing tools are TALEN, CRISPR-Cas, and ZFN [76] for structural genetic variations. These definite gene editing tools aim to develop improved microbial populations with more complex genes and to design target-specific engineered microorganisms [77, 78]. It is the cradle for altered genetic sequence makeup that differs from native variants in order to obtain targeted new microorganisms with functional genes of interest for [79, 80].

10.6.2   Genetic Variation and CRISPR Targeting

The CRISPR (Clustered Regularly Interspaced Short Palindromic Repeats)-associated (Cas) nuclease system is a highly utilized podium for genome mediated engineering [73, 74, 80]. CRIPSR-Cas mainly of Type I-III along with its subtypes provides an efficient gene editing method when applied on model organisms. CRISPRs, along other nuclease enzymes like zinc finger nucleases (ZFNs) or transcription activator-like effector nucleases (TALENs), have supplied the platform for unprecedented functional genetic research in the laboratory setup as well as the impending probability for therapeutic applications of a diverse range of genetic diseases [76, 81, 82]. At present, the *Streptococcus pyogenes* Cas9 (SpCas9) nuclease recognizing the 5'-NGG-3' PAM sequence is mostly used to study genome variations. CRISPR nuclease activity is based on Watson–Crick model of base pairing a guide RNA (gRNA) designed to pinpoint sequences with a cognate genomic DNA sequence upstream or downstream of a nuclease-recognized protospacer adjacent motif (PAM) (Fig. 10.5) [83]. As CRISPR systems actually mediate genomic cleavage at a cheap, simple and easy way, this technology of CRISPR targeting would logically affect the genetic variation by decreasing or increasing sequence homology at off-target and on-target sites or by moderating protospacer adjacent motifs.

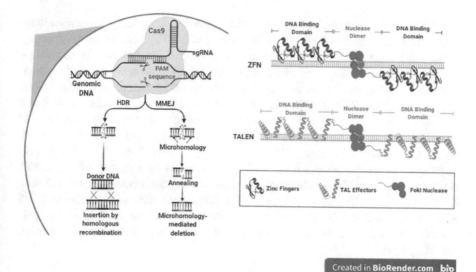

Created in **BioRender.com** bio

Fig. 10.5 Gene editing by ZFN, TALEN, CRISPR associated with Restriction enzymes like Cas 9. Mechanism of target DNA recognition in TALEN and ZFN depends on DNA–protein interaction whereas in CRISPR it is DNA–RNA interaction. DNA cleavage and repair in TALEN and ZFN take place in double-stranded DNA induced by FokI. In CRISPR, it cleaves both single- and double-stranded DNA induced by Cas9 enzyme (created in biorender.com) [76, 79, 82, 85]

Moreover, the sequential preferences of PAM sequence to study directed evolution and/or structure-guided mutagenesis can differ widely across Cas9 orthologs obtained from various bacteria such as *Campylobacter jejuni* (PAM: 5′-NNNNRYAC-3′), *Staphylococcus aureus* (PAM: 5′-NNGRRT-3′), *Neisseria meningitides* (PAM: 5′-NNNNGATT-3′), *Staphylococcus thermophilusST1* (PAM: 5′-NNAGAA-3′), *S. thermophilus A* (PAM: 5′-NGGNG-3′), and *Bacillus laterosporus* (PAM: 5′-NNNNCNDD-3′).

TALENs or Transcription activator-like effector nucleases, is a pioneering tool for gene editing and modification (Fig. 10.5) [84]. TALENs have TAL proteins initially secreted by the pathogenic bacteria *Xanthomonas*. TAL proteins are so powerful that they can combine to sequences as short as 1–2 nucleotides [85]. Each zinc finger domain recognizes a 3- to 4-bp DNA sequence, and tandem domains may be able to connect to a unique extended nucleotide sequence inside a cell's genome (generally with a length that is a multiple of 3, commonly 9 bp to 18 bp). TALENs structurally have 34 amino acid tandem repeats and show its nuclease activity after efficient binding. TALENs are now preferred for gene knock out (NHEJ-Non-homologous end joining) and gene knock in (HDR-Homology directed repair) of the desired gene or functional gene [72]. Two protein domains, one needed for sequence-specific cleavage and the other for binding and recognizing the specific binding site, combine to make the TALENs a powerful gene editing method. It is used on a diverse range including frogs, mammalian cells, rats, zebrafish, and chickens and other eukaryotic organisms [71].

ZFN or zinc finger nuclease is most widely used as artificial restriction endonuclease enzyme (Fig. 10.5) [76, 86]. ZFNs have inbuilt ZFPs (zinc finger proteins) as eukaryotic transcription factors which act as a DNA-binding domain. The nucleotide cleavage domain (Folk1) derived from *Flavobacterium okeanokoites* is also found in ZFNs [87]. Depending on the target site, the cleavage domain is surrounded by a large number of ZFPs (usually four to six). Each zinc finger domain recognizes a 3- to 4-bp DNA sequence, and tandem domains can possibly attach to an extended nucleotide sequence that is unique within a cell's genome (generally with a length that is a multiple of 3, commonly 9 bp to 18 bp). These ZFPs allow for precise target-specific unique gene editing with their eighteen base pair specificity. ZFPs structurally have an alpha-helix in place of two antiparallel running layer of sheets. ZFPs are mostly 30 amino acids long [88]. This gene editing technique is noted with knock in (HDR-Homology directed repair) and genes knock out (NHEJ-Non-homologous end joining) for flourishing prokaryotic and eukaryotic gene editing technology [89].

10.7 Microbial Glycoconjugates, Biofilm Formation, and Bioremediation of Organic Pollutants

The surfaces of all microbes are encoded with sugar molecules such as lipopolysaccharides, capsular polysaccharides, glycoproteins, secreted exopolysaccharides and lipo-oligosaccharides, in bacteria, and lipopolysaccharides in Gram-negative bacteria, lipophosphoglycan in Leishmania, and lipoarabinomannans in mycobacteria [90]. Glycoconjugates are mainly amphiphilic compounds produced on the cell membrane of the microbes. Release of the glycoconjugates depends on the exact strain of the microbes, nutritional requirements (nitrogen and mainly carbon), trace elements, and the optimal growth settings. They behave as biosurfactant during the microbial stationary growth phase [91, 92].

These molecules have hydrophobic and hydrophilic moieties that decrease the interfacial and surface tension. However, glycoconjugates, as mentioned in studies, can have wide range of structures. Glycoconjugates includes peptidoglycans, glycoproteins, glycopeptides, lipopolysaccharides, glycolipids, and glycosides. Different microbial strains also produce extracellular glycoconjugates like sophorolipids, rhamnolipids, and glycoproteins, exopolysaccharides, and glycolipopeptides. Microbial glycoconjugates amplify the bioavailability of organic pollutants, reduces surface tension, creates a solvent interface, and accelerates microbial metabolism, thus enhanced the detoxification of these harmful pollutants from the nature [92].

Organic pollutants (OPs) had adverse effects on biotic organisms present in the ecosystem, liable for diverse harmful consequences in humans, together with unfavorable mutagenic, teratogenic, and carcinogenic effects. The bioremediation of OPs usually uses physical and also chemical methods like aeration, pumping, soil washing, incineration, oxidation, etc. The rich array of metabolizing enzymes of single or mixed microbial cultures participated in the bioremediation processes through simple techniques employing both aerobic and different anaerobic metabolism [31].

The most preferred anaerobic metabolism releases parent compound like trichloroethylene (TCE), and also harmful products like vinyl chlorides (VCs) and dichloroethylene (DCE). VCs and DCE have elevated environmental toxicity than their original parent compound, i.e., TCE. Studies indicate that microbial glycoconjugates (either secreted outside or present inside the cell) and other glycolipids play a vital function in the transport of OPs across microbial membranes. Therefore, aerobic metabolism employs diverse broad spectrum catabolite enzymes (like oxygenases) to degrade Ops from contaminated sites [88, 90].

Microbial glycoconjugates accelerate the bioremediation of the OPs through biofilm formation. In environment, microorganisms interact with biotic and abiotic environmental (like synergistic and antagonistic effects) factors to produce differential glycoconjugate surfactants at polluted sites. Mixed microbial population are better adapted for biofilm production and bioremediation than single microbial strain because diverse microbe communities have more number of reporting genes and a well-bound network of diverse metabolic activities to reveal finest output within the minimal period [93]. Mixed microbial populations showed the collective result on the bioremediation of the OPs.

10.7.1   Glycoconjugates and Waste Water Treatment: A Network

Microbial glycotechnology involving the stimulated sludge process is mostly relevant for wastewater bioremediation. Microbes during aerobic digestion of waste pollutants produce flocs (floc-forming microbes) by the web of different extracellular polymeric substances (EPSs). Enzymes of microorganisms hydrolyze the sludge, releases EPSs, and recognizes glycoconjugates and polysaccharides in concert with an array of lectin [94]. Glycoconjugates can effectively decrease the surface-generated and interfacial tension of water during treatment of wastewater. Rhamnolipids reduces interfacial tension and improves solubility during removal of hydrocarbons and pretreatment of waste activated sludge [95]. The bacterial isolates of Enterobacteriaceae, Aeromonadaceae, Bacillaceae, Pseudomonadaceae, and Gordoniaceae families were efficient candidate for wastewater treatment and bioremediation [96]. The wastewater treatment candidate bacterial strains showed biofilm formation at polluted sites and antibiotic resistance due to their biosurfactant property and low surface tension values [96, 97]. Sophorolipids are employed in oil biodegradation of contaminated water and oil spill management as a glycoconjugate biosurfactant [90, 98].

10.7.2   Microbial Glycoconjugates in Pesticide Degradation

Presently, pesticides of various categories belonging to organophosphates, organochlorines, and pyrethroids group are hydrophobic, showing poor bioavailability and also low water solubility. Microbial glycoconjugates help in the process of desorption of pesticide from contaminated earth granules. They behave as biosurfactant molecules, decreases surface tension, and uplifts the biodegradation by utilizing microbial metabolic pathway [99]. These surface-active amphipathic emulsifying glycoconjugate molecules increase the partitioning of aqueous phase from the hydrophobic pesticides by releasing small emulsions at and above their actual critical micellar concentration (CMC). Thus bioavailability and mobilization of pesticides enhance their uptake microbial cells during metabolic activity [100] by making them more degradable. The commonly used glycoconjugates for pesticide bioremediation are glycolipopeptides, sophorolipids, rhamnolipids, and fructose lipids. Rhamnolipids produced by *P. aeruginosa* showed enhanced solubilization and bioavailability of endosulfan isomers of pesticides [101]. Endosulfan and hexachlorocyclohexane (HCH) are highly hydrophobic pesticides which were solubilized by thermostable rhamnolipid glycoconjugate produced by *Lysinibacillus sphaericus* strain IITR51 [102, 103]. The detoxification of persistent organochlorine like Lindane in the natural habitat continues with the glycoconjugate assemblage in a minimal salt medium by *Sphingomonas* sp. NM05, *Pseudomonas aeruginosa*, *Pseudozyma* VITJzN01, *Rhodococcus* sp. strain IITR03, *Arthrobacter globiformis*, and *Bacillus subtilis* [102].

10.7.3   Biosurfactant as Glycoconjugate

With the advancement of the humanity, various industrial materials and end products, like petroleum, pesticides, medical waste, plastic, etc., have caused a havoc pollution in spheres of air, water, and soil causing unsustainable ecosystem and dampen human well-being. Persistent pollutants go through the food chain and create various perilous results on living organisms. Microbial glycoconjugates have more applications in various industries (agricultural, textile processing, pharmaceutical, personal care, cosmetics, and food industries) and environmental relevance like hydrocarbon degradation, soil bioremediation, and oil recovery.

Bioremediation has the ability to eradicate potential pollutants through biochemical mineralization resulting in eco-friendly products or no by-products in a low operational cost, low energy requirement, economic effectiveness, and permanent bio-moderation process. Pollutants like phenols, crude oils, heavy metal, petroleum hydrocarbons, polycyclic aromatic hydrocarbons (PAHs), etc., possess high-pitched toxicity and low bioavailability to microbes, causing failure of bioremediation.

Biosurfactants, being eco-friendly, appreciably biodegradable, and multidimensional compounds, behave as additives or having surface-additive properties. They are generated by fungi, bacteria, and yeast. They are more active at very low dose and quite stable at severe environmental state like temperature (high or low), pH, and salinity. Moreover, these surfactants also increase the efficacy and/or amount of associated genes and enzymes in microbes, facilitating bioremediation of pollutants. Thus, it reduces also the toxicity generated from pollutants toward microorganisms [104].

Bacillus sp., *Pseudomonas* sp., *Candida tropicalis,* and *Citrobacter freundii* like microbes were isolated in laboratory as probable sources of biosurfactants to generate compounds like rhamnolipids, sophorolipids, and surfactin during bioremediation [105, 106]. Biosurfactants from *Pseudomonas* sp. CQ2 obtained from China (Chongqing oilfield) using ammonium nitrate and soybean oil as nitrogen and carbon sources with optimal bioleaching conditions could efficiently eliminate heavy metal Cd, Cu, and Pb from contaminated with removal efficiencies ranges from 56.9 to 78.7%. The effectiveness of biosurfactants was better than known chemical surfactants such as SDS or Tween-80 for their low-toxicity and low critical micelles concentration (CMC) [107, 108].

Bacillus nealsonii S2MT (naturally potential biosurfactant producer) was isolated from the sediment of Yanqi Lake, Beijing, China, has successful heavy engine oil-polluted soil (10–40 mg/L concentrations of contamination) remediation potential. Surfactin (powerful biosurfactants) generated from this strain has reduced surface tension, strong stability, better emulsifying abilities, therapeutic applications, and effective over wide range of environmental conditions [109].

10.8   Eco-Friendly Nanomaterials: Mitigation of Pollutants by Microbial Nanotechnology

Nanomaterials (NMs) have special physical and specific chemical properties, and scientists in different fields of environmental science, especially dealing with bioremediation, have given much attention in application of nanoparticles (nanoscale particles). When nature is exposed to elevated concentrations of pollutants (such as heavy metals and salts), that are hazardous to most microorganisms, desirable level of bioremediation may not always be achieved. Nanotechnology, a diverse field, has wide environmental benefits, including pollution prevention, remediation and treatment, pollutant exact sensing, and detection [110]. NMs employed in bioremediation have a lower toxicity to native microorganisms and boost microbial biodegradation activity (Fig. 10.6).

"Remediate" stands for the resolution of the crisis, and "bioremediation" means the utilization of various biological microorganisms (such as plants, bacteria, fungi, yeast, or their enzymes) to mineralize the pollutant, on-site or off-site remediation, and change it into non-harmful forms [111]. However, these bio-based remedial technology are extremely convenient, high throughput, and cost-effective and cause less environmental impact, and their inflexible procedures produce highly

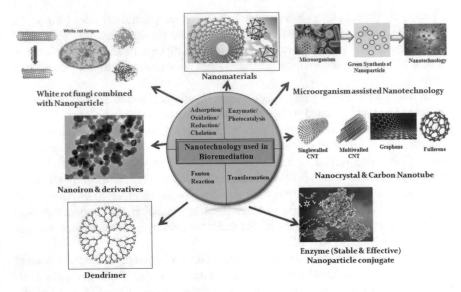

Fig. 10.6 Nanotechnology used in bioremediation

toxic by-products that cause environmental unenviable changes as well as deterioration of the microbes used in the method [112]. Bioremediation of pollutants using bio-nanomaterials is one environmentally acceptable and low cost-effective way for overcoming this barrier. In terms of environmental science, NMs have a variety of eco-friendly applications, such as substances that provide a better environment from contaminated sources in both large-scale and some portable applications.

10.8.1 Bioremediation with Nanomaterial

For in situ applications, NMs exhibit a number of desirable features or characteristics. NMs can be employed in bioremediation in waste water treatment, radioactive waste remediation, heavy metal remediation, hydrocarbon remediation, and solid waste remediation. Nanoparticles may be capable of penetrating relatively small gaps in the subsurface and remain suspended in groundwater because of their small size and exclusive surface coating property, permitting them to penetrate further and achieve greater distribution than bigger, macro-sized particles [113]. The prevalence of near sub-soil regions with constituents dissimilar from bulk regions can also motivate the chemical reactivity of materials, stimulating the involvement of interfacial free energy to the conversion of free energy of dissolution–precipitation reactions [114]. Nano-remediation methods mostly involve the application of reactive NMs (like metal oxides, carbon containing nanotubes, nanoscale size zeolites, carbon fibers, and bimetallic nanoparticles) for the detoxification and trans-degradation of contaminants.

10.8.2 Nano-Iron and Its Derivatives in Bioremediation

Iron nanoparticles behave as green nanoparticles in bioremediation owing to its redox potential property while combining with water. It is less toxic in nature and has magnetic susceptibility [115]. The elimination of As (III), a highly mobile, toxic, and prevalent arsenic species in anoxic ground water, was studied using nanoscale zero-valent iron (NZVI) [116]. Arsenic (V) has also been eradicated from ground level of water employing a colloidal reactive barrier material made of nanoscale zero-valent iron. With dissolved water and oxygen, iron also conducts "redox" reactions. Heavy metals like chromium and arsenic, pesticides (DDT, Lindane), chlorinated solvents (DCE, TCE, and PCE), and organic compounds such as nitrates have all been removed with nZVI [117]. The ability of powdered zero-valent iron to dechlorinate DDT and its linked compounds at optimum temperature has been examined [118]. Specifically, DDT, DDD [1,1-dichloro-2,2-bi's(p-chlorophenyl) ethane], and DDE [2,2-bi's(p-chlorophenyl)-1, 1-dichloroethylene] conversion by powdered zero-valent iron in buffered anaerobic aqueous solution was shown at 20 °C, in the presence and absence of nonionic surfactant like Triton X-114. *Noaea mucronata* of Chenopodiaceae family is the finest lead accumulator. This plant also accumulates for copper, zinc, and nickel as a native plant accumulator. CD accumulator plant is *Marrubium vulgare*, but the finest Fe accumulator plant is *Reseda lutea*. Nanoparticles prepared from *N. mucronata* has the best bioaccumulation ability in waste water [119]. The "ferragels" of supported zero-valent iron nanoparticles quickly dispel and immobilize Pb (II) and Cr (VI) from aqueous solution, reducing lead to Pb (0) and chromium to Cr (III) while oxidizing iron to goethite (–FeOOH). Nickel–iron nanoparticles with a large surface area have been utilized to dehalogenate trichloroethylene (TCE). At room temperature, the power of powdered zero-valent iron to dechlorinate DDT and other correlated compounds (such as DDD [1,1-dichloro-2,2-bi's(p-chlorophenyl) ethane] and DDE [2,2-bi's(p-chlorophenyl)-1, 1-dichloroethylene]) has been shown [118]. At nearly 20 °C, the outcome of powdered zero-valent iron on buffered aqueous anaerobic solution was shown in the presence and absence of nonionic surfactant Triton X-114. The role of iron nanoparticles in removal of various contaminants was shown by using Zero-valent powder iron, Iron sulfide nanoparticle, Iron nanoparticle, Fe nanocomposite, Colloidal zero-valent powder iron, and CMC4-stabilized ZVI nanoparticle in removal of Azo dye orange (II), Nitrate, arsenic (V), Herbicide: molinate, Lindane, PAHs, lead, zinc, Nickel (II), Chromium (VI), Copper (II), DDT, Cadmium, Cobalt (II), Perchlorate, etc [116–118, 120–123].

10.8.3 Carbon Nanotubes and Nanocrystals in Bioremediation

Nanomaterials based on carbon such as carbon nanotube(s) (CNT(s)) and nanocrystals allow innovative solutions to address a wide array of environmental issues [116].

Table 10.2 Removal of pollutants using carbon nanotubes (CNTs)

Types of nanoparticles	Contaminant removed	References
Carbon-based nanotubes	Lead (II), Arsenic, Atrazine, Nickel	[124]
	Organic compounds present in pesticides, dyes and several pharmaceuticals, medicines, or drugs	[111]
	Trihalomethanes	[125]
	Trichloroethane	[122]
	Methylene blue adsorption	[126]
CNTs KMnO$_4$ oxidized	Cadmium (II)	[127]
MWCNTs (multi-walled carbon tubes)	Organochlorines, Benzene, Toluene	[110]

Adsorbent particles interact with their carbon atoms on the adjoining walls of carbon nanotubes, which feature cylindrical pores. The size and structure of pores play a part in the interaction between molecules and solid surfaces. The elimination of ethyl benzene from aqueous solution has been examined using NMs such as multi-walled carbon nanotubes (MWCNTs), single-walled carbon nanotubes (SWCNTs), and hybrid carbon nanotubes (HCNTs) [124]. Organic chemicals found in colored dyes, insecticides, pharmacy medicines, and treatment medications eluted in activated sludge and wastewater were transformed by multi-walled or single-walled carbon nanotubes [122]. Relevance of carbon nanotubes was discussed in Table 10.2.

10.8.4 Enzyme Nanoparticles in Bioremediation

In bioremediation, enzymes act as biocatalysts. Enzymes' utility as cost-efficient alternatives to synthetic catalysts is restricted due to their short catalytic lifetimes and lack of stability. Binding enzymes to magnetic iron nanoparticles is a good strategy to improve their steadiness, endurance, and reusability. After attaching enzymes to magnetic iron-based nanoparticles, a magnetic field can be used to quickly separate them from reactants or end products. To make core shell magnetic nanoparticles, two separate catabolic enzymes, trypsin and peroxides, were also used as MNPs [128–131]. MNP enzyme conjugates have been discovered to be more viable, competent, and cost-effective [127]. We know that less stability of enzymes is incurred due to its oxidation property. Enzyme-conjugated nanoparticles shield the enzyme active site and thus prevent its oxidation. NMs also minimize the cell interaction of enzymes through steric hindrances and thus decrease the surface energy.

10.8.5 Dendrimers in Remediation

Dendrimers are highly branching, monodispersive macromolecules that have just lately been identified as polymer members [112]. Due to their increased reactivity, surface area, and lower toxicity, dendrimers containing NPs composites can be employed to improve catalytic activity. Polyamidoamine (PAMAM) dendrimers are a new group of nanoparticles that can be approved as water-soluble chelators [128]. Simple filtration unit for the elimination of organic pollutants based on TiO_2 (Titanium dioxide) porous ceramic filters with an alkylated poly(ethyleneimine) hyperbranched polymer, poly dendrimer (propylene imine), or cyclodextrin-impregnated pore, forming hybrid inorganic /organic filter modules with great mechanical strength and also surface area [131].

10.8.6 Microorganism-Mediated Nanotechnology

The concurrent use of microbes and biofabrication of nanomaterials makes nanotechnology more eco-friendly [132]. Because chemically generated nanoparticles may have drawbacks, green nanoparticle synthesis using fungal, plant extracts, and bacteria released enzymes could be a feasible alternative. They behave as reductive agents for the metal complex salt and produces metallic nanoparticles. *Aspergillus tubingensis* (STSP 25)-biofabricated iron oxide nanoparticles were produced from the *Avicennia officinalis* (rhizosphere) in the Sundarbans forest of India [133, 134]. With a regeneration ability of nearly five cycles, the synthesized nanoparticles were capable to eliminate higher than 90% of constituent heavy metals [copper (II), lead (II), nickel (II), and zinc (II)] from wastewater [134]. *Escherichia* sp. SINT7, a copper-resistant bacteria, was used to create copper nanoparticles. The generation of nanoparticles with the help of microbes has provided a low-cost and environment sustainable technique [134, 135].

10.8.7 Bioremediation of White Rot Fungi Combined with Nanoparticles

White rot fungi (known as WRF) are one of the most significant microbes in natural ecosystem [136]. They are recognized for having a potent enzyme system that can mineralize lignin and carbohydrates like hemicelluloses and cellulose in wood, which is important for forest biogeochemical cycles [137]. Due to the effective creation of metallic nanoparticles in bioremediation in the agricultural field, WRF has also been recognized as a possible biological reserve for biosynthesis [138]. For example, the *Trametes versicolor* was challenged with hazardous cadmium ions and in situ reduced, stable CdS nanoparticles were produced, highlighting the prospective of

WRF not only in bioremediation but also in large-scale production of metallic nanoparticles [139].

10.9 Microbial Community Proteomics: Microbial Interaction with Environment

The metaproteomic procedures offer an expensive avenue to investigate the bioremediation functions of the major dominant bacteria with other minor communities in contaminated soil, air, and water (in situ method) and better suited than traditional "artificial" laboratory setups [140]. The metaproteomic science has been extensively applied to portray microbial metabolic functions needed for bioremediation of pollutants. The metaproteome of cadmium-contaminated soil using the gel-based analysis provides information about some related proteins [141]. The biostimulation process of the dominant members of Geobacter community and their metabolic reactions to energy yield was demonstrated by metaproteomic study on the uranium-contaminated aquifer [142]. The members of the family of Betaproteobacteria (indigenous aquifer microbiota) and the Firmicutes dominate the contaminated aquifer during biostimulation with emulsified vegetable oil [143]. Metaproteomic analysis of the autochthonous bacteria was done during biodegradation of organic pollutants like chlorobenzene [144]. *Bacillus* sp. along with *Synechococcus, Sphingomonadales, Clostridium* sp., *Ralstonia solanacearum,* etc., are resistant to hydrocarbon contamination as revealed by metaproteomic survey of hydrocarbon amended soil [145]. Metaproteomic approach was employed to validate the inhibitory level of cadmium in a continuous flow of wastewater treatment in bioreactor and the biological response of an unsequenced bacterial population [146]. Metaproteomics has also happen to an imperative research field of stimulated sludge wastewater treatment using enzymes, sludge extracellular proteins in sludge digestion, and transport proteins. Both gel-based technology and non-gel-based proteomic approach were employed to identify the key players of metabolic pathways.

Metal-enriched extremely acidic (pH < 3) waters or Acid Mine Drainage (AMD) is a hazardous ecological crisis in the mining industry. The final release of water must be done after removal of associated metals and also to raise the pH. The in situ bioremediation of an AMD site was done by quantitative metaproteomic analysis (2033 proteins were identified) of the natural microbial community (Leptospirillum group II) and their related biofilm production with low complexity [147]. The Fe-oxidizing Leptospirillum (both group II with group III) chemoautotrophic bacterial communities used in AMD biofilms were employed in community genomic and metaproteomic approaches [148] to identify methyl-accepting chemotaxis proteins and methyl-independent response signal transduction.

10.10 Discussion

The application of nano-bioremediation, using better technology engineered NMs, can deliver low charge, high efficiency, minimization of chemical sludge, effective, and time cutback in situ removal processes for large-scale removal of pollutants directly and also aid in microbial degradation of waste and toxic materials. Due to the powerful potential of NMs in catalyzing biodegradation of waste and toxic materials, it is worthwhile that relevance of nanoparticles will elevate in future in sustainable development. Moreover, various health hazards or risks are also associated with NMs and have widespread ecological implications. This may reduce the relevance of NMs for environmental bioremediation. Thus, to fulfill this nanotechnology more advantageous than hazardous, regular monitoring and close intervention need to be applied sooner.

The impact of genetic structural variation on CRISPR targeting is not very specific to CRISPR as other genome editing technologies employing TALENs and ZFNs can also be used. Therapeutic genome editing is not the only tool to study impact of human genetic variation and also in treatment regime of several genetic diseases. Different variation in drug-target genes can alter drug binding property, and various genetic variants can greatly influence the rate of drug metabolism. For safer personalized, non-toxic, more effective therapeutic treatment for patients, it is necessary to optimize gRNAs by minimizing off-target potential to minimize adverse outcomes. Thus, CRISPR targeting TALENs and ZFNs should be practiced in both laboratory and clinical translation medicine setup to validate and identify the exact effects of genetic variation.

The metaproteomic science has been broadly useful to study microbial populations from different environmental habitats to provide new insights into diverse metabolic pathways, microbial diversity, metabolic potential, signal transduction cascade, ecological attributes, and microbe–environment network of interactions. Furthermore, because of the diversity and complexity of varied environmental setups, this technology still faces great drawbacks in the research of environmental microbial communities: to preserve the sample, to sort the low-abundant to high-abundant proteins, to work with low microbial populations, to critically separate the rare species, to employ the best sample collection method, and difficulty in protein extraction and proper taxonomic assignments of proteins. In environmental microbiology, the study of system biology along with biomarker discovery is needed for the precise quantitative informatics of a pre-determined group of proteins in various environmental samples. Metaproteomics in conjugation with other omics provide widespread knowledge and insight into real microbe population, their metabolic pathways, and functioning of genes and proteins.

Microbes possess many unique properties such biosurfactant production, secondary metabolites synthesis, network of biofilm formation, and other in consortium to resist the stressful environmental conditions. These properties of pollutant-resistant bacteria may be employed judicially in bioremediation. Research on multi-species biofilm producing communities has been noted for their pollutant tolerance and bio-mineralization characteristics. Thus, microbial village population showed quite promising result in bioremediation and every knots of these applied systematic need to be evaluated for better applicability.

Acknowledgements We acknowledge and show heartfelt gratitude to the department of Environmental Science, Asutosh College and the Vice Principal of Asutosh College. We also acknowledge Principal of Behala College for constant help and encouragement. We acknowledge Asutosh College and Behala College for their cooperation and help for providing the necessary infrastructure required in preparing the manuscript.

References

1. Li, Y., & Li, B. (2011). Study on fungi-bacteria consortium bioremediation of petroleum contaminated mangrove sediments amended with mixed biosurfactants. *Advanced Materials Research, 183–185*, 1163–1167.
2. Nikolopoulou, M., & Kalogerakis, N. (2018). Biostimulation strategies for enhanced bioremediation of marine oil spills including chronic pollution. *Consequences of Microbial Interactions with Hydrocarbons, Oils, and Lipids: Biodegradation and Bioremediation*, pp. 1–10.
3. Hall, E., Bernhardt, E., Bier, R., et al. (2018). Understanding how microbiomes influence the systems they inhabit: Moving from a correlative to a causal research framework. *bioRxiv*, p. 065128.
4. Montiel-Rozas, M. M., Domínguez, M. T., Madejón, E., et al. (2018). Long-term effects of organic amendments on bacterial and fungal communities in a degraded Mediterranean soil. *Geoderma, 332*, 20–28.
5. von Netzer, F., Granitsiotis, M. S., Szalay, A. R., et al. (2018). Next-generation sequencing of functional marker genes for anaerobic degraders of petroleum hydrocarbons in contaminated environments. *Anaerobic Utilization of Hydrocarbons, Oils, and Lipids*, pp. 1–20.
6. Philp, J. C., & Atlas, R. M. (2005). Bioremediation of contaminated soils and aquifers. In R. M. Atlas & J. C. Philp (Eds.), *Bioremediation: Applied microbial solutions for real-world environmental cleanup* (pp. 139–236). American Society for Microbiology (ASM) Press.
7. Maila, M. P., & Colete, T. E. (2004). Bioremediation of petroleum hydrocarbons through land farming: Are simplicity and cost-effectiveness the only advantages? *Reviews in Environmental Science and Bio/Technology, 3*, 349–360. https://doi.org/10.1007/s111157-004-6653-z
8. United States Environmental Protection Agency (USEPA). (2006). In situ and ex situ biodegradation technologies for remediation of contaminated sites. EPA/625/R-06/015.
9. Khan, F. I., Husain, T., & Hejazi, R. (2004). An overview and analysis of site remediation technologies. *Journal of Environmental Management, 71*, 95–122. https://doi.org/10.1016/j.jenvman.2004.02.003
10. Whelan, M. J., Coulon, F., Hince, G., et al. (2015). Fate and transport of petroleum hydrocarbons in engineered biopiles in polar regions. *Chemosphere, 131*, 232–240. https://doi.org/10.1016/j.chemosphere.2014.10.088
11. Delille, D., Duval, A., & Pelletier, E. (2008). Highly efficient pilot biopiles for on-site fertilization treatment of diesel oil-contaminated sub-Antarctic soil. *Cold Regions Science and Technology, 54*, 7–18. https://doi.org/10.1016/j.coldregions.2007.09.003
12. Azubuike, C. C., Chikere, C. B., & Okpokwasili, G. C. (2016). Bioremediation techniques-classification based on site of application: Principles, advantages, limitations and prospects. *World Journal of Microbiology & Biotechnology, 32*(11), 180. https://doi.org/10.1007/s11274-016-2137

13. El Fantroussi, S., & Agathos, S. N. (2005). Is bioaugmentation a feasible strategy for pollutant removal and site remediation? *Current Opinion in Microbiology, 8*, 268–275.
14. Magalhães, S. M. C., Jorge, R. M. F., & Castro, P. M. L. (2009). Investigations into the application of a combination of bioventing and biotrickling filter technologies for soil decontamination processes—A transition regime between bioventing and soil vapour extraction. *Journal of Hazardous Materials, 170*, 711–715. https://doi.org/10.1016/j.jhazmat.2009.05.008
15. Burgess, J. E., Parsons, S. A., & Stuetz, R. M. (2001). Developments in odour control and waste gas treatment biotechnology: A review. *Biotechnology Advances, 19*, 35–63. https://doi.org/10.1016/S0734-9750(00)00058-6
16. Kim, S., Krajmalnik-Brown, R., Kim, J.-O., et al. (2014). Remediation of petroleum hydrocarbon contaminated sites by DNA diagnosis-based bioslurping technology. *Science of the Total Environment, 497*, 250–259. https://doi.org/10.1016/j.scitotenv.2014.08.002
17. Kao, C. M., Chen, C. Y., Chen, S. C., et al. (2008). Application of in situ biosparging to remediate a petroleumhydrocarbon spill site: Field and microbial evaluation. *Chemosphere, 70*, 1492–1499. https://doi.org/10.1016/j.chemosphere.2007.08.029
18. Tyagi, M., da Fonseca, M. M. R., & de Carvalho, C. C. C. R. (2011). Bioaugmentation and biostimulation strategies to improve the effectiveness of bioremediation processes. *Biodegradation, 22*, 231–241. https://doi.org/10.1007/s10532-010-9394-4
19. Shah, A. A., Kato, S., Shintani, N., et al. (2014). Microbial degradation of aliphatic and aliphatic-aromatic co-polyesters. *Applied Microbiology and Biotechnology, 98*(8), 3437–3447.
20. Mohanty, A., Misra, M., & Hinrichsen, G. (2000). Biofibres, biodegradablepolymers and biocomposites: An overview. *Macromolecular Materials and Engineering, 276*, 1–24.
21. Tokiwa, Y., Calabia, B. P., Ugwu, C. U., et al. (2009). Biodegradability of plastics. *International Journal of Molecular Sciences, 10*(9), 3722–3742.
22. Sharma, M., & Dhingra, H. K. (2016). Poly-β-hydroxybutyrate: A biodegradablepolyester, biosynthesis and biodegradation. *British Microbiology Research Journal, 14*(3), 1–11.
23. Jin, H. J., Lee, B. Y., Kim, M. N., et al. (2000). Thermal and mechanicalproperties of mandelic acid-copolymerized poly (butylene succinate)and poly (ethylene adipate). *Journal of Polymer Science: Polymer Physics, 38*(11), 1504–1511.
24. Lam, C. X., Hutmacher, D. W., Schantz, J. T., et al. (2009). Evaluation of polycaprolactone scaffold degradation for 6 months in vitro and in vivo. *Journal of Biomedical Materials Research. Part A, 90*(3), 906–919.
25. Saminathan, P., Sripriya, A., Nalini, K., et al. (2014). Biodegradation of plastics by *Pseudomonas putida* isolated from garden soil samples. *Journal of Advanced Zoology, 1*(3), 34–38.
26. Muhamad, W. N. A. W., Othman, R., Shaharuddin, R. I., et al. (2015). Microorganism as plastic biodegradation agent towards sustainable environment. *Advances in Environmental Biology, 9*, 8–14.
27. Comăniță, E.-D., Hlihor, R. M., Ghinea, C., et al. (2016). Occurrence of plastic waste in the environment: Ecological and health risks. *Environmental Engineering and Management Journal, 15*, 675–685.
28. Vijaya, C., & Reddy, R. M. (2008). Impact of soil composting using municipal solid waste on biodegradation of plastics. *Indian Journal of Biotechnology, 7*, 235–239.
29. Rujnić-Sokele, M., & Pilipović, A. (2017). Challenges and opportunities of biodegradable plastics: A mini review. *Waste Management Journal, 35*(2), 132–140, 2564–2569.
30. Priyanka, N., & Archana, T. (2012). Biodegradability of polythene and plastic by the help of microorganism: A way for brighter future. *Journal of Environmental & Analytical Toxicology, 1*, 12–15.
31. Ahmed, T., Shahid, M., Azeem, F., et al. (2018). Biodegradation of plastics: Current scenario and future prospects for environmental safety. *Environmental Science and Pollution Research, 25*, 7287–7298. https://doi.org/10.1007/s11356-018-1234-9
32. Cameron, K. A., Hodson, A. J., & Osborn, A. M. (2012). Structure and diversity of bacterial, eukaryotic and archaeal communities in glacial cryoconite holes from the Arctic and the Antarctic. *FEMS Microbiology Ecology, 82*(2), 254–267.

33. Eich, A., Mildenberger, T., Laforsch, C., & Weber, M. (2015). Biofilm and diatom succession on polyethylene (PE) and biodegradable plastic bags in two marine habitats: Early signs of degradation in the pelagic and benthic zone? *PloS One, 10*, e0137201. https://doi.org/10.1371/journal.pone.0137201

34. De Tender, C. A., Devriese, L. I., Haegeman, A., et al. (2015). Bacterial community profiling of plastic litter in the Belgian part of the North Sea. *Environmental Science & Technology, 49*(16), 9629–9638.

35. Urbanek, A. K., Rymowicz, W., & Mirończuk, A. M. (2018). Degradation of plastics and plastic-degrading bacteria in cold marine habitats. *Applied Microbiology and Biotechnology*. https://doi.org/10.1007/s00253-018-9195-y

36. Maes, T., Van der Meulen, M., Devriese, L. I., et al. (2017). Microplastics baseline surveys at the water surface and in sediments of the North-East Atlantic. *Frontiers in Marine Science, 4*(135).

37. Bryant, J. A., Clemente, T. M., Viviani, D. A., et al. (2016). Diversity and activity of communities inhabiting plastic debris in the North Pacific Gyre. *mSystems, 1*(3), e00024–e00016.

38. Pauli, N. C., Petermann, J. S., Lott, C., et al. (2017). Macrofouling communities and the degradation of plastic bags in the sea: an in situ experiment. *Royal Society Open Science, 4*(10), 170549.

39. Monteiro, C., Castro, P. L., & Xavier Malcata, F. (2011). Biosorption of zinc ions from aqueous solution by the microalga *Scenedesmus obliquus. Environmental Chemistry Letters, 9*(2), 169e176.

40. Monteiro, C. M., Castro, P. M. L., & Malcata, F. X. (2012) Metal uptake by microalgae: Underlying mechanisms and practical applications. *Biotechnology Progress, 28*(2), 299e311.

41. Wilde, K. L., Stauber, J. L., Markich, S. J., et al. (2006, August). The effect of pH on the uptake and toxicity of copper and zinc in a tropical freshwater alga (*Chlorella* sp.). *Archives of Environmental Contamination and Toxicology, 51*(2), 174–85. https://doi.org/10.1007/s00 244-004-0256-0. PMID: 16583260.

42. Talebi, A. F., Tabatabaei, M., Mohtashami, S. K., et al. (2013). Comparative salt stress study on intracellular ion concentration in marine and salt-adapted freshwater strains of microalgae. *Notulae Scientia Biologicae, 5*(3), 309e315.

43. Perales-Vela, H. V., Peña-Castro, J. M., & Cañizares-Villanueva, R. O. (2006, June). Heavy metal detoxification in eukaryotic microalgae. *Chemosphere, 64*(1), 1–10. https://doi.org/10.1016/j.chemosphere.2005.11.024. PMID: 16405948.

44. Lamaia, C., Kruatrachuea, M., & Pokethitiyooka, P. (2005). *Toxicity and accumulation of lead and cadmium and lipid microbiology* (pp. 2583–2594). Springer.

45. Godlewska-żyłkiewicz, B. (2001). Analytical applications of living organisms for preconcentration of trace metals and their speciation. *Critical Reviews in Analytical Chemistry, 31*(3), 175e189.

46. Volesky, B. (2007). Biosorption and me. *Water Research, 41*(2007), 4017–4029.

47. Raza, Z. A., Abid, S., & Banat, I. M. (2018). Polyhydroxyalkanoates: Characteristics, production, recent developments and applications. *International Biodeterioration and Biodegradation, 126*(September 2017), 45–56.

48. Kasirajan, S., & Ngouajio, M. (2012). Polyethylene and biodegradable mulches for agricultural applications: A review. *Agronomy for Sustainable Development, 32*, 501–529.

49. Xu, J., & Guo, B. H. (2010). Microbial succinic acid, its polymer poly(butylene succinate), and Applications *14*, 347–388.

50. Baker, M. I., Walsh, S. P., Schwartz, Z., et al. (2012). A review of polyvinyl alcohol and its uses in cartilage and orthopedic applications. *Journal of Biomedical Materials Research - Part B Applied Biomaterials, 100B*(5), 1451–1457.

51. Hajiali, F., Tajbakhsh, S., & Shojaei, A. (2018). Fabrication and properties of polycaprolactone composites containing calcium phosphate-based ceramics and bioactive glasses in bone tissue engineering: A review. *Polymer Reviews, 58*(1), 164–207.

52. Shah, A. A., Hasan, F., Hameed, A., & Ahmed, S. (2008). Biological degradation of plastics: A comprehensive review. *Biotechnology Advances, 26*(3), 246–265.

53. Denef, V. J., VerBerkmoes, N. C., Shah, M. B., et al. (2009). Proteomics-inferred genome typing (PIGT) demonstrates inter-population recombination as a strategy for environmental adaptation. *Environmental Microbiology, 11*, 313–325.

54. Thrash, J. C., Temperton, B., Swan, B. K., et al. (2014). Single-cell enabled comparative genomics of a deep ocean SAR11 bathytype. *ISME Journal, 8*, 1440–1451.

55. Segata, N., Boernigen, D., & Tickle, T. L. (2013). Computational meta'omics for microbial community studies. *Molecular System Biology, 9*, 666.

56. Kobir, A., Shi, L., Boskovic, A., et al. (2011). Protein phosphorylation in bacterial signal transduction. *Biochimica et Biophysica Acta, 1810*, 989–994.

57. Vanacek, P., Sebestova, E., Babkova, P., et al. (2018). Exploration of enzyme diversity by integrating bioinformatics with expression analysis and biochemical characterization. *ACS Catalysis, 8*(3), 2402–2412.

58. Malla, M. A., Dubey, A., Yadav, S., et al. (2018). Understanding and designing the strategies for the microbe mediated remediation of environmental contaminants using omics approaches. *Frontiers in Microbiology*.

59. Nolte, T. M., Pinto-Gil, K., Hendriks, A. J., et al. (2018). Quantitative structure–activity relationships for primary aerobic biodegradation of organic chemicals in pristine surface waters: Starting points for predicting biodegradation under acclimatization. *Environmental Science: Processes & Impacts, 20*(1), 157–170.

60. Arora, T., Broglia, E., Thomas, G. N., et al. (2014, February). (2014) Associations between specific technologies and adolescent sleep quantity, sleep quality, and parasomnias. *Sleep Medicine, 15*(2), 240–247. https://doi.org/10.1016/j.sleep.2013.08.799 PMID: 24394730.

61. Jaiswal, S., Singh, D. K., & Shukla, P. (2019). Gene editing and system biology tools for pesticide bioremediation: A review. *Frontiers in Microbiology, 10*, 87. https://doi.org/10. 3389/fmicb.2019.00087.

62. Ravikrishnan, A., Nasre, M., & Raman, K. (2018). Enumerating all possible biosynthetic pathways in metabolic networks. *Scientific Reports, 8*(1), 9932.

63. Gonzalez-Garcia, R. A., Aispuro-Castro, R., Salgado-Manjarrez, E., et al. (2017). Metabolic pathway and flux analysis of H_2 production by an anaerobic mixed culture. *International Journal of Hydrogen Energy, 42*(7), 4069–4082.

64. Sulpice, R., & McKeown, P. C. (2015). Moving toward a comprehensive map of central plant metabolism. *Annual Review of Plant Biology, 66*, 187–210.

65. Marzorati, M., Balloi, A., De Ferra, F., et al. (2010). Identification of molecular markers to follow up the bioremediation of sites contaminated with chlorinated compounds. *Methods in Molecular Biology, 668*, 219–134.

66. Das, S., Dash, H. R., Mangwani, N., et al. (2014). Understanding molecular identification and polyphasic taxonomic approaches for genetic relatedness and phylogenetic relationships of microorganisms. *Journal of Microbiol Methods, 103*, 80–100.

67. Kong, W., Meldgin, D. R., Collins, J. J., et al. (2018). Designing microbial consortia with defined social interactions. *Nature Chemical Biology*, p. 1.

68. Borja, A. (2018). Testing the efficiency of a bacterial community-based index (microgAMBI) to assess distinct impact sources in six locations around the world. *Ecological Indicators, 85*, 594–602.

69. De Sousa, C. S., Hassan, S. S., Pinto, A. C., et al. (2018). Microbial omics: Applications in biotechnology. In *Omics technologies and bio-engineering* (pp. 3–20).

70. Khan, F., Sajid, M., & Cameotra, S. S. (2013). In silico approach for the bioremediation of toxic pollutants. *Journal of Petroleum & Environmental Biotechnology, 4*, 2.

71. Paul, D., Pandey, G., Pandey, J., et al. (2005). Accessing microbial diversity for bioremediation and environmental restoration. *Trends in Biotechnology, 23*, 135–142.

72. Pieper, D. H., & Reineke, W. (2000). Engineering bacteria for bioremediation. *Current Opinion in Microbiology, 11*, 262–270.

73. Butt, H., Jamil, M., Wang, J. Y., et al. (2018). Engineering plant architecture via CRISPR/Cas9-mediated alteration of strigolactone biosynthesis.

74. Bier, E., Harrison, M. M., O'Connor-Giles, K. M., et al. (2018). Advances in engineering the fly genome with the CRISPR-Cas system. *Genetics, 208*(1), 1–18.
75. Yadav, R., Kumar, V., Baweja, M., et al. (2018). Gene editing and genetic engineering approaches for advanced probiotics: A Review. *Critical Reviews in Food Science and Nutrition, 58*(10), 1735–1746.
76. Waryah, C. B., Moses, C., Arooj, M., et al. (2018). Zinc fingers, TALEs, and CRISPR systems: A comparison of tools for epigenome editing. In *Epigenome editing* (pp. 19–63). Humana Press.
77. Basu, S., Rabara, R. C., Negi, S., et al. (2018). Engineering PGPMOs through gene editing and system biology: A solution for phytoremediation? *Trends in Biotechnology.*
78. Dangi, A. K., Sharma, B., Hill, R. T., et al. (2018). Bioremediation through microbes: System biology and metabolic engineering approach. *Critical Reviews in Biotechnology, 39*(1), 79–98.
79. Stein, H. P., Navajas-Pérez, R., & Aranda, E. (2018). Potential for CRISPR genetic engineering to increase xenobiotic degradation capacities in model fungi. In *Approaches in bioremediation* (pp. 61–78).
80. Dai, Z., Zhang, S., Yang, Q., et al. (2018). Genetic tool development and systemic regulation in biosynthetic technology. *Biotechnology for Biofuels, 11*(1), 152.
81. Komor, A. C., Badran, A. H., & Liu, D. R. (2017). CRISPR-based technologies for the manipulation of eukaryotic genomes. *Cell, 168*, 20–36.
82. Mali, P., Yang, L., Esvelt, K. M., et al. (2013). RNA-guided human genome engineering via Cas9. *Science, 339*, 823–826.
83. Cong, L., Ran, F. A., Cox, D., et al. (2013). Multiplex genome engineering using CRISPR/Cas systems. *Science, 339*, 819–823.
84. Cox, D., Platt, R. J., & Zhang, F. (2015). Therapeutic genome editing: Prospects and challenges. *Nature Medicine, 21*, 121–131.
85. Karvelis, T., Gasiunas, G., & Siksnys, V. (2017). Harnessing the natural diversity and in vitro evolution of Cas9 to expand the genome editing toolbox. *Current Opinion in Microbiology, 37*, 88–94.
86. Niti, C., Sunita, S., Kamlesh, K., et al. (2013). Bioremediation: An emerging technology for remediation of pesticides. *Research Journal of Chemistry and Environment, 17*, 88–105.
87. Bhatt, P. (2018). Insilico tools to study the bioremediation in microorganisms. In *Handbook of research on microbial tools for environmental waste management* (pp. 389–395).
88. Covino, S., Stella, T., & Cajthaml, T. (2016). Mycoremediation of organic pollutants: Principles, opportunities, and pitfalls. In *Fungal applications in sustainable environmental biotechnology* (pp. 185–231).
89. Dvořák, P., Nikel, P. I., Damborský, J., et al. (2017). Bioremediation 3.0: Engineering pollutant-removing bacteria in the times of systemic biology. *Biotechnology Advances, 35*(7), 845–866.
90. Mengeling, B. J., & Turco, S. J. (1998). Microbial glycoconjugates. *Current Opinion in Structural Biology, 8*(5), 572–577. https://doi.org/10.1016/s0959-440x(98)80146-2
91. Varjani, S. J., & Upasani, V. N. (2016). Carbon spectrum utilization by an indigenous strain of Pseudomonas aeruginosa NCIM 5514: Production, characterization and surface active properties of biosurfactant. *Bioresource Technology, 221*, 510–516.
92. Moya, R. I., Tsaousi, K., Rudden, M., et al. (2015). Rhamnolipid and surfactin production from olive oil mill waste as sole carbon source. *Bioresource Technology, 198*, 231–236.
93. Gaur, N., Flora, G., Yadav, M., et al. (2014). A review with recent advancements on bioremediation-based abolition of heavy metals. *Environmental Sciences: Processes and Impacts, 16*, 180–193.
94. Wawrzynczyk, J., Szewczyka, E., Norrlow, O., et al. (2007). Application of enzymes, sodium tripolyphosphate and cation exchange resin for the release of extracellular polymeric substances from sewage sludge characterization of the extracted polysaccharides/glycoconjugates by a panel of lectins. *Journal of Biotechnology, 130*, 274–281.
95. Li, J. Q., Liu, W. Z., Cai, W. W., et al. (2019). Applying rhamnolipid to enhance hydrolysis and acidogenesis of waste activated sludge: Retarded methanogenic community evolution and methane production. *RSC Advances, 9*, 2034e2041.

96. Ndlovu, T., Khan, S., & Khan, W. (2016). (2016) Distribution and diversity of biosurfactant-producing bacteria in a wastewater treatment plant. *Environmental Science and Pollution Research, 23*, 9993–10004.

97. Fang, Y., Hozalski, R. M., Clapp, L. W., et al. (2002). Passive dissolution of hydrogen gas into groundwater using hollow-fiber membranes. *Water Research, 36*, 3533–3542.

98. Vogt, C., & Richnow, H. H. (2013). Geobiotechnology II. Bioremediation via *in situ* microbial degradation of organic pollutants (pp. 123–146). Springer.

99. Banat, I. M., Franzetti, A., Gandolfi, I., et al. (2010). Microbial biosurfactants production, applications and future potential. *Applied Microbiology and Biotechnology, 87*, 427–444.

100. Singh, P., Saini, H. S., & Raj, M. (2016). Rhamnolipid mediated enhanced degradation of chlorpyrifos by bacterial consortium in soil-water system. *Ecotoxicology and Environmental Safety, 134*, 156–162.

101. Karanth, N. G. K., Deo, P. G., & Veenanadig, N. K. (1999). Microbial productions of biosurfactants and their importance. *Current Science, 77*, 116–126.

102. Abdul, S. J., & Das, N. (2013). Enhanced biodegradation of lindane using oil-in-water bio-microemulsion stabilized by biosurfactant produced by a new yeast strain, *Pseudozyma* VITJzN01. *Journal of Microbiology and Biotechnology, 23*, 1598–1609.

103. Gaur, V. K., Bajaj, A., Regar, R. K., et al. (2018). Rhamnolipid from a *Lysinibacillus sphaericus* strain IITR51 and its potential application for dissolution of hydrophobic pesticides. *Bioresource Technology, 272*, 19–25.

104. Liu, L., Li, S., Garreau, H., et al. (2000). Selective enzymatic degradationsof poly (L-lactide) and poly (ε-caprolactone) blend films. *Biomacromolecules, 1*(3), 350–359.

105. Liu, Z.-F., Zeng, G.-M., Wang, J., Zhong, H., Ding, Y., & Yuan, X.-Z. (2010). Effects of monorhamnolipid and Tween 80 on the degradation of phenol by Candida tropicalis. *Process Biochemistry, 45*(5), 805–809.

106. Mishra, S., Lin, Z., Pang, S., et al. (2021). Biosurfactant is a powerful tool for the bioremediation of heavy metals from contaminated soils. *Journal of Hazardous Materials, 15*(418), 126253. https://doi.org/10.1016/j.jhazmat.2021.126253 Epub 2021 Jun 2.

107. Liang, Y. S., Yuan, X. Z., Zeng, G. M., Hu, C. L., Zhong, H., Huang, D. L., Tang, L., & Zhao, J. J. (2010). Biodelignification of rice straw by Phanerochaete chrysosporium in the presence of dirhamnolipid. *Biodegradation, 21*(4), 615–624.

108. Sun, W., Zhu, B., Yang, F., et al. (2020, February). Optimization of biosurfactant production from Pseudomonas sp. CQ2 and its application for remediation of heavy metal contaminated soil Chemosphere. *265*, 129090. https://doi.org/10.1016/j.chemosphere.129090. Epub 2020 Nov 30.

109. Phulpoto, I. A., Yu, Z., Hu, B., et al. (2020). Production and characterization of surfactin-like biosurfactant produced by novel strain Bacillus nealsonii S2MT and it's potential for oil contaminated soil remediation. *Microbial Cell Factories, 19*(1), 145. https://doi.org/10.1186/s12934-020-01402-4

110. Hua, S., Gong, J. L., Zeng, G. M., et al. (2017). Remediation of organochlorine pesticides contaminated lake sediment using activated carbon and carbon nanotubes. *Chemosphere, 77*, 65–76.

111. Xu, J., & Bhattacharyya, S. D. (2005). Membrane based bimetallic nanoparticles for environmental remediation: Synthesis and reactive properties. *Environmental Progress, 24*, 358–366.

112. Ponder, S. M., Darab, J. G., & Mallouk, T. E. (2000). Remediation of Cr(VI) and Pb(II) aqueous solutions using supported, nanoscale zero-valent iron. *Environmental Science & Technology, 34*(12).

113. Rajan, S. (2011). Nanotechnology in groundwater remediation. *International Journal of Environmental Science and Technology, 2*, 182–187.

114. Kanatzidis, M. G., & Poeppelmeier, K. R. (2007). Report from the third workshop on future directions of solid-state chemistry: The status of solid-state chemistry and its impact in the physical sciences. *Progress in Solid State Chemistry, 36*, 1–133.

115. Bolade, O. P., Williams, A. B., & Benson, N. U. (2020). Green synthesis of iron-based nanomaterials for environmental remediation: A review. *Environmental Nanotechnology, Monitoring & Management, 13*, 100279. https://doi.org/10.1016/j.enmm.2019.100279

116. Choe, S., Chang, Y. Y., Hwang, K. Y., et al. (2000). Kinetics of reductive denitrification by nanoscale zero-valent iron. *Chemosphere, 41*, 1307–1311.

117. Li, X. Q., & Zhang, W. X. (2007). Iron nanoparticles: The core shell structure and unique properties for Ni (II) sequestration. *Langmuir, 22*, 4638–4642.

118. Feitz, A. J., Joo, S. H., Guana, J., et al. (2005). Oxidative transformation of contaminants using colloidal zerovalent iron. *Colloids and Surfaces A, 265*, 88–94.

119. Mohsenzadeh, F., & Chehregani Rad, A. (2012). Bioremediation of heavy metal pollution by nano-particles of noaea mucronata. *International Journal of Bioscience, Biochemistry and Bioinformatics, 2*, 85–89.

120. Xiong, Z., Zhao, D., & Pan, G. (2007). Rapid and complete destruction of perchlorate in water and ionexchange brine using stabilized zero-valent iron nanoparticles. *Water Research, 41*, 3497–3505.

121. Feng, J., Hu, X., Yue, P. L., et al. (2003). Degradation of azo-dye orange II by a photo assisted Fenton reaction using a novel composite of iron oxide and silicate nanoparticles as a catalyst. *Industrial and Engineering Chemistry Research, 42*, 2058–2066.

122. Kshitij, C. J., Zhuonan, L., Hema, V., et al. (2016). Carbon nanotube based groundwater remediation: The case of trichloroethylene. *Molecules, 21*, 953–967.

123. Paknikar, K. M., Nagpal, V., Pethkar, A. V., et al. (2005). Degradation of lindane from aqueous solutions using iron sulfide nanoparticles stabilized by biopolymers. *Technology of Advanced Materials, 6*, 370–374.

124. Li, Y. H., Wang, S., Wei, J., et al. (2002). Lead adsorption on carbon nanotubes. *Chemical Physics Letters, 357*, 263–266.

125. Lu, C., Chung, Y. L., & Chang, K. F. (2005). Adsorption of trihalomethanes from water with carbon nanotubes. *Water Research, 39*, 1183–1189.

126. Ahsan, M. A., Jabbari, V., Imam, M. A., et al. (2020). Nanoscale nickel metal organic framework decorated over graphene oxide and carbon nanotubes for water remediation. *Science of the Total Environment, 69*, 134214. https://doi.org/10.1016/j.scitotenv.2019.134214

127. Li, Y. H., Wang, S., Luan, Z., et al. (2003). Adsorption of cadmium (II) from aqueous solution by surface oxidized carbon nanotubes. *Carbon, 41*, 1057–1062.

128. Schrick, B., Hydutsky, B. W., Blough, J. L., et al. (2004). Delivery vehicles for zerovalent metal nanoparticles in soil and groundwater. *Chemistry of Materials, 16*(11), 2187–2193.

129. Qiang, Y., Sharma, A., Paszczynski, A., et al. (2007). Conjugates of magnetic nanoparticle-enzyme for bioremediation. In *Proceedings of the 2007 NSTI Nanotechnology Conference and Trade Show* (vol. 4, pp. 656–659).

130. Ruffini-Castiglione, M., & Cremonini, R. (2009). Nanoparticles and higher plants. *Caryologia, 62*, 161–165.

131. Wang, C. B., & Zhang, W. X. (1997). Synthesizing nanoscale iron particles for rapid and complete dechlorination of TCE and PCBs. *Environmental Science & Technology, 31*(7), 2154–2156.

132. Mahanty, S., Chatterjee, S., Ghosh, S., et al. (2020). Synergistic approach towards the sustainable management of heavy metals in wastewater using mycosynthesized iron oxide nanoparticles: Biofabrication, adsorptive dynamics and chemometric modeling study. *Journal of Water Process Engineering, 37*, 101426. https://doi.org/10.1016/j.jwpe.2020.101426

133. Govarthanan, M., Jeon, C. H., Jeon, Y. H., et al. (2020). Non-toxic nano approach for wastewater treatment using Chlorella vulgaris exopolysaccharides immobilized in iron-magnetic nanoparticles. *International Journal of Biological Macromolecules, 162*, 1241–1249. https://doi.org/10.1016/j.ijbiomac.2020.06.227

134. Noman, M., Shahid, M., Ahmed, T., et al. (2020). Use of biogenic copper nanoparticles synthesized from a native Escherichia sp. as photocatalysts for azo dye degradation and treatment of textile effluents. *Environmental Pollution, 257*, 113514. https://doi.org/10.1016/j.envpol.2019.113514

135. Cheng, S., Li, N., & Jiang, L. (2019). Biodegradation of metal complex Naphthol Green B and formation of iron–sulfur nanoparticles by marine bacterium Pseudoalteromonassp CF10-13. *Bioresource Technology, 273*, 49–55. https://doi.org/10.1016/j.biortech.2018.10.082

136. Chen, G., Guan, S., Zeng, G., et al. (2013). Cadmium removal and 2,4-dichlorophenol degradation by immobilized *Phanerochaete chrysosporium* loaded with nitrogendoped TiO_2 nanoparticles. *Applied Microbiology and Biotechnology, 97*, 3149–3157.

137. Hou, J., Dong, G., Ye, Y., et al. (2014). Laccase immobilization on titania nanoparticles and titania-functionalized membranes. *Journal of Membrane Science, 452*, 229–240.

138. Xu, P., Zeng, G., Huang, D., et al. (2013). Synthesis of iron oxide nanoparticles and their application in *Phanerochaete chrysosporium* immobilization for Pb (II) removal. *Colloids and Surfaces A: Physicochemical and Engineering, 419*, 147–155.

139. Sanghi, R., & Verma, P. (2009). A facile green extracellular biosynthesis of CdS nanoparticles by immobilized fungus. *Chemical Engineering Journal, 155*, 886–891.

140. Wilmes, P., Heintz-Buschart, A., & Bond, P. L. (2015). A decade of metaproteomics: Where we stand and what the future holds. *Proteomics, 15*, 3409–3417.

141. Singleton, I., Merrington, G., Colvan, S., et al. (2003). The potential of soil protein-based methods to indicate metal contamination. *Applied Soil Ecology, 23*, 25–32.

142. Wilkins, M. J., Verberkmoes, N. C., Williams, K. H., et al. (2009). Proteogenomic monitoring of geobacter physiology during stimulated uranium bioremediation. *Applied and Environment Microbiology, 75*, 6591–6599.

143. Chourey, K., Nissen, S., Vishnivetskaya, T., et al. (2013). Environmental proteomics reveals early microbial community responses to biostimulation at a uraniumand nitrate-contaminated site. *Proteomics, 13*, 2921–2930.

144. Benndorf, D., Balcke, G. U., Harms, H., et al. (2007). Functional metaproteome analysis of protein extracts from contaminated soil and groundwater. *ISME Journal, 1*, 224–234.

145. Bastida, F., Jehmlich, N., & Lima, K. (2016). The ecological and physiological responses of the microbial community from a semiarid soil to hydrocarbon contamination and its bioremediation using compost amendment. *Journal of Proteomics, 135*, 162–169.

146. Carla, M. R., Lacerda, L. H. C., & Kenneth, F. R. (2007). Metaproteomic analysis of a bacterial community response to cadmium exposure. *Journal of Proteome Research, 6*, 1145–1152.

147. Ram, R. J., VerBerkmoes, N. C., & Thelen, M. P. (2015). Community proteomics of a natural microbial biofilm. *Science, 308*, 1915–1920.

148. Goltsman, D. S., Denef, V. J., Singer, S. W., et al. (2009). Community genomic and proteomic analyses of chemoautotrophic iron-oxidizing "leptospirillum rubarum" (group II) and "leptospirillum ferrodiazotrophum" (group III) bacteria in acid mine drainage biofilms. *Applied and Environment Microbiology, 75*, 4599–4615.

Chapter 11
Internet of Things (IoT): Emphasizing Its Applications and Emergence in Environmental Management—The Profound Cases

Mustafa Kayyali

Abstract Internet of Things (IoT) is a modern technical method that aims to relate things represented by devices and sensors and connect them to the Internet to transfer and receive data with each other without any kind of human intervention or human supervision. It is a virtual network that combines different things classified within electronics, software, sensors, and actuators. It connects them through the Internet, which allows these things to exchange data with each other, and it was introduced at the end of 1999 by Professor Kevin Ashton. It is worth mentioning that the term "thing" in the Internet of Things is not limited to inanimate objects and small devices only, the "thing" in IoT may be a woman who is carrying a heart rate device, for example, a girl carrying a tracker device, and a car equipped with sensors and indicators. Environmental applications are also considered one of the most important areas and current and future trends of Internet of Things technology. Through the use of environmental applications of technology, human life can be greatly improved such as desalination of salty sea water, weather forecasting, wastewater refining, obtaining the highest level of agricultural crops, and many other beneficial uses of technology.

Keywords IoT · Internet of Things · Green management · Environmental management · Environmental monitoring

11.1 Introduction

Simply speaking, "things" in Internet of Things are all devices that can communicate and correspond with each other using the Internet (refer Fig. 11.1). All those devices are allowed and have the ability to collect and exchange information, generate and process data, with the aim of improving decision-making and automating processes, as do emergency medical systems based on health monitors, which may be sent to the doctor or call the emergency automatically when it detects sudden changes in the patient's body.

M. Kayyali (✉)
Azteca University, Chalco, Mexico
e-mail: kayyali@heranking.com

© The Author(s), under exclusive license to Springer Nature Singapore Pte Ltd. 2022　　201
P. K. Paul et al. (eds.), *Environmental Informatics*,
https://doi.org/10.1007/978-981-19-2083-7_11

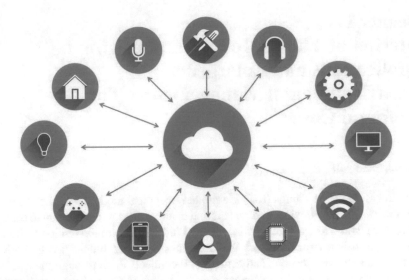

Fig. 11.1 IoT supported and associated tools and systems

The idea of the Internet of Things reflects the impact of the World Wide Web in the near future. Thanks to the Internet of Things, most electronic devices are now autonomous and do not require human intervention. These devices will be able to report malfunctions and repair themselves, and the concept will be expanded to include everything you can imagine, from the cars you drive to the lights in your home. Even your wardrobe gives you the opportunity to dress up depending on whether you feel hot or cold.

A notable example is the control of heating appliances within a building from operation and shutdown. As a result, the device notifies and repairs in advance of the possibility of an accident or malfunction. It should be noted that the Internet of Things lists many everyday objects that can be combined according to and applied via its systems, such as industrial machines, wearable devices, and many others, and it must be noted that this period when, and while we are witnessing IoT will be buzzing with it as the best way to make life more efficient than ever.

11.2 Objectives of the Work

IoT has a wide networks of applications and one of them is scientific applications, while other applications were related to academia. This chapter aims to identify the applications of Internet of Things in environmental management, and in improving the environmental awareness of people.

11.3 The History of IoT

To talk about the development and emergence of the Internet of Things, we can go back in history to the development and emergence of the Internet as we know it, which took place in stages. From WWII, until the early nineties of the twentieth century, the network was the preserve of military uses and applications, particularly with the US Army.

Then, there was a strategic decision to open the door for civilian applications in the late 1980s and early 1990s. Many of the militaries admit that they did not expect the massive spread of the Internet and its services worldwide, and they did not expect the applications to reach all walks of life. With the spread of mobile phones as a new method of technology, (with the access rate of the Internet reaching 100% in a large number of countries in the world, the emergence of smartphone and tablet technology, and generations of data transmission services), the door opened very wide for the expansion of the phenomenon of social platforms communication (both audio and visual). All this developments and advances in the Technogym led to the emergence and appearance of the third generation of the Internet.

The Internet of Things or what is referred to as (IoT) has become one of the technologies that we often hear about, and it is one of the things that the most famous and largest companies have given an important space. This is due to the importance of this high-tech at the present time in various sectors.

Although the term has not been heard by us before, its real presence in life dates back to 1974, which witnessed the emergence of ATM machines, which is one of the models of the Internet of Things devices. Regardless of the aforementioned information, 87% of people in 2015 have never heard of IoT, or its real meaning. In 2008, it was indicated according to different studies that number of devices connecting to IoT is bigger than number of people in all countries of the world [7]. And this number reached nearly five billion devices in 2015. There are many very many examples of devices under the category of the Internet of Things today, and we use them daily from computers, smartphones, smartwatches, smart glasses, and TVs, they are all practical examples of Internet of Things devices, The same applies to some cars, such as Tesla cars, which are connected to the Internet and communicate with each other to improve the quality of automated driving and data exchange. The innovation of the Internet of Things was not a matter of the moment, but rather it was based on a series of previous developments and achievements.

In 1999, MIT technology expert Kevin Ashton presented his IoT proposal in a lecture at his workplace at Proctor and Gamble, where he had the idea of using radio-frequency identification (RFID) wireless chips in the supply system. Kevin connected those products by lining them with a wireless receiver that could monitor sales and inventory and signal when the additional market was needed, and he assumed that such collected data and information would and could help solve and find solutions too many real-world urgent problems. In 2010, the concept of the Internet of Things began to gain some popularity. Information leaked had shown that a service that was provided from Google, namely "StreetView", was not only providing 360-degree images, but also had saved and stored a lot of data and information from people's Wi-Fi and Internet networks. This service was seen as the beginning of a new strategy of not only indexing the Internet but also indexing the physical world. In the same year, the Chinese government indicated that IoT would be its priority for the next years [4].

The concept of the Internet of Things was launched in 1991, and its development continued through the 1990s until Reza Raji described how it worked in the IEEE Spectrum in 1994. Since then, several companies have started to propose solutions based on the principle of the Internet of Things system. The ideal coining of an expression related to genius/smart devices has been around since 1982 by Coca Cola machine. The machine was able to report the stock of drinks and whether the drinks that had just entered the machine were cold. In 1990, a scientist called John Rumpke was able to connect the toasted bread heating machine to the Internet, and this connection allowed him to turn it ON or OFF using the Internet.

In June 2002 [1], a research chapter indicating IoT was submitted for use at the Nordic researchers in logistics conference in Norway. An article in Finnish was previously published in January 2002.

Although these dates and applications are mentioned as bright signs in the history of technology, the most prominent and important date in 2008, which scientists believe is the date of the birth of the Internet of Things, and this is due to the fact that 2008 was the year in which the numbers of people were equal with the volume of devices connected to the Internet for the first time in history (refer Fig. 11.1).

11.4   How IoT Works?

An ideal IoT system combines four different parts: sensors/devices, communications, data processors, and user platform. Devices with sensors and embedded objects connected to IoT, collecting information from different devices and using analytics to present the most important information with applications designed to specific needs. These powerful IoT platforms and systems can discover useful information and information that are very helpful for useful operation. This information can be used to identify patterns, make suggestions, and detect potential problems before they arise. For example, if I own a car designing firm, I might want to be familiar with all components that are important. By using IoT, I can:

- Use the sensors to find out the most popular showroom areas and where customers are late.
- Examine the available sales data to see which components are selling the most.
- Automatic comparison of sales data with inventory levels for consumption of popular products.

The information provided by the connected devices allows us to make genius decisions based on authentic information about the components I need to equip, saving me time and money. Smart objects mean that you can automate certain tasks, especially if they are repetitive, repetitive, time consuming, or even dangerous. Let us look at some examples to see what it looks like in real life.

IoT solutions work by integrating networked computers and sensors into everyday gadgets and products. Then, information is collected and exchanged between tools automatically, and then these tools can communicate with each other through their connection to an ever-expanding virtual network.

In 2022, mobiles are playing a great and main role in the IoT. This is due to the fact that IoT devices can be changed, controlled, and assigned by mobile phones. You can use your mobile to communicate with your smart monitor, for example, to select the perfect temperature for your workplace and for your home. Doing that is very helpful for international economy and environment because it saves energy and avoids unneeded heating or cooling while you are away from home or workplace. This can save your money and time. IoT devices have sensors and mini-computer processing units that act on the data collected by the sensors by using machine learning or AI technologies. Essentially, IoT devices are smart small-computers, connected to the Internet. Machine learning process occur when computers learn in a similar way to human beings by collecting data and information from their surrounding environments. This is what make those IoT devices act in a smart way. Deep learning process at those devices happen in this structure.

Machine learning is an integrated part of computer science that attempts to make devices capable of learning without having to be programmed literally. In general, simply, machine learning can be defined as teaching machines to perform tasks on their own without having to program them literally every time to perform a particular task, but rather programming them in a way that they are able to learn to perform many different tasks. That does not mean your laptop will discuss the key points of last night's big political discussion with you. But at least, your connected refrigerator may send you a message or an SMS on your mobile that you are low on butter or meat because it knows by detection you are near a market.

The establishment of the Internet of Things system depends on the various main components, which are smart devices that support the web for the purpose of harnessing them to process data and sensors, as well as connecting devices of many kinds for the goal of collecting information from their environment and transmitting them to be used, where the secret of working in them lies in the extent of interdependence and communication of all devices The Internet of Things with special sensors to attract and analyze the required data, so that important data from various parts is obtained, without the need for any human being interaction, and special protocols must be present to connect this great network and devices that are connected to the web.

IoT network consists of a very large network of interconnected and related devices. These devices transfer, process, and collect very big and huge amounts of data and information about how they operate and details about how they store data and information. This stored data is sent to large cloud servers located all over the world. The cloud sends relevant instructions, orders, and directions based on the information received.

11.5   How Big IoT is?

Internet of Things (IoT) prospects and goals had grown dramatically from a theoretical expression to a major priority for many companies and institutions in the
last 15 years. This is due to the fact that companies integrate, and implement IoT
devices into their network, and structure infrastructures. By the help of IoT, those
giant companies are looking for new methods to use and manage the data they are
collecting.

IoT-enabled devices and models can connect to international networks, and
operate in international arenas. It is very difficult to clarify to people how much
big the IoT is. Maybe, this can be shown in the following points:

- In the year 2021, there were more than ten billion active Internet of Things devices
  all over the world.
- It is estimated that the number of active Internet of Things devices will be much
  more than twenty-five billion in 2030. A number that is a way bigger than number
  of people all over the world.
- By 2025, it is expected that there will be approximately 152,200 Internet of Things
  devices connecting to the Internet every 60 s.
- It is expected that IoT solutions and investments have the potential to generate
  nearly 7 trillion dollars in economic value by 2025.
- 83% of organizations, companies, and high-tech firms have improved their
  efficiency by applying IoT technology.
- It is estimated that global spending on Internet of Things gadgets and techs will
  reach $15 trillion in period between 2019 and 2025.
- The amount of data and information generated by IoT devices and tools is expected
  to reach 73.1 ZB zettabytes in 2025. The number will multiply each three years.
- By the end of 2020, 5.8 billion automotive and enterprise gadgets were depending
  and applying IoT.
- High-tech firms and companies saved 54% of their costs by using IoT.
- Banking and financial services solutions depending on IoT are expected to grow
  to $2.03 billion by 2023.
- Every second, 127 devices new devices are logging to the Internet for the first
  time.
- The number of mobile phones depending on IoT connections is expected to reach
  3.5 billion in 2023.
- Investments in IoT by international companies is expected to reach to $15 trillion
  by 2025.

11.6 Future of IoT

The future of Internet of Things has the abilities to be without any limit. Advantages to the Internet will be enhanced through increased network abilities, and well-built artificial intelligence (AI) systems and protocols.

Although most of us do not have a smart home, IoT applications are becoming very popular. According to a study conducted by the IEEE International Conference on Consumer Electronics, the number of devices connected to the Internet is expected to reach 50 billion devices by the end of 2020 [8]. Experts estimate that the number will rise to 200 billion devices by 2025. This means that the number of Internet users and connected devices The Internet is constantly increasing and our relationship with things will change, and the ways of communication between humans and devices will become smarter.

Looking at how many devices were connected to IoT in the last few years gives us the future prospects of the tech, and were it is going to. In 2016, more than 4.7 billion things were connected to the Internet. In 2022, the market will rise to nearly 11.6 billion IoT devices.

Consumers will not be the only ones using IoT devices. Cities and businesses are with no time embracing genius high-techs to save time and money.

This means cities will be able to automate and using IoT at every place [3]. Those cities will manage and collect data through things like visitor registration, video camera monitoring systems, bike rental stations and taxis. IoT will be here and everywhere.

11.7 Green Applications of IoT

The level of science and technology has improved greatly, and more and more advanced technologies have appeared and are widely used in various industries of society. IoT technology plays a very prominent role in environmental monitoring, especially in the context of the current severe environmental pollution, which brings urgent challenges to the long-term development of human society. IoT technology has also become an important means of environmental protection, it can monitor changes in environment quality in real time, and once serious pollution occurs, it will give early and effective warning and take appropriate measures to deal with the situation (refer: Fig. 11.2).

The application of Internet of Things technology in the field of environmental monitoring should give way to the advantages of this technology. Play a more solid guarantee of environmental protection. The application of Internet of Things technology in environmental monitoring can monitor and manage environmental information in real time, help improve the quality and efficiency of environmental information collection, and assist in environmental protection work.

Fig. 11.2 Environmental
monitoring applications of
IoT

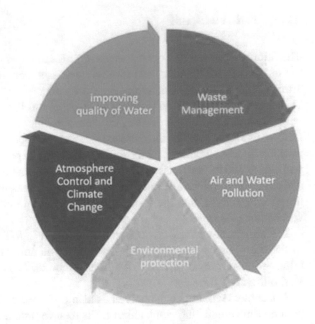

11.7.1 Environmental Protection Using Internet of Things

The main advantage of IoT technology is the realization of the connection of things, which can collect the information of the same objects through advanced sensors and intelligent devices, and transmit it to the information platform for unified analysis and management. Environmental Internet of Things has a wide range of practical applications, covering all parts of the country, and is an advanced information system for monitoring and managing pollution sources. The Internet of Things for environmental protection has gradually become the main means of environmental pollution control. Through the application of a large number of advanced technologies, the environmental management business model has undergone a great change. In general, the application of the Internet of Things to protect the environment in environmental monitoring is of far-reaching importance, and it has played a great role in promoting the implementation of the concept of scientific development and environmental protection [5].

11.7.2 Communication Between IoT and Environmental Monitoring

Nowadays, scientific research on things technology is gradually deepening and multiplying, which provides a strong guarantee for the sustainable development and improvement of environmental protection cause, and is of great significance to the sustainable development of human society in the future [6]. In order to more effectively improve the current environmental pollution problems, it is necessary to increase the importance of environmental monitoring and apply more advanced technologies to enhance environmental monitoring to become more streamlined.

11.7.3   Emerging Internet of Things Environmental Protection Application: The Scenario

In the current era, Internet of Things technology has become one of the most important means of environmental monitoring, and its role in social, and culture development and improvement has become very distinguished. The Internet of Things for environmental protection mainly emphasizes the further integration of advanced technologies on the basis of the traditional environmental protection industry, and promotes environmental protection work to become more systematic and standardized [10]. Accordingly, many countries has promulgated a series of regulations and regulations related to environmental protection, but it has not achieved significant results in environmental protection, and has attracted widespread attention. In the field of environment saving and the practical application of IoT, it can realize the collection, transmission, analysis, and storage of environmental information, which will help improve the effectiveness and quality of environmental protection work, promote the sustainable development of environmental management work, and contribute to the future environmental development [9].

11.8   Atmosphere Control and Climate Change

Atmospheric monitoring is an essential part of environment monitoring. Monitoring experts are required to monitor and study pollutant to determine whether the content of pollutants in the atmosphere are under the expected and accepted rates and limits, or exceeds the quality limits. The use of the Internet of Things contributes to monitoring global warming factors and measures the level of greenhouse gas emissions in the atmosphere, and the level of risk that this change can cause in specific population areas. In addition, the use of the Internet of Things contributes to controlling the atmosphere and creating the appropriate weather conditions for a specific region and contributing to artificial rain projects that can contribute to solving the problems of desertification and drought in some affected areas around the world. The usages of IoT sensor high-tech in environment monitoring can monitor toxic substances in places where sensors can be installed, or the sensors can be installed in densely populated areas, so that the sensor monitoring range is bigger.

11.9   Controlling and Improving Quality of Water via IoT

The use of the Internet of Things in water monitoring systems contributes to improving the quality of potable water, and identifying the most important factors that contribute to water purification and removal of impurities. The application of this modern and contemporary technology contributes to identifying problems related to water sources and measuring their pollution levels and the levels through which this water can be suitable for human consumption. The application of this technology contributes to the analysis of data and information for evaluating the extent, level, and degree of water purity. It also contributes to protecting clean and pure water wells from pollution and sending warnings and danger signals in the event that this pure and fresh water reaches high levels of pollution.

11.10   Wastewater Treatment Monitoring

With the construction and development of industrialization and the acceleration of urbanization, people's corresponding demand for water resources is increasing. At the same time, people's interest in the protection and reuse of water resources is also gradually increasing, especially with regard to the quality of the aquatic environment. Water quality monitoring has become the main means of suppressing the problem of water pollution, and it has far-reaching impact and significance, and the quality and efficiency of water quality monitoring has significant limitations, and the effect of sewage treatment has become unsatisfactory. However, the application of IoT sensor technology in sewage treatment monitoring can lead to real-time monitoring of sewage treatment conditions, which can effectively reduce personnel labor intensity and enhance the originality and comprehensiveness of sewage treatment technology to ensure its effectiveness.

11.11   Air Monitoring and Internet of Things: The Emergence

The Internet of Things (IoT) application contributes to monitoring air currents and wind direction around the world. By studying data related to air speed and direction, it is possible to predict the locations and locations of hurricanes and weather disasters, in addition to identifying areas that are likely and expected to declare a state of emergency due to difficult weather conditions. In addition, by determining the movement of winds, it is possible to determine the areas and time periods expected for the occurrence of dust storms in desert areas, and the backup effects and protection methods that countries can undertake to protect against these very harsh weather conditions. The Internet of Things contributes to determining air pollution levels as a result of emissions from factories and companies. Therefore, there are significant economic effects that can result from data resulting from studies related to the impact of the Internet of Things on the atmosphere, and may cause the enactment of new policies that reduce the release of these emissions [2].

There are many other applications of IoT in green management such as tracking irrigation systems, fertilizer schedules, monitoring weather conditions, monitoring crops from pests and dangers such as fires, monitoring salt and pesticide percentages, smart grazing and herd follow-up, measuring environmental pollution of water, especially for fish farms. Of course, we do not forget here to monitor forests and forest resources through surveillance devices and drones.

11.12 Conclusion

It is clear that Internet of Things tech is developing in a rapid manner and in a way that enhances the well-being and development of mankind and helps to improve the services provided. It is also clear that the Internet of Things is evolving in a way that enables and includes new applications that were not used before. This encourages companies, factories, investors, and capital to make additional and new investments in this new field. It also encourages scientists, researchers, and programmers to present new research and ideas that simulate the optimal mechanisms for developing this technology. Internet of Things tech can be applied in most scientific, research, social, and environmental fields and sectors. This tech can also be invested in climatic, environmental, and atmospheric conditions, weather forecasting, wind speed and rainfall, and forecasting the occurrence of floods, hurricanes, volcanoes, and tornadoes. In addition, it is possible to create simulation models that simulate the atmospheric and climatic weather of a region and predict its weather.

11.13 Recommendations and Suggestions

We recommend holding extensive studies and research to demonstrate the applications of the Internet of Things in the field of environmental management and public administration. We also recommend making additional investments to support this technology and support its great benefits.

References

1. Cheng, Y., & Zhao, L. (2019). Application and research of GIS intelligent environmental protection internet of things system construction technology. In *2019 International Conference on Information Technology and Computer Application (ITCA)*. https://doi.org/10.1109/itca49 981.2019.00014
2. Environmental Protection Agency. (n.d.). *How to use air sensors: Air sensor guidebook.* EPA. Retrieved January 22, 2022, from https://www.epa.gov/air-sensor-toolbox/how-use-air-sensors-air-sensor-guidebook
3. Kyriazis, D., Varvarigou, T., Rossi, A., White, D., & Cooper, J. (2013). Sustainable smart city IOT applications: Heat and electricity management & eco-conscious cruise control for public transportation. In *2013 IEEE 14th International Symposium on "A World of Wireless, Mobile and Multimedia Networks" (WoWMoM)*. https://doi.org/10.1109/wowmom.2013.6583500
4. Leonardos, K. (n.d.). *Analysis of the applicability of internet of things projects.* Retrieved January 22, 2022, from http://monografias.poli.ufrj.br/monografias/monopoli10024147.pdf
5. Maroli, A., Narwane, V. S., & Gardas, B. B. (2021). Applications of IOT for achieving sustainability in agricultural sector: A comprehensive review. *Journal of Environmental Management, 298*, 113488. https://doi.org/10.1016/j.jenvman.2021.113488
6. Hassan, M. N., Islam, M. R., Faisal, F., Semantha, F. H., Siddique, A. H., & Hasan, M. (2020). An IoT based environment monitoring system. In *2020 3rd International Conference on Intelligent Sustainable Systems (ICISS)*, 2020 (pp. 1119–1124). https://doi.org/10.1109/ICISS49785. 2020.9316050
7. Marques, G., Miranda, N., Kumar Bhoi, A., Garcia-Zapirain, B., Hamrioui, S., & de la Torre Díez, I. (2020). Internet of things and enhanced living environments: Measuring and mapping air quality using cyber-physical systems and mobile computing technologies. *Sensors, 20*(3), 720. https://doi.org/10.3390/s20030720
8. Vivek, S., Verma, D., & Krishnan, P. (2018). Towards solving the IoT standards gap. In *2018 International Conference on Advances in Computing, Communications and Informatics (ICACCI)*, 2018 (pp. 1441–1447). https://doi.org/10.1109/ICACCI.2018.8554506
9. Zhang, Z., & Zhang, S. (2021). Application of internet of things and Naive Bayes in public health environmental management of government institutions in China. *Journal of Healthcare Engineering, 2021*, 1–7. https://doi.org/10.1155/2021/9171756
10. Zhao, Y. J., Du, B., & Liu, B. K. (2013). Smart environmental protection: The new pathway for the application of the internet of things in environmental management. *Applied Mechanics and Materials, 411–414*, 2245–2250. https://doi.org/10.4028/www.scientific.net/amm.411-414.2245

Chapter 12
LiFi-Based Energy-Efficient Traffic Sensing and Controlling System Management for Smart City Application

Abhinandan Sarkar, Rajdeep Chakraborty, and Hoimanti Dutta

Abstract Alternative research efforts have been conducted for the last few years to overcome the deficit spectrum of electromagnetic waves which can release the network traffic from the saturated radio frequency domain. With some of the culture and development, it can be stated that optical wireless communication can lead to a new spectrum of data sharing. Consumption of data and its usage has increased more than 100 times in the last 10 years. Development of around 80 billion IOT systems has also contributed to the crisis of radio spectrum availability and increased traffic. It is forecasted that by 2022 there will be use of around 50 zettabytes of data, that can be imagined as many bits of stars and planets in the universe. With the recent contribution from researches, a new domain has been termed as visible light communication (VLC) and will reinforce the communication protocol. In this domain, the architectural transformation resulted in light fidelity (Li-Fi), replacing the wireless fidelity (WiFi) with added security and unrestricted bandwidth allocation.

There are 300 Tetra Hz unused bandwidth (1000 times 300 Giga Hz of radio frequency spectrum) available at higher frequencies in the visible light spectrum. LiFi uses the visible light spectrum for communication, which is much faster than radio frequency, and can be easily used in near field communication. Many researches have served with multiple conceptions and misconceptions in this experimental area. The technology is advancing with the speed of its own concept, visible light. Establishing the liable system and computing the data, parametrical diversity is in progress. Professor Harald Haas of Edinburgh University has provided many clear out reach for the proposed system. Many tech-giants have configured their own. But in our daily life, bringing ease to our society has been a concept and dream till now. Converging all the thought at a point with the recent development and implementation, in this chapter we claim to successfully design a light fidelity-based system, which can be used for traffic signal sensing and managing, and it will be energy efficient. As it is studied that LiFi system transmits data through LED and receives through photocell,

A. Sarkar
Shree Ramkrishna Institute of Science and Technology, Kolkata, India

R. Chakraborty (✉) · H. Dutta
Netaji Subhash Engineering College, Kolkata, India
e-mail: rajdeep.chak@gmail.com

LED is a clear source of energy-efficient output and photocell (solar cells) which absorbs light energy and converts into electrical energy is a proven energy-efficient receptor.

Our system proposes a long-range LED, which transmits light spectrum of over 1.5 km and reflects back after incident on the surface of the vehicles. Solar cells used as receptors of this light beam track the congestion of the beams. The implementation of the transmitter and receiver can be done in affordable and reliable using microcontroller like ATmega 328. The data can be retrieved and processed, and the density of vehicles can be traced, with easy controlling of the traffic signals. With amplification of the incident light beam on solar cells, it can be reused for power support to the circuitry, making it less biasing potential consuming device. The emergency vehicles can also be traced with some deviation of visible light spectrum wavelengths. This design opens up path for many researches on the field of smart city application, environment and its different parameter monitoring systems.

As the system uses LED as a source and solar panel as a receptor, it is very much energy efficient. The solar panel can charge the inbuilt battery itself and can operate in day and night condition. The conventional traffic management system uses mainly camera-oriented system, so this proposed system is quite energy efficient. After feeding the count in the edge level gateways in each of the traffic points, the traffic light may be controlled upon the policy of the city. Thus, this proposed system is quite relevant for smart city applications or smart traffic management system. This proposed system is also cost-effective as it is only using LED, small photocell (here it is solar receptor) and microcontroller ATmega 328 for deployment, and traffic control can be done through edge computing using Raspberry Pi.

Keywords Light fidelity (LiFi) · Energy efficient · Traffic sensing · Traffic controlling · Smart city · IoT

12.1 Introduction

Communication, more precisely development of the communication system or also can be termed as the ways of transferring data, was one of the dominant research cultures from the prehistoric years of human evolution. The species evolved with the symbolic representation on cave walls to communicate with each other. This develop different sound frequencies in various directions. With the separation of the land masses due to the movement of the tectonic plates or the lithosphere area, the culture and vocables became different. Written data transfer came into existence, and with the help of pigeon, it was studied that long distance communication also took place. This whole development happened in the early or medieval era. In the modern times, we can see a rapid growth of communication system, and we had dedicated system for letter distribution and then came the telegrams, trunk calls, computers and land phones. A noticeable time arrived with the invention of wireless communication system and Internet [1]. Life became so easy to connect with distant persons. Another remarkable evolution came with fax machine, where letter can be sent instantly to distant places. With the emergence and development of Internet, email came, land phones got changed to mobile phones, wireless communication developed more with time, and we can see speeding up of data transfer. We can see development of devices; the separated world came to our palm. But with these evolutions, complexity of the systems also increased [2].

When we talk about wireless communication, we mean communication through waves, at a fixed bandwidth and wavelength. Though we are at the verge of expanded communication, we feel unsafe because our life data has come to a risk. Data from different devices get hacked while transferring through unsafe communication ports like WiFi. Though it is studied that data transfer through wireless fidelity technology is quite fast, usage does not limit its risk or insecurity with data transmission and receive. WiFi ports get easily cloned, through which connected devices lose their data. As well we can see that the bandwidth allocation of the propagated wave is too congested, due to which it is becoming slower with days passing. With the development of Internet of Things technology, around 80 million devices are getting connected in the world around. Collision and congestion during data transfer are prevailing problems nowadays.

As a lifesaver, a new technology emerged, invented by Professor Harald Hass of University of Edinburgh, communication through visible light communication. The spectrum frequency ranges from 4.3×10^{14} Hz to 7.5×10^{14} Hz and 400 nm to 800 nm in wavelength [3]. This band is readily available in our indoor environment, which is quite fast and secured for usage. The data is transmitted through the visible light, through a commonly used solid-state device, LED. The question comes why LED? and why cannot we use fluorescents or incandescent light for transmitting data. As we study communication technology, we are clear with the concept of data transmission, and we know that modulation and demodulation of the signal are the spine of the technology. The current intensity of the LED can be easily modulated when compared with other sources. The lifetime of the light emitting diode (LED) is longer, with more efficiency and durability. So what is an LED? LED is a p–n junction semiconductor which emits narrow spectrum light when electric current flows through it, and this can be termed as electroluminescenes [4]. Wavelength of the LED can be determined by its material used to manufacture and the energy gap. For visible light communication (VLC), the use of white light is dominant over others. Some of the unique characteristics of VLC are as follows: the transmission of data is a mathematical representation of channel impulse response or it can be said as transmission matrix [3]. The modulation is on the intensity factor, so phase and frequency are not dependant. The received data is real and positive as the modulation is on the instantaneous power [4]. Another feature is multipath fading, which can be easily omitted in case of transmission through visible light spectrum and use of LED [5]. Reduced cost and security are dominant features for this technology. As visible light does not penetrate through opaque object, we have a considerable parameter of line of sight for data transmission. If we can modulate the signal, decode and transmit in a proper way, this spectrum would prove to be the most safest region of communication. A VLC system can be called as a non-coherent and non-negative signal transmitter.

When we compare radio waves and VLC, both use electromagnetic spectrum for communication but due to the wavelength we know visible light cannot pass any opaque barrier, so data security can be sustained in VLC system. In case of radio waves, though there is loss of signal strength when it passes through thick walls, it passes, and there can be loss of data or theft of it. So in visible light communication, it is restricted to a specified place, range is short, data security is excellent, and speed of data transfer is also good, with reduction of data fading. And as intensity modulation and direct detection method are used to communicate in VLC system, it also gives a considerable advantage over RF. The development of VLC is not to eliminate and reduce the use of radio waves, but to complement its usage [5].

Before going in depth with our proposed model or system, an overall subjective study of the VLC and LiFi systems, its advantage over the contemporary technologies and its various applications need to be studied.

Let us know about the VLC system and its different blocks of functionality. The basic system of the VLC consists of three sections. The transmitter, channel and the receiver make up this three sections. The transmitter block consists of modulator and the transmitting LED; the receiver block consists of a photodiode for capturing the light emitted from the LED, a demodulator; and finally, we get the data. We can also have a cloud data storing facility for remote data access. As our application is based on edge computing, it is beneficial to have remote data serving. The channel in the system is the space between the transmitter and the receiver, which can get affected by interference, noise and attenuation [3]. The VLC system for practical orientation can be single output or multi-output device. But it can be easily seen that, as modulation is dependent on intensity of light, and it can be termed as illumination of that place, or the illumination created by the transmitter, we need enough lexus to transmit data when we are using it for indoor networking. When compared with intensity of fluorescent light or incandescent light, the intensity or power of illumination of LEDs is much more. If we look around us and compare it with our naked eyes also, it can be seen that the illumination created by a 60w fluorescent bulb is much less when compared with using a 18w LED bulb. As discussed earlier, modulation of intensity becomes easy as LED is current-driven semiconductor illuminator. In the VLC system, we can include another block, signal conditioning, that will amplify the signal, reduce noise, filter the signal and provide a more specific and relevant data transmission. In the picture below, we can see a basic block diagram of the VLC system.

In every communication system, there will be losses during data transmission and the received signal carries noise; the signal conditioning block shown in Fig. 12.1 will help in minimizing the noise and other attenuations. Another block that can be added for more optimization is the line decoder, which will maintain the sequence of the data received and also decode the signal in a desirable output.

Let us also understand about the layered architecture of the LiFi system. The diagram of the architecture is given in Fig. 12.2. The basic architectural flow has three layers, physical layer, MAC layer and the application layer. The physical layer

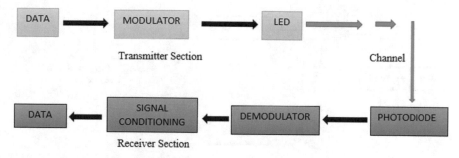

Transmitter Section

Channel

Receiver Section

Fig. 12.1 Basic block diagram of a VLC system

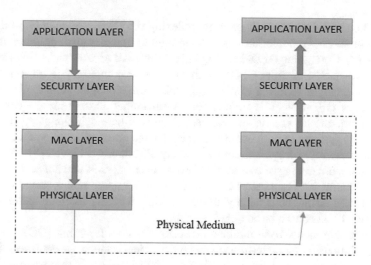

Fig. 12.2 Layered architecture of LiFi system [7]

connects between the device and the medium, and in simple words, it acts as the transmitter and the receiver of the LiFi system. Different modulation techniques are followed in different systems, and according to the layer modularity and construction, mainly it is studied that on–off keying, orthogonal frequency division multiplexing and quadrature amplitude modulation, colour shift keying and pulse modulation techniques are used. The medium access control layer is concerned with visibility support, dimming and mobility support, VPAN disassociation and networking beacon generation for coordinating device. As data security is one of the important aspects in today's communication ports, we propose another layer in the architecture for security purpose, and lastly, we have the application layer. So proposing three important layer, physical layer, MAC layer and the security layer or can be said as added security layer [6].

As stated above about the modulation technique, let us know in brief about the mentioned techniques, for the better understanding of the proposed concept. Starting

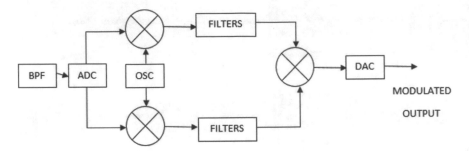

Fig. 12.3 Building blocks of QAM

with orthogonal frequency division multiplexing (OFDM), it is mainly done when conventional serial modulation schemes are used like quadrature amplitude modulation (QAM). Combining OFDM with QAM minimizes the inter-symbol interference. Another advantage of using OFDM is that the signal complexity in transmitter and receiver is easily transferred to digital domain from analogue domain. The implementation of OFDM is quite simple, data is transmitted in multiple frequencies on parallel path, and hence, the symbol period tends to be much elongated compared to the serial system but with same data rate. So, the effect of inter-symbol interference is maximum to 1 symbol, and hence, equalisation is achieved [8]. In a QAM system, there is a mixing of two amplitude modulated signals, and hence, it increases effective bandwidth. For a better-optimized interference free signal, less noisy signal OFDM-QAM technique is widely used in multiple wireless communication system [9]. Figure 12.3 shows a basic QAM system.

Another commonly used modulation scheme is on–off keying (OOK). Though there are limitations like interference and spectral inefficiency, we achieve enhanced energy per bit and lower power levels than 802.15.4 [10]. In this scheme, the on state is represented with 1 and 0 is used to represent off state, but the LED is not completely off, and it is dimmed to reduce the intensity level and trigger the off state [7].

Pulse modulation (PM) technique and colour shift keying (CSK) are other few generally used modulation schemes. As previously discussed, for LiFi, the intensity of light or modulation of it is the working factor, and we do not need to modulate the frequency or the phase, so in case of pulse modulation schemes, pulse amplitude modulation or pulse width modulation is used to vary the intensity according to the data bits [11]. By the name colour shift keying, we can identify the theory behind CSK, that modulation is done by variance of 3 LED, red green and blue, and it is dependent on the colour space chromaticity diagram [7].

Though there are misconceptions for LiFi as a sustainable communication system, like LiFi is line of sight scheme, it cannot be dimmed, it does not work in sunlight, it flickers a lot, and it is for downlink only, these limitations can be recovered by using variety of modulation techniques, among the above stated few like OFDM intensity modulations and uni-polar OFDM [12].

LiFi with varied applications, in different domain, comes encapsulated with new age computing techniques and gives a new path for data communication, with very little power consumption and simplified networking theories. Our article concept is based on such an application of visible light, where it is established in the centre of a multi-junction road. The LED is fixed in this way that it does not hit the driver sight, but is long range enough (around 1.5 km) to travel and incident on vehicles at an angle, so that it reflects back to origin, to get absorbed and detected by the implemented solar panels or photodiodes at the junction. The intensity of the reflected ray produces a current, more the intensity of reflection, more the current generation, and we can predict this as more traffic in that path. If there was less traffic, the pitch used in making roads would not reflect back much light, as maximum of it will get absorbed by the deep ash colour of it. The reflected ray would come from the shiny metallic bodies of the cars or vehicles. To implement this project, we tend to get through some minute details, how the circuit gets its power in the middle of the road? Who gets the data? How it can be fetched from remote places? The sunlight during the day will interfere with LEDs, how to optimise the concept during day time? How to process the signal against bright sunlight? Will there be flickering with ageing of the light? Is it cost-efficient? How to implement edge computing techniques on a lamp post? Will the design prioritize emergency vehicles? All this queries will be discussed in our methodology section to draw the concept clear to the expected readers. Section 12.1 discusses the related studies, Sect. 12.2 gives the detailed methodology, and Sect. 12.3 draws the conclusion. References are listed at last.

12.2 Related Studies

In this section of the article, we discuss about some of the related works in this field. This section will provide a more insight to the design concept.

The technology got its inception in 1990, from Japan, Germany and Korea. In [13], they discuss about the scarcity of radio waves, its capacity, congestion, efficiency, availability and security. The deficiency in radio waves gives an alternate solution of light fidelity. The chapter proposes in establishing a LiFi-based system to prioritize the vehicles and differ the signals automatically. As traffic light uses LED, establishing such a system would not be of great challenge. Emergency vehicles can also transmit the signal of priority through the on top mounted signal lights, which makes the traffic movement easy and broadcasts the signals or delays up ahead. The chapter also shares some applications of LiFi in different fields, like airways, medicals, underwater vehicles and hotspots.

The article [14] states the visible light communication to be bi-directional, that it can be used for both uplink and downlink. The connectivity is in large area, with high speed data transfer and security. The articles propose a system that has a LiFi-enabled taillight and headlight of vehicles and also the traffic signals. This will help in road safety and traffic management by establishing a vehicle-to-vehicle communication drive, hence reducing road accidents and optimizing the traffic signals.

Article [15] also describes the LiFi system like the previous article, but the theory of the chapter is based on the different technical standards and modulation techniques used and is going to be used in near future for establishing a LiFi domain for traffic management. It shares the use of RFID in traffic management with its advantages and disadvantages with the future of LiFi and how much workable it is going to become.

In articles [16] and [17], the proposed system is of vehicle-to-vehicle communication, where the head lights are made as LiFi transmitter and tail lights are made as LiFi receiver. This concept helps in alerting the traffic signal about emergency vehicles, and once the signal is received, immediately the signal turns green, decreasing the wait time of the vehicle in congested roads. This V2V communication system helps in communicating between the vehicles and sharing required information in the road.

12.3   Methodology

In this section, we discuss about the detailed description of the project. Till now, we have discussed in detail about the visible light communication, the backbone of our design. We have said how VLC works and what are the different modulation schemes used to transmit data. Though there was some shortcomings and misconceptions about LiFi, it can also be sorted using specific multiplexing and modulation techniques.

Our design is a complement to the latest technologies used in traffic control. Today, we have a pre-defined counter, that defines how much a car has to wait for the green signal. We have camera which is very costly, and it monitors the traffic inflow and outflow, the density of vehicles and traffic rules. But it has a problem that it is not remotely monitored, the traffic outpost is very near, and maximum number of cameras are wired to the display devices. In metro cities, some WiFi-enabled cameras are established for a certain distant monitoring of the vehicles, but are too costly. It has been noticed that emergency vehicles like ambulance and fire brigade vehicles do not get enough space in tight traffic to go out. Once a road becomes congested, it has to be immediately evacuated, to maintain a smooth traffic. When manual control or fixed timing is set, the traffic cannot be smoothen immediately. With costly visual capturing devices, it is not affordable to establish it in every corner of the city or town.

Throwing some light on another aspect, remote monitoring and automatic monitoring of traffic become difficult when WiFi devices are not connected. We intend to provide a one stop solution to the issues arising in everyday life. With the implementation of edge computing, the data is processed in real time on the location and the traffic is monitored.

As previously discussed, a VLC transmitter is a LED, and the receptor is a solar cell/panel or photodiodes. The cost is very less in comparison to the camera vision. We need a microcontroller to process the data. We prefer using Arduino UNO developmental board, which has AtMega328 microcontroller, inbuilt ADC to connect

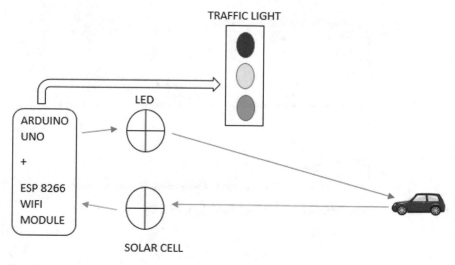

Fig. 12.4 Basic concept of the project

digital and analogue devices. It also has a embedded crystal oscillator to trigger the machine states for proper timed operation. The LED and the solar cell are attached to the board for light transmission, receiving of reflected back rays, processing of the intensity received and taking action on the traffic signal. We can connect a ESP8266 Arduino compatible WiFi module to upload the traffic data on the cloud, which will give a further detailed study of the traffic timings and controls and from where the characteristic of the solar cell can also be studied. The figure below shows the basic concept of the project.

As it can be seen in the basic block in Fig. 12.4, Arduino board is connected to the LED, and the intensity of the light in LED is modulated by the Arduino. LED transmits the light towards the road, incidence on the cars, gets reflected back towards the solar cell. The cell captures the light and produces current accordingly. More the intensity of light, more will be the current electricity generation. This generation is in analogue form, and the ADC converts the data into digital bits. More the intensity, more number of higher bits flow in the processor. Once the processor fetches the data after processing, if found to be a constant flow of higher bits, it can be followed as congested road, or still traffic. Immediately, the processor sends signal to the traffic light to change from red signal to green signal. So that heavy traffic is minimized in that road, giving a smooth movement of vehicles. In Fig. 12.5, we can see the analogue or continuous change of current with the flow of different vehicles on the road. There is a fall in the level with minimum vehicles, and there is rise in the signal with increasing vehicles. It can also be seen a approximately constant signal when the signal is red and the vehicles are still, giving a brief insight of a congested road. This signal is converted into digital signal by feeding it to the analogue to digital converter of the UNO board.

The relationship between light intensity and current can be given as:

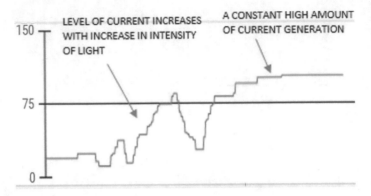

Fig. 12.5 Variation of solar cell signal (current) with light intensity

$$L_{\text{intensity}} \infty I_{\text{generation}} \tag{12.1}$$

From Eq. (12.1), we can understand that as the intensity of light increases, the flow of electron in the circuit increases because more photons strike the panel, so more generation of current. It can also be traced about the congestion of road, with increase in the current flow from the solar cell. A digital output of the solar cell is shown in Fig. 12.6.

If we notice the curve of Fig. 12.6, it shows a square wave. From the curve, it is visualized that the bit pattern is tentatively 11,110,011,111,111 and so on; when the current generation is not above the threshold value or the margin, the Arduino processes the data as 0; as it crosses the threshold value, the ADC gives 1 as output; and accordingly, the traffic signals are manipulated automatically. With connection of ESP 8266 WiFi module with Arduino, the whole data and its variation can be uploaded in the cloud storage for remote accessing and post-processing of other parameters.

In the introduction section, we have raised some questions, some of the queries we have already discussed, and in this paragraph, we will discuss the rest in brief. As

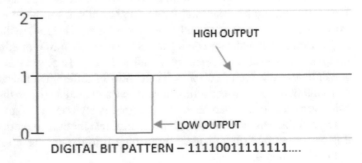

Fig. 12.6 Digital output of Solar cell

we see, we have a solar cell or panel attached as receptor, and the battery which is providing the power will get charged from the current generated in the cell. During daytime, due to the presence of sunlight, the battery will get charged, which will continue to supply power during the night time, with a small amount of charging from the light intensity during night. Though there will be sunlight with OFDM-QAM modulation techniques, we can reduce the interference or minimize it to such level that we get a prominent data. With implementation of solid-state device, the system does not get affected by flickering with ageing. And the system can prioritize emergency vehicles, as emergency vehicles have signal lights on top; with colour detection schemes, we can detect the wavelength and frequency of the red light; and special action can be taken to bypass the emergency vehicles.

The processing becomes challenging when we come across a junction point like 4 points or 5 points. Detecting the prominent congestion can incur a little delay of few seconds. But comparing with a camera system, it is way cheaper and a better way of data communication, as it does not use radio waves for this process. When we make the hybrid model [18] for cloud interfacing, we require a WiFi connected to the LiFi system for continuous updation. The graphical representation Fig. 12.7 below shows the output of receptor for a 4-point junction. We will also highlight the decision-making process of the microcontroller against the signal generated.

From the above representation, we see the variation of the current in the solar panels due to the different intensity of light at the 4 signal points. We assume all the four signals are red, so the signal becomes constant after a certain time, and we will see which signal will be green first. And the sample is taken for few minutes after the signals turn red. If we see for lane 3, the signal value is very low near about 20–50. In lane 4, the value is near about 190–200, so we can predict that the lane is quite full. The value of lane 2 is little bit higher than lane 4, just crossing 200, so we can say lane 2 will be green before lane 4 at this instant and then lane 3. Now, from lane 1, the value is reaching to 350–390, and when compared to the other lanes, it is highly congested. So lane 1 will get the green signal first and then lane 2 followed by

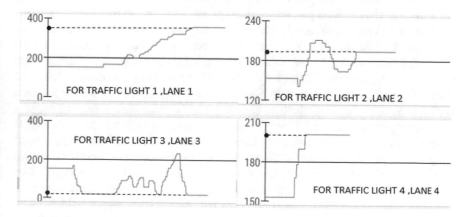

Fig. 12.7 Signal variation at different points

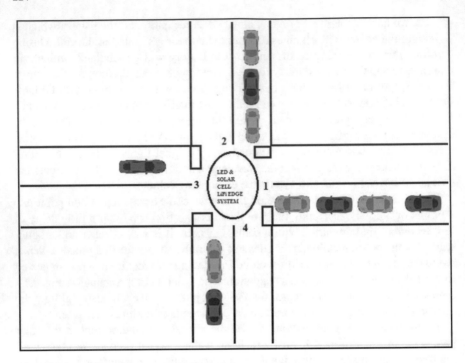

Fig. 12.8 Structural concept of the traffic management system

lane 4 and lane 3. After analysing the data from the graph, now we can visualize the system established on the 4-point traffic crossing. In Fig. 12.8, we give the structural concept of the traffic management system.

In the above figure, we have shown the deployment format of the system, and the figure also relates to the graphical outputs in Fig. 12.7. The lanes have been marked accordingly, and the density of vehicles is also picturized for a better understanding. As defined in the previous paragraph, the signalling will work from lane 1 and then 2 followed by lane 4 and lane 3. The LiFi system will be mounted on a tall tower in the centre of the junction, with four sensor-receptor system to get data from four directions. We can use only one Arduino UNO as it has enough ports to interface with four input–output devices. Though it seems that there may be lag in data fetching, but giving a appropriate delay with the four processes, the Arduino gets enough space for the previous execution cycle to get complete. So data over-writing or lagging in delivering the output can be minimized. The microcontroller compares the signal generated in the receptors and gives the signal change command towards the most congested lane. Giving it a Go! Signal for the lane to become decently filled and moving towards the next highest signal and so on. Once it executes the signal from all the panels, it waits for the next congestion signal to act upon. There may be a situation where, once the traffic signal goes red, the vehicles get packed up, while the other lanes are moving, so the controller checks the signal for few seconds, till it gets a good intensity on the panel and executes the signal on that lane. The whole process can be defined in multiple ways the programmers wants it to be.

So from this design of the LiFi system, and applying some basic calculations, we can predict the congestion of traffic on the road and can establish an automatic traffic management system giving a path for development in smart cities.

12.4 Conclusion

Though the inception is studied to be in 1990 [13], but the practical demonstration of the technology was done in 2011, in Ted talk series by Professor Harald Haas, he demonstrated the working capabilities of LiFi. He also showed how in foggy weather the data will get transmitted, and he showed a transmission of video through a led lamp, which gets played in the receiving side laptop. He proves the transmission by blocking the path of visible light. Many researches and development are in process for the light fidelity. Many applications are developed. We propose for one of the application of visible light data transfer concept for traffic management. The article describes about the advantage of LiFi over WiFi. We can use the techniques to get the appropriate data. We describe how intensity of light affects the current generation in the solar cells, thus varying the signals in the microcontroller board. There are multiple future scopes from this article for further development, like pedestrian management, accident monitoring and alarming system and data communication between vehicles. Attaching with GSM module, we can predict the location of the vehicle giving a complementing solution to GPS. Thus, we can conclude that light fidelity will take over the new generation of industries, transportation, smart cities and many more real-world applications.

References

1. Khandal, D., & Jain, S. (2014). Li-fi (light fidelity): The future technology in wireless communication. *International Journal of Information & Computation Technology, 4*(16), 1687–1694.
2. Leydesdorff, L. (1994, March 21). The evolution of communication systems. The evolution of communication systems. *International Journal of Systems Research and Information Science, 6*, 219–230. Available at SSRN: https://ssrn.com/abstract=2236680
3. Webster, J. G., Ndjiongue, A. R., Ferreira, H. C., & Ngatched, T. M. N. (2015). Visible light communications (VLC) technology. In J. G. Webster (Ed.), *Wiley Encyclopedia of Electrical and Electronics Engineering*. https://doi.org/10.1002/047134608X.W8267
4. Gancarz, J., Elgala, H., & Little, T. D. C. (2013). Impact of lighting requirements on VLC systems. *IEEE Communications Magazine, 51*(12), 34–41. https://doi.org/10.1109/MCOM.2013.6685755
5. Burchardt, H., Serafimovski, N., Tsonev, D., Videv, S., & Haas, H. (2014). VLC: Beyond point-to-point communication. *IEEE Communications Magazine, 52*(7), 98–105. https://doi.org/10.1109/MCOM.2014.6852089

6. Harald, H. (2018). LiFi is a paradigm-shifting 5G technology. *Reviews in Physics, 3*, 26–31. ISSN 2405-4283, https://doi.org/10.1016/j.revip.2017.10.001

7. Ullah Khan, L. (2017). Visible light communication: Applications, architecture, standardization and research challenges. *Digital Communications and Networks, 3*(2), 78–88, ISSN 2352-8648, https://doi.org/10.1016/j.dcan.2016.07.004

8. Armstrong, J. (2009). OFDM for optical communications. *Journal of Lightwave Technology, 27*(3), 189–204. doi: https://doi.org/10.1109/JLT.2008.2010061

9. Li, X. (2008). Simulink-based simulation of quadrature amplitude modulation (QAM) system. In *Proceedings of the 2008 IAJC-IJME International Conference.*

10. Daly, D. C., & Chandrakasan, A. P. (2007). An energy-efficient OOK transceiver for wireless sensor networks. *IEEE Journal of Solid-State Circuits, 42*(5), 1003–1011. https://doi.org/10.1109/JSSC.2007.894323

11. Deloraine, E. M. (1949). Pulse modulation. *Proceedings of the IRE, 37*(6), 702–705. https://doi.org/10.1109/JRPROC.1949.233950

12. Haas, H. (2016). LiFi: Conceptions, misconceptions and opportunities. *IEEE Photonics Conference (IPC), 2016*, 680–681. https://doi.org/10.1109/IPCon.2016.7831279

13. Kalaiselvi, V. K. G., Sangavi, A., & Dhivya. (2017). Li-Fi technology in traffic light. In *2017 2nd International Conference on Computing and Communications Technologies (ICCCT)*, 2017, pp. 404–407. doi: https://doi.org/10.1109/ICCCT2.2017.7972307

14. Yeasmin, N., Zaman, R., & Mouri, I. J. (2016). Traffic control management and road safety using vehicle to vehicle data transmission based on Li-Fi technology. *International Journal of Computer Science, Engineering and Information Technology, 6*(3/4), 1–7.

15. Al_Barazanchi, I., Sahy, S. A., & Jaaz, Z. A. (2021, February). Traffic Management with Deployment of Li-Fi Technology. In *Journal of Physics: Conference Series* (Vol. 1804, No. 1, p. 012141). IOP Publishing.

16. Shanmughasundaram, R., Vadanan, S. P., & Dharmarajan, V. (2018, February). Li-Fi based automatic traffic signal control for emergency vehicles. In *2018 Second International Conference on Advances in Electronics, Computers and Communications (ICAECC)* (pp. 1–5). IEEE.

17. Mugunthan, S. R. (2020). Concept of Li-Fi on smart communication between vehicles and traffic signals. *Journal: Journal of Ubiquitous Computing and Communication Technologies, 2*, 59–69.

18. Abuella, H., et al. (2021). Hybrid RF/VLC systems: A comprehensive survey on network topologies, performance analyses, applications, and future directions. *IEEE Access, 9*, 160402–160436. https://doi.org/10.1109/ACCESS.2021.3129154

19. Albraheem, L. I., Alhudaithy, L. H., Aljaser, A. A., Aldhafian, M. R., & Bahliwah, G. M. (2018). Toward designing a Li-Fi-based hierarchical IoT architecture. *IEEE Access, 6*, 40811–40825. https://doi.org/10.1109/ACCESS.2018.2857627

Chapter 13
Technology-Based Management and Monitoring System to Combat COVID-19: An Environmental Disaster of the Century

Sandeep Poddar

Abstract COVID-19 and environmental factors were connected to an elevated risk of pre-existing diseases associated with illness severity, immune system impairment, viral survival and transport, and viral exposure-inducing behaviours. The global health calamity caused by the COVID-19 epidemic has a wide range of negative consequences in every aspects of life, which includes the environment, economy, social, political, and cultural realms. COVID-19 infestation, in the first wave with more severity and a severe disaster in our human society. Advancement in technology especially in information technology (IT) applications in disaster management has become an essential part. COVID-19 a sudden epidemic disaster in all sectors of our human society. Strategic preparedness and response plan we have taken at national and international levels to overcome the ongoing challenges in the response to COVID-19. To identify, manage, track, and care for new cases of COVID-19, all countries enhanced their preparedness, alertness, and response. The present study focussed on information technology (IT)-based control over human and monitoring of COVID-19 environment calamity.

Keywords Environmental disaster · COVID-19 · Mobile apps · Communication

13.1 Introduction

Pollution of the air and water is examples of environmental factors. Acute respiratory syndrome is linked to chemical exposure, climate, and the built environment. The spread of coronavirus-2 (SARS-CoV-2) and coronavirus disease 2019 (COVID-19) is a severe threat to this century. Environmental variables and COVID-19 were connected to a higher chance of previous diseases related to illness severity, immune system impairment, viral survival and transport, and viral exposure-inducing behaviours.

S. Poddar (✉)
Deputy Vice Chancellor (Research & Innovation), Lincoln University College, 47301 Petaling Jaya, Selangor, D. E, Malaysia
e-mail: sandeeppoddar@lincoln.edu.my

© The Author(s), under exclusive license to Springer Nature Singapore Pte Ltd. 2022 227
P. K. Paul et al. (eds.), *Environmental Informatics*,
https://doi.org/10.1007/978-981-19-2083-7_13

Coronaviruses (CoVs) having an envelope and belong to positive stranded RNA virus family with a huge number of members. Non-severe acute respiratory syndrome (SARS)-like human CoVs (HCoV 229E, NL63, OC43, and HKU1) are globally prevalent and account for a significant portion of upper respiratory tract infections [1]. SARS-CoV-2 is a new type of coronavirus that was found recently and is the cause of the COVID-19 outbreak. Prevention and treatment of this disease have been made extremely challenging due to COVID-19 mutations. The ability of mutations to produce new variants with high transmissibility, disrupt viral fitness, and enhance virus reproduction is a necessary trait. In light of the foregoing, an attempt is made to determine the epidemiological pattern of the COVID-19 pandemic, if any, in terms of case incidence and death in various geo-climatological regions of the planet. The global health calamity caused by the COVID-19 epidemic has a wide range of negative consequences in all aspects of life, including the environment, economy, social, political, and cultural realms. The first phase was more severe and resulted in a major catastrophe in our human civilisation.

Global environment life on Earth and life below water as per United Nations Sustainable Development Goals are extensively affected by COVID-19. The quick spread of this virus is one of the main reasons of breaking the human social activity. The virus's spread triggered a shift in people's lifestyles, resulting in widespread job losses and a threat to people's survival as businesses collapsed. To minimise the spread of the virus, millions of individuals have been placed under lockdown. As a precaution, people were advised to use masks, gloves, and hand sanitizer on a regular basis, resulting in the development of a vast volume of medical waste in the environment. The transportation system, particularly international and interstate, has been shut down, and there are still restrictions in place. Overall, economic activity has ceased, and stock markets have fallen in lockstep with declining carbon emissions. "Social distancing" is the one major way to fight this novel disease [2]. But, people do not want to maintain the social distancing.

With an increase in the number of infected cases, there is also an increase in the number of patients who have recovered. Not everyone who has recovered from SARS-CoV-2 infection is symptom-free. Symptoms are also subject to change. Bone and joint pain, weariness, disorientation, sleeplessness, palpitations, and headaches have been reported in some cases. Other signs of irreversible lung scarring and malfunction occurred, especially in patients with severe pulmonary illness [3].

There are several research work carried out by our group in different aspects related to COVID-19 [4–9] Vaccination as one of the major reliefs to combat this pandemic. Chatki and Tabassum (2021) described different vaccines used in COVID-19 [10].

Advancement in technology especially in information technology (IT) applications in disaster management has become an essential part. COVID-19 a sudden epidemic disaster in all sectors of our human society. Strategic preparedness and response plan we have taken at national and international levels to overcome the ongoing challenges in the response to COVID-19. To identify, manage, track, and care for new cases of COVID-19, all countries enhanced their level of preparedness, alertness, and reaction.

The mobile-based communication cloud-based operating system also worked out [11].

In this present study, IT software-based control over human and monitoring of COVID-19 environment calamity has been discussed. The mobile-based apps listed in Tables 13.1 and 13.2 with their release dates to combat the COVID-19 disaster.

The RFID-based tracking system also imposed in several countries, e.g. Malaysia which is linked with Home quarantine Management system (refer Figs. 13.1 and 13.2).

To combat the spread of the current pandemic, various countries are employing a variety of digital tools and artificial intelligence (AI) for a variety of objectives, ranging from digital contact tracing to diagnostics, health information management, and outbreak prediction. WHO encouraging all developing countries to include the IT-based application in the National Health Policy [12].

Case management and real-time health surveillance are possible using mobile phone apps which track individual movements. Without laws and regulations in place to combat the stigma surrounding the pandemic and respect the rights of people

Table 13.1 Android and harmony OS-based contact-tracing apps

No	App name	Country	Release date
1	Aarogya Setu	India	11 April2020
2	Alhosn UAE	United Arab Emirates	07 April2020
3	Aman	Jordan	06 June2020
4	BeAware Bahrain	Bahrain	28 March2020
5	Bluezone	Vietnam	16 April 2020
6	Care 19 Alert	North Dakota (United States)	12 August2020
7	Coronavírus-SUS	Brazil	05 July2020
8	Corona-Warn-App	Germany	16 June 2020
9	COVID Alert	Canada	21 September 2020
10	COVIDSafe	Australia	04 June 2020
11	COVIDWISE	Virginia (United States)	04 August 2020
12	CRUSH COVID RI	Rhode Island (United States)	18 May 2020
13	Druk Trace	Bhutan	02 May 2020
14	GuideSafe	Alabama (United States)	16 August 2020
15	MySejahtera	Malaysia	16 April 2020
16	NHS COVID-19	England and Wales (United Kingdom)	12 August 2020
17	NZ COVID-19 Tracer	New Zealand	19 May 2020
18	SwissCOVID	Switzerland	25 Jun 2020
19	Tabaud	Saudi Arabia	14 June 2020
20	TousAntiCOVID	France	10 June 2020
21	TraceTogether	Singapore	09 March 2020

Table 13.2 iOS-based contact-tracing apps

No	App name	Country	Release date
1	AarogyaSetu	India	02 April 2020
2	Alhosn UAE	United Arab Emirates	07 April 2020
3	Aman	Jordan	06 June 2020
4	BeAware	Bahrain	28 March 2020
5	Bluezone	Vietnam	18 April 2020
6	Care19 Alert	North Dakota (United States)	31 July 2020
7	Coronavírus-SUS	Brazil	18 March 2020
8	Corona-Warn-App	Germany	16 June 2020
9	COVID Alert	Canada	31 July 2020
10	COVIDSafe	Australia	26 April 2020
11	COVIDWISE	Virginia (United States)	5 August 2020
12	CRUSH COVID RI	Rhode Island (United States)	20 May 2020
13	Druk Trace	Bhutan	20 April 2020
14	GuideSafe	Alabama (United States)	17 August 2020
15	MySejahtera	Malaysia	16 April 2020
16	NHS COVID-19	England and Wales (United Kingdom)	24 September 2020
17	NZ COVID Tracer	New Zealand	19 May 2020
18	SwissCOVID	Switzerland	25 Jun 2020
19	Tabaud	Saudi Arabia	25 June 2020
20	TousAntiCOVID	France	22 October 2020
21	TraceTogether	Singapore	20 March 2020

Source Min-Allah et al. 2021[24]

who are most marginalised, digital contact-tracing risks undercutting the purpose of epidemic reduction [13].

Molecular materials have unparalleled features that make them ideal for high-density computer system integration. Organic semiconductor materials for low-cost circuits to genetically engineered proteins for commercial imaging equipment are all examples of current molecular applications. Molecular computing should be viewed as an opportunity for new applications rather than a rival to traditional computing. Advances in molecular computing are projected to benefit sectors such as micro-robotics and bioimmersive computing. Both fresh computing concepts and material advances will be required to make progress [14].

Fig. 13.1 RFID-based
digital tracker wrist band.
Source Author

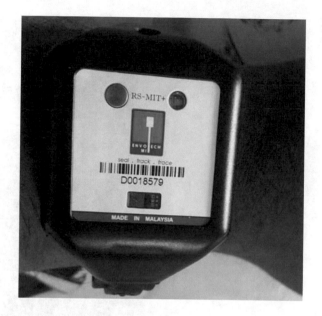

13.2 Objective

COVID-19 pandemic is the major threat of environment in today's scenario. The present study focusses on strategic preparedness and response plan we have taken at national and international levels to overcome the ongoing challenges in the response to COVID-19.

To identify, manage, track, and care for new cases of COVID-19, all countries enhanced their preparedness, alertness, and response.

Advancement and novel applications of information technology (IT)-based control over human and monitoring of COVID-19 environment calamity.

13.3 Detection

The antibody test and the RT-PCR test are the two most common types of tests used to identify COVID-19 all over the world. The antibody test, which is an indirect method of testing, can determine whether or not the immune system has contacted the virus. After an infection has taken root, antibodies can take up to 9–28 days to produce, which is a long time, and by that time, the infected individual could have transmitted the disease if not carefully isolated. In contrast, RT-PCR testing is rapid and can detect COVID-19 in 4–6 h. Given the scale of the pandemic, this is also too soon, and RT-PCR testing has its own set of restrictions. The high expense of importing the reagents and other elements utilised in the kits is one such barrier.

Fig. 13.2 RFID-based digital tracker (CTU) for COVID-19. *Source* Author

Researchers of different fields, medical, biotechnology as well as artificial intelligence, data science, machine learning, actively participating to develop tools to detect and prevent this pandemic with their technical skills and expertise in respective area described a deep learning algorithm for detecting COVID-19 in X-rays using nCOVnet tracking [15–20].

The coronavirus disease (COVID-19) reaction has been greatly aided by mobile health (mHealth) apps [21].

The Home Quarantine Management System (HQMS) is integrated with GPS technology for real-time monitoring of individuals subject to Ministry of Health home quarantine surveillance and observation orders, which will be implemented on 8 December 2021 [22].

The OxCGRT has focussed on following four indices that group different families of policy indicators:

GRI (all categories)

Stringency index (containment and closure policies)
CHI (containment and closure and health policies)
ESI (economic support measures).
All the above index is having a series of individual policy response indicators.

13.4   Gathering

The mobile apps help to find gathering, etc. but still having loopholes. As people concern is not want to disclose the gatherings.

Ordinal scales in the programme OxCGRT distinguish between, say, a restriction on meetings of more than ten people and a prohibition on gatherings of more than 100 individuals. The disadvantage of an ordinal approach is that it categorises varied observations into pre-determined categories. For example, during spring 2020, the United Kingdom and France both had essentially identical stay-at-home orders, and both were designated as the second-highest ordinal point on that indication. To deliver the highest level of accuracy and consistency, OxCGRT depends on human judgement and contextual understanding rather than automated data gathering or coding. Individual contributors must carefully examine several regulations within each area in order to use the ordinal scales to assign a code that best fits each indicator [23].

13.5   Communication

Medical workers working in sample collection centres across the country can use the app as a hand-held tool. The app cannot be used by the person or patient who is undergoing the test. RT-PCR findings are not provided via this app or the portal https://covid19cc.nic.in. The sample collection facility will send samples of various types of specimens to ICMR laboratories for COVID-19 confirmation by RT-PCR. To provide advance notice, mobile-based applications are deployed.

Different mobile-based applications Aarogya Setu (India), MySejahtera (Malaysia), COVID-19-aware MN (State of Minnesota), etc. have been mentioned in Tables 13.1 and 13.2 with their release dates to combat the COVID-19 disaster. For MySejahtera App, refer Fig. 13.3.

13.6   Social Stigma and Apps:

The harmful impact of societal stigma on the emotional, mental, and physical well-being of COVID-19 survivors and their families has been documented [26–28].

Fig. 13.3 MySejahtera app in Malaysia displaying different status of persons. *Source* https://forum. lowyat.net/topic/5190199 [25]

By using mobile apps with helpline facility in some countries, e.g. in Malaysia, the necessary food and other items have been delivered at doorstep. It not only maintains the spreading of disease but also keep the persons identity confidential.

13.7 Conclusion

Global environment life on Earth and life below water as per United Nations Sustainable Development Goals are extensively affected by COVID-19. The quick spread of this virus is one of the main reasons of breaking the human social activity. The researchers of different sectors struggle hard to overcome this environmental calamity. The medical health workers, biotechnologists, life science researchers, computer application specialists all working together to overcome this disaster and threat of the century. The advancement and use of technology is proven to support the fight against this environmental threat. We have no choice, but to take the help of technology to maintain physical distancing and combat the disease.

References

1. Raoult, D., Zumla, A., Locatelli, F., Ippolito, G., & Kroemer, G. (2020). Coronavirus infections: Epidemiological, clinical and immunological features and hypotheses. *Cell Stress, 4*(4), 66–75. https://doi.org/10.15698/cst2020.04.216

2. Shereen, M. A., Khan, S., Kazmi, A., Bashir, N., & Siddique, R. (2020). COVID-19 infection: Origin, transmission, and characteristics of human coronaviruses. *Journal of Advanced Research, 24*, 91–98. https://doi.org/10.1016/j.jare.2020.03.005

3. Huang, C., Huang , L., Wang, Y., Li, X., Ren, L., Gu, X., et al. (2021). 6-month consequences of COVID-19 in patients discharged from hospital: A cohort study. *The Lancet, 397*(10270), 220–232.https://doi.org/10.1016/S0140-6736(20)32656-8

4. Bhattacharya, S., Basu, P., & Poddar, S. (2020). Changing epidemiology of SARS-CoV in the context of COVID-19 pandemic. *Journal of Preventive Medicine and Hygiene, 61*(2), E130. https://doi.org/10.15167/2421-4248/jpmh2020.61.2.1541

5. Bhattacharya, S., Sinha, S., Bhattacharya, S., & Poddar, S. (2021). Studies on the awareness, apprehensions and aspirations of the university students of West Bengal, India in the context of COVID-19 pandemic. *Malaysian Journal of Medical Research (MJMR), 5*(2), 29–33.

6. Bhowmik, D., & Poddar, S. (2021). Nexus between unemployment and mental health. In: Poddar et al. (Eds.), *Human resource management and mental health-a psychosocial aspect*. Lincoln Research and Publications Limited, Australia and Lincoln University College, Malaysia.https://doi.org/10.46977/book.2021.hrmmh

7. Dioso, R. I., Poddar, S., Abdullah, B. F., Hassan, H. C. (2021). Effectiveness of online health education on healthy diets and regular exercises in achieving a health related quality of life during this pandemic era. *Malaysian Journal of Medicine & Health Science*, 67–72.

8. Rahayu, S., Pertiwi, W. E., Meidina, L., & Poddar, S. (2021). The roles of knowledge and perceptions in Covid-19 transmission prevention behavior. *Malaysian Journal of Medicine and Health Sciences, 17*, 62–66.

9. Ansar, W. (2020). Periscopic view on COVID 19 infection: A review. *International Journal of Advancement in Life Sciences Research, 3*(3), 1–15. https://doi.org/10.31632/ijalsr.20.v03 i03.001

10. Chatki, P. K., & Tabassum, S. (2021). Juxtaposition on discrete Covid-19 vaccines: For rudimentary and pivotal cognizance. *International Journal of Advancement in Life Sciences Research, 4*(4), 1–6. https://doi.org/10.31632/ijalsr.2021.v04i04.001

11. Bexci, M. S., Poddar, S., & Athithan, A. (2019). Cloud based convergent unrestricted mobile communication assistance via evolutionary adaptive ad hoc strategies for disaster relief. International Journal of Engineering and Advanced Technology (IJEAT) 8(6), 3460–3462. https://doi.org/10.35940/ijeat.F9518.088619

12. Davis, S. (2020). The Trojan Horse: Digital Health, human rights, and global health governance. *Health and Human Rights, 22*(2), 41–47.

13. Davis, S. L. (2020). Contact tracing apps: Extra risks for women and marginalized groups. *Health and Human Rights Journal, 4*.

14. Zauner, K. P. (2005). Molecular information technology. *Critical Reviews in Solid State and Materials Sciences, 30*(1), 33–69. https://doi.org/10.1080/10408430590918387

15. Alimadadi, A., Aryal, S., Manandhar, I., Munroe, P. B., Joe, B., & Cheng, X. (2020). Artificial intelligence and machine learning to fight COVID-19. *Physiological Genomics, 52*(4), 200–202. https://doi.org/10.1152/physiolgenomics.00029.2020

16. Gozes, O., Frid-Adar, M., Greenspan, H., Browning, P. D., Zhang, H., Ji, W., et al. (2020). Rapid ai development cycle for the coronavirus (covid-19) pandemic: Initial results for automated detection & patient monitoring using deep learning ct image analysis. arXiv preprint arXiv: 2003.05037

17. Fanelli, D., & Piazza, F. (2020). Analysis and forecast of COVID-19 spreading in China, Italy and France. *Chaos, Solitons & Fractals, 134*, 109761. https://doi.org/10.1016/j.chaos.2020. 109761

18. Hall, L. O., Paul, R., Goldgof, D. B., & Goldgof, G. M. (2020). Finding covid-19 from chest x-rays using deep learning on a small dataset. arXiv preprint arXiv:2004.02060
19. Panwar, H., Gupta, P. K., Siddiqui, M. K., Morales-Menendez, R., & Singh, V. (2020). Application of deep learning for fast detection of COVID-19 in X-Rays using nCOVnet. *Chaos, Solitons & Fractals, 138*, 109944. https://doi.org/10.1016/j.chaos.2020.109944
20. Xu, X., Jiang, X., Ma, C., Du, P., Li, X., Lv, S., ... & Li, L (2020). A deep learning system to screen novel coronavirus disease 2019 Pneumonia, *Engineering, 6*(10) 1122–129. https://doi.org/10.1016/j.eng.2020.04.010
21. Singh, J. L. H., Couch, D., & Yap, K.(2020). Mobile health apps that help with COVID-19 management: Scoping review. *JMIR Nursing, 3*(1), e20596. https://doi.org/10.2196/20596
22. Uem Edgenta. (2022). Home quarantine management system, Available at https://www.uemedgenta.com/404.aspx?aspxerrorpath=/media/news-announcements/home-quarantine-management-s.aspx
23. Hale, T., Angrist, N., Goldszmidt, R., Kira, B., Petherick, A., Phillips, T., et al. (2021). A global panel database of pandemic policies (Oxford COVID-19 Government Response Tracker). *Nature Human Behaviour, 5*(4), 529–538. https://doi.org/10.1038/s41562-021-01079-8
24. Min-Allah, N., Alahmed, B. A., Albreek, E. M., Alghamdi, L. S., Alawad, D. A., Alharbi, A. S., et al. (2021). A survey of COVID-19 contact-tracing apps. *Computers in Biology and Medicine, 137*, 104787. https://doi.org/10.1016/j.compbiomed.2021.104787
25. MySejahtera application to assist in monitoring COVID-19. Bernama. 21 April 2020. Archived from the original on 22 June 2021. Retrieved June 22, 2021.
26. Cheng, H.-Y., Jian, S.-W., Liu, D.-P., Ng, T.-C., Huang, W.-T., & Lin, H.-H. (2020). Contact tracing assessment of COVID-19 transmission dynamics in Taiwan and risk at different exposure periods before and after symptom onset. *JAMA Internal Medicine, 180*, 1156–1163. https://doi.org/10.1001/jamainternmed.2020.2020
27. Chew, C. C., Lim, X. J., Chang, C. T., et al. (2021). Experiences of social stigma among patients tested positive for COVID-19 and their family members: A qualitative study. *BMC Public Health, 21*, 1623. https://doi.org/10.1186/s12889-021-11679-8
28. Hasinuddin, M., Rusana, R., Noviana, U., Ekawati, H., Aini, N., & Poddar, S. (2021). Parents' psychological aspect in caring for the child during COVID-19 pandemic. *Jurnal Ners, 16*(2), 193–197. https://doi.org/10.20473/jn.v16i2.30637

Chapter 14
Drone Applications in Wildlife Research—A Synoptic Review

Subhendu Mazumdar

Abstract Unmanned aerial vehicles (popularly known as drones) are being increasingly used in wildlife research. These remote-controlled devices are often used to collect information from areas which are otherwise very difficult to access by the researcher. Therefore, drones are promisingly advantageous over various traditional research techniques. Application of these technologies not only increases the accessibility but also reduces human effort and collect accurate information by causing minimal disturbance to the habitat. For these reasons, application of UAVs has gained popularity over last few decades in various domains of ecological research across the globe. In wildlife research, the UAVs are used in population surveys, dispersion studies, nest monitoring, radio telemetry studies, habitat assessment, etc. Undoubtedly, UAVs show great prospect of its application in many ecological and wildlife research in future. In such a scenario, this review is an attempt to provide the researchers with a glimpse of the remarkable progress along with the benefits and future scopes of this modern technology in various wildlife researches.

Keywords Unmanned aerial vehicles · Wildlife · Biodiversity · Conservation

14.1 Introduction

Unprecedented technological development, improvement, and expansion have been noticed over last few decades. Widespread digital facilities, greater connectivity, rise in novel ways of open source and collaborative technology development initiatives by various manufacturers and DIY/maker communities since 4th industrial revolution [12] have made many technologies cheaper and better available for the scientific and research communities across the globe. Consequently, main such technologies have also trickled down into various types of ecological researches. Some of these technologies enabled the scientific communities to either overcome certain hurdles in monitoring wildlife in inaccessible terrains, while others helped to collect

S. Mazumdar (✉)
Department of Zoology, Shibpur Dinobundhoo Institution (College), 412/1, G.T. Road (South), Shibpur, Howrah 711102, India
e-mail: subhendumazumdar@gmail.com

scientific data in a more efficient way. In addition to the well-established technological applications, new technologies as well as novel applications are often being tried by the wildlife practitioners [112]. Data thus collected ultimately contribute in adopting better conservation initiatives for many species and ensure their continued existence by alleviating the threats. Therefore, all these technologies useful in conservation of wildlife and biodiversity are collectively considered as "conservation technology", which is progressively emerging as a new discipline [12, 79]. One such conservation technology includes drones (Fig. 14.1), which are also known as unmanned/unoccupied aerial vehicles (UAVs), unmanned/unoccupied aerial systems (UASs), or remotely piloted aircraft systems (RPASs).

14.2   Drones—An Overview

Drones are unmanned aerial vehicles (UAVs) or remotely piloted aircraft systems (RPASs) with remote manual control or with programmed GPS-controlled navigation path [80]. These unmanned aerial vehicles are available in different shapes and sizes (Table 14.1) and offer great versatility in carrying different types of devices and sensors [90]. However, the type and size of the drone are mainly determined by the payload which in turn depends on the nature of sensor to be used to achieve specific goals [90]. Fixed-wing UAVs (unmanned airplanes) and rotary-wing UAVs (unmanned multicopters) are the most popularly used drones in wildlife research. Among the rotary-wing UAVs, the ones with single wing are called "helicopter UAV," while the ones with multiple wings are called "multicopter UAV." The fixed-wing and rotary-wing types of drones are different in various characteristic features (Table 14.2), [3].

Commonly, still and video cameras, thermal/infrared sensors, acoustic devices, loudspeakers, sprayers, etc., are mounted on the drones based on the purpose for which it is used [124]. The time and distance of flight are usually determined by the amount of such payloads. For instance, AeryonScout tricopter is able to fly at an altitude of 330 m with 300 g payload within a range up to 3 km [145].

These devices were first designed during Second World War and were exclusively used for military purposes till twentieth century [40]. During twenty-first century, there have been significant improvement in drone technology along with reduction in their size and cost [4]. So, apart from their military uses, drones are now being used by the civilians for various purposes [3, 1, 19, 24]. Approximately, two million drones are sold annually every year worldwide, and this astounding number is increasing every year [1]. In recent times, drones are applied in various fields (Fig. 14.2), of which some of the non-military applications include aerial photography and videography, product and medicine delivery [6, 77, 145], law enforcement and patrolling [40], fire control [101], disaster management, search and rescue operations, geographic mapping, weather forecasting, precision agriculture [66, 141], and wildlife monitoring and management. Nevertheless, owing to its sophisticated

(a) Quadcopter drone in operation (Photo courtesy: Ayan Dutta)

(b) High resolution aerial image taken by the quadcopter drone (Photo courtesy: Kanad Roy)

Fig. 14.1 **a** Quadcopter drone and **b** aerial image captured by the drone

nature and associated benefits over the traditional methods, drones are now increasingly being adopted for novel applications in many other domains. Over the last decade, such drones have gained immense popularity among the scientific communities [100, 148]. In the present article, discussion will be restricted on its use in various wildlife researches. The main aim of this study was to identify the areas of wildlife research where drones were used based on the systematic review of

Table 14.1 Categories of drones as per DGCA

Types of drones	Weight	Size
Nano	≤250 g	<30 mm
Micro	250 g–2 kg	30–100 mm
Small	2–25 kg	300–500 mm
Medium	25–150 kg	300–500 mm
Large	≥150 kg	>2 m

Table 14.2 Major trade-offs between fixed-wing and multi-rotor drones

Fixed-wing UAVs	Rotary-wing UAVs
Fly longer distances and operate beyond the line of sight (either autonomously or by remote piloting from ground), but unable to hover or flying at very low speed	Fly shorter distances and operate within line of vision of ground operator. These are able to hover or fly at very low speed
Relatively less energy consumption as they can glide during flight. So, continuous flight time is more (>1 h)	High energy consumption as they are unable to glide resulting in short flight duration (<1 h)
Larger in size	Smaller in size
Carry heavier payloads	Payload capacity is comparatively less
Relatively less noisy	Relatively more noisy
Require large open spaces and linear strips for takeoff and landing	Vertical takeoff and landing possible. Hence, any linear takeoff/landing strip is not required
Require more training to operate	Require less training to operate
Price is relatively high	Price is relatively low
Less popular in wildlife studies	Very popular in wildlife studies

published literature in order to provide the researchers with the snapshot of the remarkable progress along with the benefits and future scopes of application of this modern technology in wildlife research. This review might be useful for the wildlife researchers, forest managers, biodiversity conservationists, as well as others working in different branches of applied ecology who are willing to adopt this technology in their respective domains of research.

14.3   Methodology

Comprehensive and systematic search was carried out in Google Web (www.goo gle.com), Google Scholar (www.scholar.google.co.in), and ResearchGate (www. researchgate.net) to find out published peer-reviewed articles that used drones in wildlife research. "Wildlife research," along with any of the following terminologies, viz., drones, unmanned/unoccupied aerial vehicles (UAVs), unmanned/unoccupied

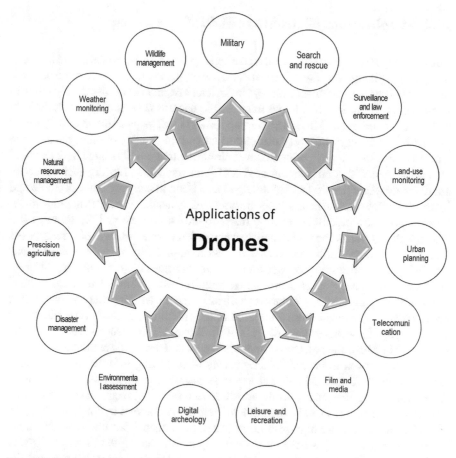

Fig. 14.2 Some applications of drones

vehicle systems (UVSs), unmanned/unoccupied aircraft system (UAS), remotely piloted aerial systems (RPAS) were used as search phrases.

Among the outputs, the articles which (i) dealt on various aspects of wildlife (excluding socioeconomic and veterinary studies), (ii) published in peer-reviewed journals, and (iii) where drones were used for data collection, were downloaded. Then, the downloaded articles along with the cross-references were thoroughly scanned. Any article not satisfying any one of the above-mentioned three inclusion criteria were removed. Finally, only, the articles published in English were retained for this review. Despite of strictly adhering to this systematic search criteria and due to unavailability of access to full texts, some of the articles might be inadvertently left out. Still, to the best of my understanding and belief, adequate number of wildlife studies have been considered in this review to provide a synoptic view on the areas of wildlife research where drones are largely being used.

14.4 Applications of Drones in Wildlife Research

Aerial surveys (from manned aircrafts like helicopters) have been an age-old methodology in wildlife monitoring and census. However, these surveys were often adopted
in open habitats (e.g., herbivore surveys in African savannah), and also during wildlife
surveys carried out in difficult and inaccessible terrains (like surveys of medium to
large sized faunal species thriving in alpine snows or on steep mountain cliffs), where
ground-based surveys was very difficult to carry out. Studies in other parts of the
world showed that surveys using manned aircrafts often requires huge budget and is
also associated with a fair amount of risk of fatal accidents [126]. In such scenario,
small off-the-shelf drones (<2 kg) emerged as a boon in carrying out repeatable and
intensive surveys. These instruments are now widely available, easy to operate, allow
larger spatial coverage in lesser time than ground-based surveys [22, 21, 69, 146].
These tools also allow the researchers to carry out finer scale surveys of smaller areas
and obtain high-resolution and accurate spatiotemporal data from various locations
(including those from remote areas and challenging terrains) and during different
weather conditions. Repeat surveys also require less additional time and effort.
Application of drones also reduce the disturbance to wildlife due to less human
presence and/or reduced noise compared to the traditional ground, vessel-based or
aerial surveys in occupied aircrafts [60, 103]. Besides, it reduces the risk of a close
encounter between the researcher and wild species. Hence, these unmanned aerial
vehicles are proven to be very efficient tools in collecting information on various
aspects of wildlife and its habitat at low-cost and with minimal risks [123]. Therefore, these drones are often considered better than satellite images in studies carried
out at low to medium spatial extent due to their flexible revisit option at lesser operational cost and for the high-resolution images obtained from low altitude flights
[3, 80]. All these benefits have ultimately helped in reducing the overall budget of
research by minimizing the time and manpower. For these reasons, there have been
an upsurge in the use of drones in wildlife, biodiversity, and associated ecological
and environmental research is across the globe [21, 24, 77, 87, 104, 112].

Drones are being increasingly used in monitoring and conservation of flora, fauna,
and their habitats, both by the protected area managers and communities [2, 90] which
provide more accurate data [62]. Nevertheless, application of drones in wildlife
conservation and management is fairly recent [87]. All studies reviewed here have
been grouped under two broad categories—(i) drones in wildlife research and (ii)
drones in wildlife conservation [124].

14.4.1 Drones in Wildlife Research

In wildlife research, drones are often envisaged for the purpose of detecting wild
animals and to generate data on their abundance (wildlife census); [57, 86, 120, 144]
in order to model and map their density distribution (species distribution model),

[47, 77, 110, 135, 138], for tracking any focal species (wildlife telemetry), [105], monitoring them in the wild (wildlife monitoring), to assess various animal behavior (ethological studies), [11, 42], along with collecting data on several other biological parameters of wild animals and their habitat condition [26, 77, 96, 124]. Drones are also used to minimize human–animal conflicts, to detect and combat forest fires, to prevent wildlife crime, and for better law enforcement. Researchers have used drones to study wildlife thriving in different habitats ranging from terrestrial to freshwater lakes, rivers, estuaries, and oceans [68, 67, 84, 85, 88].

In terrestrial habitats, drones have been used to survey a wide range of taxa and generated various novel informations. Drones are used in spatial and volumetric analysis [29, 30,151, 31] as well as for carbon content assessments [102] in dense forest habitats. Deep unstable mud, impenetrable tree stands and the presence of venomous snakes, crocodiles, and hippopotamus in swampy mangrove forests of east Africa make those habitats practically inaccessible for ground-based surveys, where quadcopters fitted with optical and IR cameras serve as only viable alternative [38]. LiDAR mounted on drones is used to calculate leaf area density and crown structure of Brazilian Amazon forests [28]. For vegetation mapping across landscapes, finer spatial resolution of drone data is often preferred over satellite images with relatively course spatial resolution [81, 118]. Such drone-based surveys are particularly useful to generate vegetation indices as an index of forest health, to measure habitat heterogeneity, to assess canopy density and also to and identify canopy-dwelling species [38, 136]. Researches can also locate the nests of orangutans and chimpanzees [14, 138] in the drone images with excellent spatial resolution by studying the distinct features in the canopy structures. Small UAVs were used for mapping orangutan habitats [135] and also to locate many other canopy-dwelling primates [57, 131]. Drone-based surveys were particularly carried out to study species that are found in remote and "difficult to access" terrains throughout the globe [73]. For instance, drones are used to observe nests of raptors [5], canopy breeding birds [111, 147], cliff nesting birds [15, 52], to study polar bears in Arctic [8], Tibetan antelopes in cold deserts of Tibetan plateau [64], detecting animals in African Savannah [122], to survey Penguin colonies in Antarctica [120], so on and so forth. Drones with thermal sensors are also used to locate small, secretive, and cryptic species [71, 96]. Drones are useful in habitat mapping as they generate excellent high-resolution orthographic photographs and other images useful in 3D mapping modeling and projection. Drone images have also been found to provide more useful information on the structure composition of flora and fauna, over ground biomass as well as about various other crucial ecological components in deserts [27] and grasslands [43]. Drone-based thermal infrared remote sensing has been done to count whitetailed deer [25] and to locate ground nest of birds [125, 129]. In coastal habitats, drones are often used to generate geomorphological data of sand dunes [37, 93] and assessment of shoreline vegetation [142]. One interesting study using drone revealed the impact of the sea-level rise and subsequent squeezing of coastal width on turtle nesting [139].

In marine ecosystems, drones are used to identify benthic habitats [38] as well as to study coral reef morphology and diversity [20]. Drones have often been used by researchers to study marine megafauna [41, 45]. The marine taxa which have already been monitored using UAVs include agglomerations of jelly fishes [127], elasmobranchs [58, 76, 75], turtles [44, 134], pinnipeds [78], cetaceans [46, 49], seals [115], manatees [83], dugongs [60, 95, 116], and whales [49, 51]. Gray et al. [49] were able to distinguish between humpback whales, minke whales, and blue whales with 98% accuracy from drone imageries using neural network and photogrammetric techniques. Morphometric features (like length, height, and volume of the body) are measured from drone images using 3D models which help to assess their body conditions [17, 23, 78, 113]. Rajpurkar et al. [116] used lightweight quadcopter drone to study dugongs in marine waters adjoining Andaman and Nicobar Islands and documented marine litters, boats, and high-speed vessels to be the potential threats. Live recordings of several threatened marine taxa (like Endangered Zebra shark *Stegostoma fasciatum*, Vulnerable Spotted eagle ray *Aetobatus ocellatus*) we also done using drones. Besides, drones are also used in ocean observation system [72, 88] to identify the location of seagrass [56] and coastal fish nursery grounds [143], monitoring floating marine litter [45]. This clearly shows that the lightweight UAVs can serve as excellent tools to study the distribution and threats of a wide range of marine taxa (especially those with serious conservation concern) and also to efficiently identify the threats and other illegal activities inside the marine-protected areas [116]. In freshwater ecosystems, drone images are used to study channel bathymetry, topography, and various other dimensions of rivers [68, 67, 84, 85, 149]. Better estimation of two Amazon River dolphin (tucuxi and pink dolphins) populations were obtained using small off-the-shelf multirotor UAVs [106]. However, Joyce et al. [70] opined that adequate care need to be taken to reduce sunshine glares on surface during drone-based surveys carried out in aquatic habitats.

Interestingly, I found that the drone-based studies in terrestrial habitats involved maximum species of birds (in breeding and non-breeding stages) followed by mammals and reptiles, while in aquatic habitats drones were mostly used to study mammals, followed by birds and reptiles.

14.4.2   Drones in Wildlife Conservation

Although the application of drone in direct conservation actions are limited as compared to their wide-scale applications for research purposes [124], yet UAVs are increasingly being used in various types of wildlife conservation initiatives throughout the globe [114, 137]. Inadequate resources (fund, vehicles, manpower, etc.) and difficult terrains are major challenges in many protected areas situated in the developing tropical countries with dense forests. Shortage of frontline staff in protected areas, as well as in protected areas located in difficult terrains, often result in a large area being infrequently patrolled as ground-based surveillance is labor intensive and more time-consuming. In these areas, drones might empower the frontline patrolling staff to monitor the vast and difficult to reach areas more frequently and thus enable the park managers to efficiently conserve the wildlife and other resources inside those protected areas [290]. Such increased vigilance using this cutting edge conservation technology has empowered the patrolling staff and ground-based law enforcement officials to efficiently locate and combat various illegal activities like poaching [18, 54, 94, 104, 109], illegal logging and mining [77] and thereby apprehend the offenders [77, 104, 128]. Moreover, such drone-based patrolling data might be modeled to predict spatiotemporal patterns of these crimes [128, 130], thereby enabling the authorities to prioritize areas for conservation action and/or to identify the gap areas. The high-resolution images taken by the drones also serve as important evidence in the court of law to prosecute the convicts [130]. Such use of drones by the officials might serve as deterrent for the wildlife criminals [104, 128]. However, there have been reports from US and Germany where the drones used by animal welfare groups were shot down by the hunters [18]. In addition, these tools might also help in early detection and quick response in forest fire incidents [39, 101], efficient search and rescue operations [16, 48], and to reduce human–animal conflicts [53, 128]. These unmanned aerial vehicles might also be used in forest restoration programs to sprinkle seeds in the deforested areas or over the degraded forest patches [133]. The major prey species of endangered black-footed ferrets, the prairie dogs, were inoculated against deadly plague by dropping peanut butter pellets with vaccine in front of their burrows from customized octocopter drones [33]. To assess the grazing pressure inside Yellow River Source National Park in China, drones were deployed to effectively collect abundance data of large herbivores [150]. In India, a project titled "E-Bird Technology for Tiger Conservation" is going to be implemented in joint collaboration of Wildlife Institute of India and National Tiger Conservation Authority, wherein UAVs will be used as a surveillance and monitoring tool for tiger conservation. Initially such UAV based tiger surveys will be implemented in representative tiger reserves and eventually in all tiger reserves of the country in near future. In addition to the monitoring of tiger populations, this project also aims to collect visual data on animal movements, forest fires, habitat data, and illegal activities inside the tiger reserves. Now-a-days, many animal documentaries are prepared using drone clips. Such documentaries in public domains create interest among the common people toward wildlife and also for the location. This, in turn, will boost wildlife tourism by increasing the tourist footfall in various protected areas and other wilderness areas.

14.5 Challenges of Drone Application in Wildlife Research

Despite of many advantages, there exists some technological, operational, and legal challenges in the application of drones. Many low-cost drones have limited flight time (around 30 min) and low operational distance. Also, there are strict legal restrictions in drone flying in many areas [55, 80].

Census of wildlife using drones can only be carried out in open landscapes devoid of much canopy cover where the individual animals are easy to count. Moreover, manual counting of the focal species in a large number of aerial images obtained by the drone cameras requires a lot of expertise, patience, and time. Also, due to the high spatial resolution of the aerial images obtained from the drone platforms, the present algorithms associated with pixel-based analysis of the satellite images are inefficient to produce land cover and vegetation classification as well as in habitat suitability analyses [107, 148]. In such circumstances, deep learning techniques and object-based image analysis (OBIA) as well as standardized sampling protocol and newer statistical analytical techniques need to be adopted in the analysis of wildlife data procured through drones [59, 91]. Such automated image processing will undoubtedly improve the acceptability and usefulness of such drone imageries among the scientific communities [32].

Security of equipment and data are some of the other major challenges. Although comparative assessments revealed that application of drones is more accurate and incur lesser cost than the labor intensive and more time-consuming traditional aerial, boat based, or ground surveys at least for studies carried out in small to medium spatial extents [62, 90, 87, 147], yet many researchers carrying out researches on a shoestring budget (in countries like India) find it very difficult to afford these technologies. So, any loss of equipment and its associated data stored onboard is a major setback for these researchers. In addition, skilled research personnel and sufficient other IT infrastructural facilities (hardware and software) are also needed for efficient storage and analysis of large amount of data collected by drones.

Also, there are certain other social and ethical issues that need to be considered with due importance if drones are to be used in wildlife conservation. Support of the local people and their active involvement is the key to success for any long-term conservation program [65, 92]. If the communities do not accept drones hovering overhead, then it is necessary to convince them about the need for using drones. Otherwise, forceful use of drones might result in various retaliation, which will ultimately jeopardize the entire conservation scenario. Privacy and psychological well-being of the local inhabitants need to be considered with due consideration and empathy. Some of the negative social impacts might be averted if some members from the local communities are given the responsibility of drone flying after proper training and under the supervision of experts [108]. There are reports of initiatives taken by several wildlife practitioners and NGOs to provide training to tribal and indigenous communities and successfully involving them in wildlife monitoring and patrolling, as well as in combating forest fires and deforestation using this new-age conservation technologies [38].

Several other negative aspects and challenges of drone application on wildlife have also been pointed out by several authors [40, 55, 61, 90, 82, 135]. Drones are operational at a height, which are also used by many other volant organisms like as birds and bats [35]. Such coexistence in similar aerial space poses a risk of collision (leading to the injury of the animal and/or damage of the instrument) and also causes disturbances and other negative effects for the species. The breeding pairs are particularly sensitive to the presence of any novel objects (such as UAVs) in close proximity. Such disturbances are often manifested by subtle changes in their behavioral patterns [121] like increased vigilance, reduced foraging rates, and frequent display of threat postures and/or calls. Sensitive species might also experience certain physiological changes (like increased heart rate and greater secretion of stress hormones.). It is pertinent to mention here that American black bears manifested rise in heart rate during their first encounter with UAVs, which restored within normal range after getting habituated by subsequent exposures [34]. Variable behavioral responses were also noticed in different seal species [115]. Particularly, disturbances to the breeding pairs might result in abandoning the newborn offspring by the parents, thereby affecting their reproductive success. Nesting lesser snow geese are disturbed by drones [9]. Frequent drone encounters might even result in avoidance of the area by the sensitive species and breeding individuals, which again will jeopardize the local population and overall community composition. Therefore, enough care should be taken while carrying out research on these sensitive species using drones. Flying of drones should be avoided too close to any sensitive species and breeding pairs, especially birds. Flying birds may consider UAVs as predators or potential competitors invading their flying space and might either attack and get injured, or manifest neophobic response and escape [99]. On the other hand, the colony breeding birds are known to leave their nest and attack any novel object approaching their nest in groups considering it to be a threat for their chicks, thereby wasting energy. Such panicked individuals may collide with each other or with the UAVs resulting in the injury of adults. While the adults are busy chasing away UAVs, the eggs/chicks remain exposed in the nests. Sometimes, the wary parents even knock over the eggs/chicks from the nests. Holldorf [63] showed the effects of drone on wildlife behavior. Rebolo-Ifran et al. [121] reported that drones elicited negative behavioral response in at least 50 species of animals living in all types of habitat and from all countries across the globe. Undoubtedly, these negative impacts of drones on reptiles [13], birds [147], and mammals [33, 115, 117] are matters of concern. On the contrary, minimal aggressive or alarm behavior was manifested by colony nesting water birds [10] and raptors [89, 140]. This clearly indicates that the response toward drones varies in different species. I assume that such responses may even vary in different individuals and during different life stages (juvenile vs adult, male vs female, breeding vs non-breeding, etc.) of the same species. This assumptions need to be tested through in-depth scientific research. Thereafter, appropriate policies, guidelines, recommendations, and drone flying protocols may be formulated based on robust scientific findings. Urgent attention need to be given on negative impacts of recreational drone flying on wildlife without much delay [1, 82, 103]. US National Parks Service has already taken initiatives to prohibit private drone flying inside US National Parks to minimize the negative impacts of drone on wildlife [50].

14.6 Conclusion

The overall benefit of drones in overcoming the hindrances of carrying out wildlife research, particularly those done in difficult and inaccessible terrains undoubtedly outweighs the drawbacks. Compared to the traditional methodologies, drone surveys might serve as improved non-invasive and less disturbing methodology for acquiring high-resolution data at frequent intervals. Nevertheless, these surveys are to be done after proper testing and minimizing the negative effects, and also, the ethics and guidelines are to be strictly adhered [61]. Wildlife researcher should thoroughly go through the scientific articles and reviews on the ethical issues [97, 98, 140] to have a fair idea on potential disturbances of drones on wildlife [103, 121, 124] strictly follow the guidelines [61] and protocols [7, 36, 119, 120]. They also need to be aware of the existing laws, regulations, and restrictions regarding drone flying in their respective study areas. Finally, with all its benefit and with the ongoing advancements to tackle the challenges in a more efficient way, it may be clearly assumed that drones are going to become one of the major state-of-art conservation technology in wildlife researches in near future. Likewise, India will also witness an emergence in the use of drones in newer regimes of wildlife, biodiversity, and applied ecological research in the days to come.

Acknowledgements I would like to thank Mr. Kanad Roy, who provided me with the photographs of drones and helped me in formatting the reference list. Thanks are also due to Smt. Bidisa Mazumdar and Mr. Rishan Mazumdar for their assistance during manuscript preparation. I express my sincere gratitude for Dr. P.K.Paul for encouraging me to write this review article. The Principal of Shibpur Dinobundhoo Institution (College) is gratefully acknowledged for encouragement and infrastructural support.

References

1. APO-100 FAA. (2018). FAA aerospace forecast: Fiscal years 2018–2038 [www document]. URL www.faa.gov/data_research/aviation/aerospace_forecasts/media/FY2018-38_FAA_Aerospace_Forecast.pdf
2. Ancin-Murguzur, F., Munoz, L., Monz, C., & Hausner, V. (2019). Drones as a tool to monitor human impacts and vegetation changes in parks and protected areas. *Remote Sensing in Ecology and Conservation, 6*(1), 105–113. https://doi.org/10.1002/rse2.127
3. Anderson, K., & Gaston, K. (2013). Lightweight unmanned aerial vehicles will revolutionize spatial ecology. *Frontiers in Ecology and the Environment, 11*(3), 138–146. https://doi.org/10.1890/120150
4. Anderson, C. (2012). How I accidentally kickstarted the domestic drone boom (Wired). Retrieved from http://www.wired.com/2012/06/ff_drones/
5. Andrew, M., & Shephard, J. (2017). Semi-automated detection of eagle nests: An application of very high-resolution image data and advanced image analyses to wildlife surveys. *Remote Sensing in Ecology and Conservation, 3*(2), 66–80. https://doi.org/10.1002/rse2.38
6. Arthur, C. (2014). *Amazon seeks US permission to test Prime Air delivery drones.* The Guardian. Retrieved January 23, 2022, from https://www.theguardian.com/technology/2014/jul/11/amazon-prime-air-delivery-drones
7. Barnas, A., Chabot, D., Hodgson, A., Johnston, D., Bird, D., & Ellis-Felege, S. (2020). A standardized protocol for reporting methods when using drones for wildlife research. *Journal of Unmanned Vehicle Systems, 8*(2), 89–98. https://doi.org/10.1139/juvs-2019-0011
8. Barnas, A., Felege, C., Rockwell, R., & Ellis-Felege, S. (2018). A pilot(less) study on the use of an unmanned aircraft system for studying polar bears (*Ursus maritimus*). *Polar Biology, 41*(5), 1055–1062. https://doi.org/10.1007/s00300-018-2270-0
9. Barnas, A., Newman, R., Felege, C. J., Corcoran, M. P., Hervey, S. D., Stechmann, T. J., Rockwell, R. F., & Ellis-Felege, S. N. (2018). Evaluating behavioral responses of nesting

lesser snow geese to unmanned aircraft surveys. *Ecology and Evolution, 8*(2), 1328–1338. https://doi.org/10.1002/ece3.2018.8.issue-210.1002/ece3.3731

10. Barr, J., Green, M., DeMaso, S., & Hardy, T. (2020). Drone surveys do not increase colony-wide flight behaviour at waterbird nesting sites, but sensitivity varies among species. *Scientific Reports, 10*(1). https://doi.org/10.1038/s41598-020-60543-z

11. Basu, C., Deacon, F., Hutchinson, J., & Wilson, A. (2019). The running kinematics of free-roaming giraffes, measured using a low cost unmanned aerial vehicle (UAV). *PeerJ, 7*, e6312. https://doi.org/10.7717/peerj.6312

12. Berger-Tal, O., & Lahoz-Monfort, J. J. (2018). Conservation technology: The next generation. *Conservation Letters, 11*, e12458. https://doi.org/10.1111/conl.12458

13. Bevan, E., Whiting, S., Tucker, T., Guinea, M., Raith, A., & Douglas, R. (2018). Measuring behavioral responses of sea turtles, saltwater crocodiles, and crested terns to drone disturbance to define ethical operating thresholds. *PLoS ONE, 13*(3), e0194460. https://doi.org/10.1371/journal.pone.0194460

14. Bonnin, N., van Andel, A. C., Kerby, J. T., Piel, A. K., Pintea, L., & Wich, S. A. (2018) Assessment of chimpanzee nest detectability in drone-acquired images. *Drones, 2,* 17. https://doi.org/10.3390/drones2020017

15. Brisson-Curadeau, É., Bird, D., Burke, C., Fifield, D., Pace, P., Sherley, R., & Elliott, K. (2017). Seabird species vary in behavioural response to drone census. *Scientific Reports, 7*(1). https://doi.org/10.1038/s41598-017-18202-3

16. Burke, C., Rashman, M. F., Longmore, S. N., McAree, O., Glover-Kapfer, P., Ancrenaz, M., & Wich, S.A. (2019) Successful observation of orangutans in the wild with thermal-equipped drones. *Journal of Unmanned Vehicle Systems, 99,* 1–25. https://www.mdpi.com/2504-446X/3/2/34/pdf

17. Burnett, J., Lemos, L., Barlow, D., Wing, M., Chandler, T., & Torres, L. (2018). Estimating morphometric attributes of baleen whales with photogrammetry from small UASs: A case study with blue and gray whales. *Marine Mammal Science, 35*(1), 108–139. https://doi.org/10.1111/mms.12527

18. CABS. (2014). Model aircraft films bird trappers on Malta—Drone shot down by hunters Police seize nets and protected birds. Committee Against Bird Slaughter website. http://www.komitee.de/en/actions-and-projects/malta/spring-bpc-2012/model-aircraftf ilms-bird-trappers

19. Canal, D., & Negro, J. J. (2018) Use of drones for research and conservation of birds of prey. In: *Birds of prey* (pp. 325–337). Springer.

20. Casella, E., Collin, A., Harris, D., Ferse, S., Bejarano, S., Parravicini, V., Hench, J. L., & Rovere, A. (2017). Mapping coral reefs using consumer-grade drones and structure from motion photogrammetry techniques. *Coral Reefs, 36,* 269–275. https://doi.org/10.1007/s00338-016-1522-0

21. Chabot, D., & Bird, D. (2015). Wildlife research and management methods in the 21st century: Where do unmanned aircraft fit in? *Journal of Unmanned Vehicle Systems, 3*(4), 137–155. https://doi.org/10.1139/juvs-2015-0021

22. Chabot, D. (2009). Systematic evaluation of a stock unmanned aerial vehicle (UAV) system for small-scale wildlife survey applications. MSc. thesis. Montreal, QC: Mcgill University.

23. Christiansen, F., Sironi, M., Moore, M., Di Martino, M., Ricciardi, M., Warick, H., et al. (2019). Estimating body mass of free-living whales using aerial photogrammetry and 3D volumetrics. *Methods in Ecology and Evolution, 10*(12), 2034–2044. https://doi.org/10.1111/2041-210x.13298

24. Christie, K., Gilbert, S., Brown, C., Hatfield, M., & Hanson, L. (2016). Unmanned aircraft systems in wildlife research: Current and future applications of a transformative technology. *Frontiers in Ecology and the Environment, 14*(5), 241–251. https://doi.org/10.1002/fee.1281

25. Chrétien, L., Théau, J., & Ménard, P. (2016). Visible and thermal infrared remote sensing for the detection of white-tailed deer using an unmanned aerial system. *Wildlife Society Bulletin, 40*(1), 181–191. https://doi.org/10.1002/wsb.629

26. Cleguer, C., Kelly, N., Tyne, J., Wieser, M., Peel, D., & Hodgson, A. (2021). A novel method for using small unoccupied aerial vehicles to survey wildlife species and model their density distribution. *Frontiers In Marine Science, 8*.https://doi.org/10.3389/fmars.2021.640338

27. Cunliffe, A., Brazier, R., & Anderson, K. (2016). Ultra-fine grain landscape-scale quantification of dryland vegetation structure with drone-acquired structure-from-motion photogrammetry. *Remote Sensing of Environment, 183*, 129–143. https://doi.org/10.1016/j.rse.2016.05.019

28. d'Oliveira, M., Broadbent, E., Oliveira, L., Almeida, D., Papa, D., Ferreira, M. E., et al. (2020). Aboveground biomass estimation in amazonian tropical forests: A comparison of aircraft- and GatorEye UAV-borne LiDAR data in the chico mendes extractive reserve in Acre, Brazil. *Remote Sensing, 12*(11), 1754. https://doi.org/10.3390/rs12111754

29. Dandois, J., & Ellis, E. (2010). Remote sensing of vegetation structure using computer vision. *Remote Sensing, 2*(4), 1157–1176. https://doi.org/10.3390/rs2041157

30. Dandois, J., & Ellis, E. (2013). High spatial resolution three-dimensional mapping of vegetation spectral dynamics using computer vision. *Remote Sensing of Environment, 136*, 259–276. https://doi.org/10.1016/j.rse.2013.04.005

31. Dandois, J., Olano, M., & Ellis, E. (2015) Optimal altitude, overlap, and weather conditions for computer vision UAV estimates of forest structure. *Remote Sens., 7*, 13895–13920.

32. Dell, A., Bender, J., Branson, K., Couzin, I., de Polavieja, G., Noldus, L., et al. (2014). Automated image-based tracking and its application in ecology. *Trends in Ecology & Evolution, 29*(7), 417–428. https://doi.org/10.1016/j.tree.2014.05.004

33. Ditmer, M., Vincent, J., Werden, L., Tanner, J., Laske, T., Iaizzo, P., et al. (2015). Bears show a physiological but limited behavioral response to unmanned aerial vehicles. *Current Biology, 25*(17), 2278–2283. https://doi.org/10.1016/j.cub.2015.07.024

34. Ditmer, M., Werden, L., Tanner, J., Vincent, J., Callahan, P., & Iaizzo, P. et al. (2019). Bears habituate to the repeated exposure of a novel stimulus, unmanned aircraft systems. *Conservation Physiology, 7*(1). https://doi.org/10.1093/conphys/coy067

35. Dolbeer, R. (2006). Height distribution of birds recorded by collisions with civil aircraft. *Journal of Wildlife Management, 70*(5), 1345–1350. https://doi.org/10.2193/0022-541x(2006)70[1345:hdobrb]2.0.co;2

36. Doukari, M., Batsaris, M., Papakonstantinou, A., & Topouzelis, K. (2019). A protocol for aerial survey in coastal areas using UAS. *Remote Sensing, 11*(16), 1913. https://doi.org/10.3390/rs11161913

37. Duffy, J., Cunliffe, A., DeBell, L., Sandbrook, C., Wich, S., Shutler, J., et al. (2018). Location, location, location: Considerations when using lightweight drones in challenging environments. *Remote Sensing in Ecology and Conservation, 4*(1), 7–19. https://doi.org/10.1002/rse2.58

38. Duffy, J. P., Anderson, K., Shapiro, A. C., Spina Avino, F. L. DeBell, & Glover-Kapfer, P. (2020). Drone technologies for conservation. *WWF Conservation Technology Series, 1*(5). *WWF*. Retrieved from: https://space-science.wwf.de/drones/WWF_CT_Drones_2020_web.pdf

39. Ferreira, S., & van-Aarde, R. (2009). Aerial survey intensity as a determinant of estimates of African elephant population sizes and trends. *South African Journal of Wildlife Research, 39*(2), 181–191.https://doi.org/10.3957/056.039.0205

40. Finn, R., & Wright, D. (2012). Unmanned aircraft systems: Surveillance, ethics and privacy in civil applications. *Computer Law and Security Review, 28*(2), 184–194. https://doi.org/10.1016/j.clsr.2012.01.005

41. Fiori, L., Doshi, A., Martinez, E., Orams, M., & Bollard-Breen, B. (2017). The use of unmanned aerial systems in marine mammal research. *Remote Sensing, 9*(6), 543. https://doi.org/10.3390/rs9060543

42. Fiori, L., Martinez, E., Bader, M., Orams, M., & Bollard, B. (2020). Insights into the use of an unmanned aerial vehicle (UAV) to investigate the behavior of humpback whales (Megaptera novaeangliae) in Vava'u, Kingdom of Tonga. *Marine Mammal Science, 36*(1), 209–223. https://doi.org/10.1111/mms.12637

43. Forsmoo, J., Anderson, K., Macleod, C., Wilkinson, M., & Brazier, R. (2018). Drone-based structure-from-motion photogrammetry captures grassland sward height variability. *Journal of Applied Ecology, 55*(6), 2587–2599. https://doi.org/10.1111/1365-2664.13148

44. Fuentes, M., Bell, I., Hagihara, R., Hamann, M., Hazel, J., Huth, A., et al. (2015). Improving in-water estimates of marine turtle abundance by adjusting aerial survey counts for perception and availability biases. *Journal of Experimental Marine Biology and Ecology, 471*, 77–83. https://doi.org/10.1016/j.jembe.2015.05.003

45. Garcia-Garin, O., Aguilar, A., Borrell, A., Gozalbes, P., Lobo, A., & Penadés-Suay, J., et al. (2020). Who's better at spotting? A comparison between aerial photography and observer-based methods to monitor floating marine litter and marine mega-fauna. *Environmental Pollution, 258*, 113680. https://doi.org/10.1016/j.envpol.2019.113680

46. Gill, L., D'Amelio, P., Adreani, N., Sagunsky, H., Gahr, M., & Maat, A. (2016). A minimum-impact, flexible tool to study vocal communication of small animals with precise individual-level resolution. *Methods In Ecology And Evolution, 7*(11), 1349–1358. https://doi.org/10.1111/2041-210x.12610

47. Goebel, M., Perryman, W., Hinke, J., Krause, D., Hann, N., Gardner, S., & LeRoi, D. (2015). A small unmanned aerial system for estimating abundance and size of Antarctic predators. *Polar Biology, 38*(5), 619–630. https://doi.org/10.1007/s00300-014-1625-4

48. Goodrich, M., Morse, B., Gerhardt, D., Cooper, J., Quigley, M., Adams, J., & Humphrey, C. (2008). Supporting wilderness search and rescue using a camera-equipped mini UAV. *Journal Of Field Robotics, 25*(1–2), 89–110. https://doi.org/10.1002/rob.20226

49. Gray, P., Fleishman, A., Klein, D., McKown, M., Bézy, V., Lohmann, K., & Johnston, D. (2019). A convolutional neural network for detecting sea turtles in drone imagery. *Methods In Ecology And Evolution, 10*(3), 345–355. https://doi.org/10.1111/2041-210x.13132

50. Guardian. (2014). US officials move to ban drones from national parks. The Guardian, June 20. Retrieved January 23, 2022, from https://www.theguardian.com/environment/2014/jun/20/national-park-service-ban-drones

51. Guirado, E., Tabik, S., Rivas, M., Alcaraz-Segura, D., & Herrera, F. (2019). Whale counting in satellite and aerial images with deep learning. *Scientific Reports, 9*(1), 14259. https://doi.org/10.1038/s41598-019-50795-9

52. Hadjikyriakou, T., Kassinis, N., Skarlatos, D., Charilaou, P., & Kirschel, A. (2020). Breeding success of Eleonora's Falcon in Cyprus revisited using survey techniques for cliff-nesting species. *The Condor, 122*(4), duaa045. https://doi.org/10.1093/condor/duaa045

53. Hahn, N., Mwakatobe, A., Konuche, J., de Souza, N., Keyyu, J., Goss, M., et al. (2016). Unmanned aerial vehicles mitigate human–elephant conflict on the borders of Tanzanian Parks: A case study. *Oryx, 51*(3), 513–516. https://doi.org/10.1017/s0030605316000946

54. Hambrecht, L., Brown, R., Piel, A., & Wich, S. (2019). Detecting 'poachers' with drones: Factors influencing the probability of detection with TIR and RGB imaging in miombo woodlands, Tanzania. *Biological Conservation, 233*, 109–117. https://doi.org/10.1016/j.biocon.2019.02.017

55. Hardin, P., Lulla, V., Jensen, R., & Jensen, J. (2018). Small Unmanned Aerial Systems (sUAS) for environmental remote sensing: Challenges and opportunities revisited. *Giscience and Remote Sensing, 56*(2), 309–322. https://doi.org/10.1080/15481603.2018.1510088

56. Hays, G. C., Alcoverro, T., Christianen, M. J., Duarte, C. M., Hamann, M., Macreadie, P. I., et al. (2018). New tools to identify the location of seagrass meadows: Marine grazers as habitat indicators. *Frontiers in Marine Science, 5*, 9. https://doi.org/10.3389/fmars.2018.00009

57. He, G., Yang, H. T., Pan, R. L., Sun, Y. W., Zheng, P. B., Wang, J. H., Jin, X. L., Zhang, J. J., Li, B. G., & Guo, S. T. (2020). Using unmanned aerial vehicles with thermal-image acquisition cameras for animal surveys: A case study on the Sichuan snub-nosed monkey in the Qinling Mountains. *Integrative Zoology, 15*, 79–86. https://doi.org/10.1111/1749-4877.12410

58. Hensel, E., Wenclawski, S., & Layman, C. A. (2018). Using a small, consumer grade drone to identify and count marine megafauna in shallow habitats. *Latin American Journal of Aquatic Research, 46*, 1025–1033. https://doi.org/10.3856/vol46-issue5-fulltext-15

59. Hodgson, J., Baylis, S., Mott, R., Herrod, A., & Clarke, R. (2016). Precision wildlife monitoring using unmanned aerial vehicles. *Scientific Reports, 6*, 22574. https://doi.org/10.1038/srep22574

60. Hodgson, A., Kelly, N., & Peel, D. (2013). Unmanned aerial vehicles (UAVs) for surveying marine fauna: A dugong case study. *PLoS ONE, 8*, e79556. https://doi.org/10.1371/journal.pone.0079556

61. Hodgson, J. C., & Koh, L. P. (2016). Best practice for minimizing unmanned aerial vehicle disturbance to wildlife in biological field research. *Current Biology, 26*, R404–R405. https://doi.org/10.1016/j.cub.2016.04.001

62. Hodgson, J. C., Mott, R., Baylis, S. M., Pham, T. T., Wotherspoon, S., Kilpatrick, A. D., et al. (2018). Drones count wildlife more accurately and precisely than humans. *Methods in Ecology and Evolution, 9*, 1160–1167. https://doi.org/10.1111/2041-210X.12974

63. Holldorf, E. (2018). Avifauna ethological response to unmanned aircraft systems. MS Thesis, University of San Francisco, CA. May. https://repository.usfca.edu/capstone/771/.

64. Hu, J., Wu, X., & Dai, M. (2020). Estimating the population size of migrating Tibetan antelopes Pantholops hodgsonii with unmanned aerial vehicles. *Oryx, 54*(1), 101–109. https://doi.org/10.1017/S0030605317001673

65. Hulme, D., & Murphree, M. W. (2001). *African wildlife and livelihoods: The promise and performance of community conservation.* James Currey.

66. Hunt, E. R., Hively, W. D., Fujikawa, S. J., Linden, D. S., Daughtry, C. S. T., & McCarty, G. W. (2010). Acquisition of NIRgreen-blue digital photographs from unmanned aircraft for crop monitoring. *Remote Sensing, 2*, 290–305.

67. Husson, E., Hagner, O., & Ecke, F. (2014). Unmanned aircraft systems help to map aquatic vegetation. *Applied Vegetation Science, 17*, 567–577.

68. Husson, E. (2016). Images from unmanned aircraft systems for surveying aquatic and riparian vegetation. *Acta Universitatis Agriculturae Sueciae, 115*(53).

69. Jones, G. P., Pearlstine, L. G., & Percival, H. F. (2006). An assessment of small unmanned aerial vehicles for wildlife research. *Wildlife Society Bulletin, 34*, 750–758. https://doi.org/10.2193/0091-7648(2006)34[750:aaosua]2.0.co;2

70. Joyce, K., Duce, S., Leahy, S., Leon, J., & Maier, S. (2018). Principles and practice of acquiring drone-based image data in marine environments. *Marine and Freshwater Research, 70*(7), 952–963. https://doi.org/10.1071/mf17380

71. Karp, D. (2020). Detecting small and cryptic animals by combining thermography and a wildlife detection dog. *Science and Reports, 10*, 5220. https://doi.org/10.1038/s41598-020-61594-y

72. Kay, S., Hedley, J., & Lavender, S. (2009). Sun glint correction of high and low spatial resolution images of aquatic scenes: A review of methods for visible and near-infrared wavelengths. *Remote Sensing, 1*(4), 697–730. https://doi.org/10.3390/rs1040697

73. Kays, R., Crofoot, M., Jetz, W., & Wikelski, M. (2015). Terrestrial animal tracking as an eye on life and planet. *Science, 348*(6240). https://doi.org/10.1126/science.aaa2478

74. Kelaher, B., Colefax, A., Tagliafico, A., Bishop, M., Giles, A., & Butcher, P. (2020). Assessing variation in assemblages of large marine fauna off ocean beaches using drones. *Marine and Freshwater Research, 71*(1), 68–77. https://doi.org/10.1071/mf18375

75. Kiszka, J. J., Mourier, J., Gastrich, K., & Heithaus, M. R. (2016). Using unmanned aerial vehicles (UAVs) to investigate shark and ray densities in a shallow coral lagoon. *Marine Ecology Progress Series, 560*, 237–242. https://doi.org/10.3354/meps11945

76. Kiszka, J. J., & Heithaus, M. R. (2018). Using aerial surveys to investigate the distribution, abundance, and behavior of sharks and rays, In: J. C. Carrier, M. R. Heithaus, & C. A. Simpfendorfer (Eds.), *Shark research: Emerging technologies and applications for the field and laboratory.* CRC Press.

77. Koh, L., & Wich, S. (2012). Dawn of drone ecology: Low-cost autonomous aerial vehicles for conservation. *Tropical Conservation Science, 5*(2), 121–132. https://doi.org/10.1177/194008291200500202

78. Krause, D., Hinke, J., Perryman, W., Goebel, M., & LeRoi, D. (2017). An accurate and adaptable photogrammetric approach for estimating the mass and body condition of pinnipeds using an unmanned aerial system. *PLoS ONE, 12*(11), e0187465. https://doi.org/10.1371/jou rnal.pone.0187465

79. Lahoz-Monfort, J., Chadès, I., Davies, A., Fegraus, E., Game, E., Guillera-Arroita, G., et al. (2019). A call for international leadership and coordination to realize the potential of conservation technology. *BioScience, 69*(10), 823–832. https://doi.org/10.1093/biosci/biz090

80. Lahoz-Monfort, J., & Magrath, M. J. L. (2021). A comprehensive overview of technologies for species and habitat monitoring and conservation. *BioScience, 71*(10), 1038–1062. https:// doi.org/10.1093/biosci/biab073

81. Laliberte, A., & Rango, A. (2009). Texture and scale in object-based analysis of subdecimeter resolution unmanned aerial vehicle (UAV) imagery. *IEEE Transactions On Geoscience and Remote Sensing, 47*(3), 761–770. https://doi.org/10.1109/tgrs.2008.2009355

82. Lambertucci, S., Shepard, E., & Wilson, R. (2015). Human-wildlife conflicts in a crowded airspace. *Science, 348*(6234), 502–504. https://doi.org/10.1126/science.aaa6743

83. Landeo-Yauri, S., Castelblanco-Martínez, D., Hénaut, Y., Arreola, M., & Ramos, E. (2021). Behavioural and physiological responses of captive Antillean manatees to small aerial drones. *Wildlife Research.* https://doi.org/10.1071/wr20159

84. Lejot, J., Delacourt, C., Piégay, H., Fournier, T., Trémélo, M., & Allemand, P. (2007). Very high spatial resolution imagery for channel bathymetry and topography from an unmanned mapping controlled platform. *Earth Surface Processes and Landforms, 32*(11), 1705–1725. https://doi.org/10.1002/esp.1595

85. Lin, J., Shu, L., Zuo, H., & Zhang, B. (2012). Experimental observation and assessment of ice conditions with a fixed-wing unmanned aerial vehicle over Yellow River, China. *Journal of Applied Remote Sensing, 6*(1), 063586. https://doi.org/10.1117/1.jrs.6.063586

86. Linchant, J., Lhoest, S., Quevauvillers, S., Lejeune, P., Vermeulen, C., Semeki Ngabinzeke, J., et al. (2018). UAS imagery reveals new survey opportunities for counting hippos. *PLoS ONE, 13*(11), e0206413. https://doi.org/10.1371/journal.pone.0206413

87. Linchant, J., Lisein, J., Semeki, J., Lejeune, P., & Vermeulen, C. (2015). Are unmanned aircraft systems (UASs) the future of wildlife monitoring? A review of accomplishments and challenges. *Mammal Review, 45*(4), 239–252. https://doi.org/10.1111/mam.12046

88. Lomax, A., Corso, W., & Etro, J. (2005). Employing unmanned aerial vehicles (UAVS) as an element of the integrated ocean observing system. *Proceedings of Oceans 2005 MTS/IEEE.* https://doi.org/10.1109/oceans.2005.1639759

89. Lyons, M., Brandis, K., Callaghan, C., McCann, J., Mills, C., Ryall, S., & Kingsford, R. (2018). Bird interactions with drones, from individuals to large colonies. *Australian Field Ornithology, 35,* 51–56. https://doi.org/10.20938/afo35051056

90. López, J. J., & Mulero-Pázmány, M. (2019). Drones for conservation in protected areas: Present and future. *Drones, 3*(1), 10. https://doi.org/10.3390/drones3010010

91. Ma, Q., Su, Y., & Guo, Q. (2017) Comparison of canopy cover estimations from airborne LiDAR, aerial imagery, and satellite imagery. *IEEE Journal of Selected Topics in Applied Earth Observations and Remote Sensing, 10*(9), 4225–4236. https://doi.org/10.1109/JSTARS. 2017.2711482

92. Maffey, G., Homans, H., Banks, K., & Arts, K. (2015). Digital technology and human development: A charter for nature conservation. *Ambio, 44*(S4), 527–537. https://doi.org/10.1007/ s13280-015-0703-3

93. Mancini, F., Dubbini, M., Gattelli, M., Stecchi, F., Fabbri, S., & Gabbianelli, G. (2013). Using unmanned aerial vehicles (UAV) for high-resolution reconstruction of topography: The structure from motion approach on coastal environments. *Remote Sensing, 5*(12), 6880–6898. https://doi.org/10.3390/rs5126880

94. Marks, P. (2014). Elephants and rhinos benefit from drone surveillance. *New Scientist, 221*(2956), 24. https://doi.org/10.1016/s0262-4079(14)60323-7

95. Marsh, H., Lawler, I., Kwan, D., Delean, S., Pollock, K., & Alldredge, M. (2004). Aerial surveys and the potential biological removal technique indicate that the Torres Strait dugong

fishery is unsustainable. *Animal Conservation, 7*(4), 435–443. https://doi.org/10.1017/s13679
43004001635
96. Martin, J., Edwards, H., Burgess, M., Percival, H., Fagan, D., Gardner, B., et al. (2012).
    Estimating distribution of hidden objects with drones: From tennis balls to manatees. *PLoS
    ONE, 7*(6), e38882. https://doi.org/10.1371/journal.pone.0038882
97. Marx, G. (1998). Ethics for the new surveillance. *The Information Society, 14*(3), 171–185.
    https://doi.org/10.1080/019722498128809
98. Marx, G. (2004). What's new about the "new surveillance"?: Classifying for change and
    continuity. *Knowledge, Technology & Policy, 17*(1), 18–37. https://doi.org/10.1007/bf0268
    7074
99. McEvoy, J., Hall, G., & McDonald, P. (2016). Evaluation of unmanned aerial vehicle shape,
    flight path and camera type for waterfowl surveys: Disturbance effects and species recognition.
    PeerJ, 4, e1831. https://doi.org/10.7717/peerj.1831
100. Melesse, A., Weng, Q., Thenkabail, P., & Senay, G. (2007). Remote sensing sensors and
    applications in environmental resources mapping and modelling. *Sensors, 7*(12), 3209–3241.
    https://doi.org/10.3390/s7123209
101. Merino, L., Caballero, F., Martínez-de-Dios, J., Maza, I., & Ollero, A. (2011). An unmanned
    aircraft system for automatic forest fire monitoring and measurement. *Journal Of Intelligent &
    Robotic Systems, 65*(1–4), 533–548. https://doi.org/10.1007/s10846-011-9560-x
102. Mlambo, R., Woodhouse, I., Gerard, F., & Anderson, K. (2017). Structure from motion (SfM)
    photogrammetry with drone data: A low cost method for monitoring greenhouse gas emissions
    from forests in developing countries. *Forests, 8*(3), 68. https://doi.org/10.3390/f8030068
103. Mulero-Pázmány, M., Jenni-Eiermann, S., Strebel, N., Sattler, T., Negro, J., & Tablado, Z.
    (2017). Unmanned aircraft systems as a new source of disturbance for wildlife: A systematic
    review. *PLoS ONE, 12*(6), e0178448. https://doi.org/10.1371/journal.pone.0178448
104. Mulero-Pázmány, M., Stolper, R., van Essen, L., Negro, J., & Sassen, T. (2014). Remotely
    piloted aircraft systems as a rhinoceros anti-poaching tool in Africa. *PLoS ONE, 9*(1), e83873.
    https://doi.org/10.1371/journal.pone.0083873
105. Muller, C., Chilvers, B., Barker, Z., Barnsdale, K., Battley, P., French, R., et al. (2019). Aerial
    VHF tracking of wildlife using an unmanned aerial vehicle (UAV): Comparing efficiency
    of yellow-eyed penguin (*Megadyptes antipodes*) nest location methods. *Wildlife Research,
    46*(2), 145. https://doi.org/10.1071/wr17147
106. Oliveira, R., Näsi, R., Niemeläinen, O., Nyholm, L., Alhonoja, K., & Kaivosoja, J. et al.
    (2019). Assessment of RGB and hyperspectral UAV remote sensing for grass quantity and
    quality estimation. In *The international archives of the photogrammetry, remote sensing and
    spatial information sciences, XLII-2/W13*, (pp. 489–494). https://doi.org/10.5194/isprs-arc
    hives-xlii-2-w13-489-2019
107. Pande-Chhetri, R., Abd-Elrahman, A., Liu, T., Morton, J., & Wilhelm, V. (2017). Object-
    based classification of wetland vegetation using very high-resolution unmanned air system
    imagery. *European Journal Of Remote Sensing, 50*(1), 564–576. https://doi.org/10.1080/227
    97254.2017.1373602
108. Paneque-Gálvez, J., McCall, M., Napoletano, B., Wich, S., & Koh, L. (2014). Small drones
    for community-based forest monitoring: An assessment of their feasibility and potential in
    tropical areas. *Forests, 5*(6), 1481–1507. https://doi.org/10.3390/f5061481
109. Paul, J. K., Yuvaraj, T., & Gundepudi, K. (2020). Demonstrating low-cost unmanned aerial
    vehicle for anti-poaching. In: *2020 IEEE 17Th India Council International Conference
    (INDICON)*. https://doi.org/10.1109/indicon49873.2020.9342131
110. Perryman, W., Goebel, M. E., Ash, N., LeRoi, D. J., & Gardner, S. (2014). Small unmanned
    aerial systems for estimating abundance of krill-dependent predators: A feasibility study with
    preliminary results. In J. G. Walsh (Ed.), *AMLR 2010–2011 field season report* (pp. 64–72).
    U.S. Department of Commerce, NOAA Technical Memorandum NMFS-SWFSC-524.
111. Pickett, M., Taggart, B., Rivers, J., Adrean, L. & Nelson, S. (2017). Utilizing UAS to locate
    nesting seabirds within the canopy of mature forests. *Auvsi Xponential 2017*, Dallas.

112. Pimm, S., Alibhai, S., Bergl, R., Dehgan, A., Giri, C., Jewell, Z., et al. (2015). Emerging technologies to conserve biodiversity. *Trends in Ecology & Evolution, 30*(11), 685–696. https://doi.org/10.1016/j.tree.2015.08.008

113. Pirotta, V., Smith, A., Ostrowski, M., Russell, D., Jonsen, I., Grech, A., & Harcourt, R. (2017). An economical custom-built drone for assessing whale health. *Frontiers in Marine Science, 4*.https://doi.org/10.3389/fmars.2017.00425

114. Platt, J. (2012). *Eye in the Sky: Drones help conserve Sumatran orangutans and other wildlife.* Scientific American Blog Network. Retrieved January 23, 2022, from https://blogs.scientificamerican.com/extinction-countdown/drones-help-conserve-sumatran-orangutans-wildlife/.

115. Pomeroy, P., O'Connor, L., & Davies, P. (2015). Assessing use of and reaction to unmanned aerial systems in gray and harbor seals during breeding and molt in the UK. *Journal Of Unmanned Vehicle Systems, 3*(3), 102–113. https://doi.org/10.1139/juvs-2015-0013

116. Rajpurkar, S., Pande, A., Sharma, S., Gole, S., Dudhat, S., Johnson, J. A., & Sivakumar, K. (2021). Light-weight unmanned aerial vehicle surveys detect dugongs and other globally threatened marine species from the Andaman and Nicobar Islands, India. *Current Science, 121*(2), 195–197. https://www.currentscience.ac.in/Volumes/121/02/0195.pdf

117. Ramos, E., Maloney, B., Magnasco, M., & Reiss, D. (2018). Bottlenose dolphins and Antillean manatees respond to small multi-rotor unmanned aerial systems. *Frontiers in Marine Science, 5*.https://doi.org/10.3389/fmars.2018.00316

118. Rango, A., Laliberte, A., Steele, C., Herrick, J., Bestelmeyer, B., Schmugge, T., et al. (2006). Research article: Using unmanned aerial vehicles for rangelands: current applications and future potentials. *Environmental Practice, 8*(3), 159–168. https://doi.org/10.1017/s1466046606060224

119. Raoult, V., Colefax, A., Allan, B., Cagnazzi, D., Castelblanco-Martínez, N., & Ierodiaconou, D., et al. (2020). Operational protocols for the use of drones in marine animal research. *Drones, 4*(4), 64.https://doi.org/10.3390/drones4040064

120. Ratcliffe, N., Guihen, D., Robst, J., Crofts, S., Stanworth, A., & Enderlein, P. (2015). A protocol for the aerial survey of penguin colonies using UAVs. *Journal Of Unmanned Vehicle Systems, 3*(3), 95–101. https://doi.org/10.1139/juvs-2015-0006

121. Rebolo-Ifrán, N., Graña Grilli, M., & Lambertucci, S. (2019). Drones as a threat to wildlife: YouTube complements science in providing evidence about their effect. *Environmental Conservation, 46*(3), 205–210. https://doi.org/10.1017/s0376892919000080

122. Rey, N., Volpi, M., Joost, S., & Tuia, D. (2017). Detecting animals in African Savanna with UAVs and the crowds. *Remote Sensing Of Environment, 200*, 341–351. https://doi.org/10.1016/j.rse.2017.08.026

123. Rodríguez, A., Negro, J., Mulero, M., Rodríguez, C., Hernández-Pliego, J., & Bustamante, J. (2012). The eye in the sky: Combined use of unmanned aerial systems and GPS data loggers for ecological research and conservation of small birds. *PLoS ONE, 7*(12), e50336. https://doi.org/10.1371/journal.pone.0050336

124. Sandbrook, C. (2015). The social implications of using drones for biodiversity conservation. *Ambio, 44*(S4), 636–647. https://doi.org/10.1007/s13280-015-0714-0

125. Santangeli, A., Chen, Y., Kluen, E., Chirumamilla, R., Tiainen, J., & Loehr, J. (2020). Integrating drone-borne thermal imaging with artificial intelligence to locate bird nests on agricultural land. *Scientific Reports, 10*(1). https://doi.org/10.1038/s41598-020-67898-3

126. Sasse, D. B. (2003). Job-related mortality of wildlife workers in the United States, 1937–2000. *Wildlife Society Bulletin*, 1015–1020.

127. Schaub, J., Hunt, B., Pakhomov, E., Holmes, K., Lu, Y., & Quayle, L. (2018). Using unmanned aerial vehicles (UAVs) to measure jellyfish aggregations. *Marine Ecology Progress Series, 591*, 29–36. https://doi.org/10.3354/meps12414

128. Schiffman, R. (2014). Drones flying high as new tool for field biologists. *Science, 344*(6183), 459–459. https://doi.org/10.1126/science.344.6183.459

129. Scholten, C., Kamphuis, A., Vredevoogd, K., Lee-Strydhorst, K., Atma, J., Shea, C., et al. (2019). Real-time thermal imagery from an unmanned aerial vehicle can locate ground nests of a grassland songbird at rates similar to traditional methods. *Biological Conservation, 233*, 241–246. https://doi.org/10.1016/j.biocon.2019.03.001

130. Snitch, T. (2014). Poachers kill three elephants an hour. Here's how to stop them. *The telegraph.* Retrieved January 23, 2022, from http://www.telegraph.co.uk/news/earth/environment/conservation/10634747/Poachers-kill-three-elephantsan-hour.-Heres-how-to-stop-them.html

131. Spaan, D., Burke, C., McAree, O., Aureli, F., Rangel-Rivera, C., Hutschenreiter, A., et al. (2019). Thermal infrared imaging from drones offers a major advance for spider monkey surveys. *Drones, 3*(2), 34. https://doi.org/10.3390/drones3020034

132. Stöcker, C., Bennett, R., Nex, F., Gerke, M., & Zevenbergen, J. (2017). Review of the current state of UAV regulations. *Remote Sensing, 9*(5), 459. https://doi.org/10.3390/rs9050459

133. Sutherland, W., Bardsley, S., Clout, M., Depledge, M., Dicks, L., Fellman, L., et al. (2013). A horizon scan of global conservation issues for 2013. *Trends in Ecology & Evolution, 28*(1), 16–22. https://doi.org/10.1016/j.tree.2012.10.022

134. Sykora-Bodie, S., Bezy, V., Johnston, D., Newton, E., & Lohmann, K. (2017). Quantifying nearshore sea turtle densities: Applications of unmanned aerial systems for population assessments. *Scientific Reports, 7*(1). https://doi.org/10.1038/s41598-017-17719-x

135. Szantoi, Z., Smith, S., Strona, G., Koh, L., & Wich, S. (2017). Mapping orangutan habitat and agricultural areas using Landsat OLI imagery augmented with unmanned aircraft system aerial photography. *International Journal Of Remote Sensing, 38*(8–10), 2231–2245. https://doi.org/10.1080/01431161.2017.1280638

136. Tian, J., Wang, L., Li, X., Gong, H., Shi, C., Zhong, R., & Liu, X. (2017). Comparison of UAV and WorldView-2 imagery for mapping leaf area index of mangrove forest. *International Journal of Applied Earth Observation and Geoinformation, 61*, 22–31. https://doi.org/10.1016/j.jag.2017.05.002

137. UNEP. (2013a). A new eye in the sky: Eco-drones. Global Environment Alert Service Bulletin, May 2013. United Nations Environment Programme. http://www.unep.org/geas/

138. van Andel, A., Wich, S., Boesch, C., Koh, L., Robbins, M., Kelly, J., & Kuehl, H. (2015). Locating chimpanzee nests and identifying fruiting trees with an unmanned aerial vehicle. *American Journal Of Primatology, 77*(10), 1122–1134. https://doi.org/10.1002/ajp.22446

139. Varela, M., Patrício, A., Anderson, K., Broderick, A., DeBell, L., Hawkes, L., et al. (2018). Assessing climate change associated sea-level rise impacts on sea turtle nesting beaches using drones, photogrammetry and a novel GPS system. *Global Change Biology, 25*(2), 753–762. https://doi.org/10.1111/gcb.14526

140. Vas, E., Lescroël, A., Duriez, O., Boguszewski, G., & Grémillet, D. (2015). Approaching birds with drones: First experiments and ethical guidelines. *Biology Letters, 11*(2), 20140754. https://doi.org/10.1098/rsbl.2014.0754

141. Velusamy, P., Rajendran, S., Mahendran, R. K., Naseer, S., Shafiq, Md., & Choi, J. -G. (2022). Unmanned Aerial Vehicles (UAV) in precision agriculture: Applications and challenges. *Energies, 15*(1), 217. https://doi.org/10.3390/en15010217

142. Ventura, D., Bonifazi, A., Gravina, M., Belluscio, A., & Ardizzone, G. (2018). Mapping and classification of ecologically sensitive marine habitats using Unmanned Aerial Vehicle (UAV) imagery and Object-Based Image Analysis (OBIA). *Remote Sensing, 10*(9), 1331. https://doi.org/10.3390/rs10091331

143. Ventura, D., Bruno, M., Jona Lasinio, G., Belluscio, A., & Ardizzone, G. (2016). A low-cost drone based application for identifying and mapping of coastal fish nursery grounds. *Estuarine, Coastal And Shelf Science, 171*, 85–98. https://doi.org/10.1016/j.ecss.2016.01.030

144. Vermeulen, C., Lejeune, P., Lisein, J., Sawadogo, P., & Bouché, P. (2013). Unmanned aerial survey of elephants. *PLoS ONE, 8*(2), e54700. https://doi.org/10.1371/journal.pone.0054700

145. Watts, A., Ambrosia, V., & Hinkley, E. (2012). Unmanned aircraft systems in remote sensing and scientific research: Classification and considerations of use. *Remote Sensing, 4*(6), 1671–1692. https://doi.org/10.3390/rs4061671

146. Watts, A., Perry, J., Smith, S., Burgess, M., Wilkinson, B., Szantoi, Z., et al. (2010). Small unmanned aircraft systems for low-altitude aerial surveys. *Journal Of Wildlife Management, 74*(7), 1614–1619. https://doi.org/10.2193/2009-425

147. Weissensteiner, M., Poelstra, J., & Wolf, J. (2015). Low-budget ready-to-fly unmanned aerial vehicles: An effective tool for evaluating the nesting status of canopy-breeding bird species. *Journal Of Avian Biology, 46*(4), 425–430. https://doi.org/10.1111/jav.00619

148. Whitehead, K., & Hugenholtz, C. (2014). Remote sensing of the environment with small unmanned aircraft systems (UASs), part 1: A review of progress and challenges. *Journal Of Unmanned Vehicle Systems, 02*(03), 69–85. https://doi.org/10.1139/juvs-2014-0006

149. Woodget, A., Austrums, R., Maddock, I., & Habit, E. (2017). Drones and digital photogrammetry: From classifications to continuums for monitoring river habitat and hydromorphology. *WIREs Water, 4*(4), e1222. https://doi.org/10.1002/wat2.1222

150. Yang, F., Shao, Q., & Jiang, Z. (2019). A population census of large herbivores based on UAV and its effects on grazing pressure in the Yellow-River-Source National Park, China. *International Journal Of Environmental Research And Public Health, 16*(22), 4402. https://doi.org/10.3390/ijerph16224402

151. Zahawi, R., Dandois, J., Holl, K., Nadwodny, D., Reid, J., & Ellis, E. (2015). Using lightweight unmanned aerial vehicles to monitor tropical forest recovery. *Biological Conservation, 186*, 287–295. https://doi.org/10.1016/j.biocon.2015.03.031

Chapter 15
An Integrated Application of AHP and GIS-Based Model to Identify the Waterlogged Zones Susceptible for Water-Borne Diseases

Pranay Paul (ID) **and Rumki Sarkar** (ID)

Abstract Flood is considered as the annual event in the Lower Ganga Plain. It causes adverse impact on the population in terms of economic and social life. Floodwater hampers the physical life both directly (drowning, snake bite, hypothermia, etc.) and indirectly (water and vector-borne diseases). The abundance of cutoffs and marshy lands in Uttar Dinajpur district leads to the outbreak of water-borne diseases during the monsoon and post-monsoon session. Therefore, an attempt has been made to classify the region into index-based potential waterlogging zone using the multi-criteria decision making analysis. Total eight geo-hydrological parameters were taken into consideration such as elevation, slope, relative relief, geomorphology, groundwater depth, drainage density, soil, and MNDWI. The normalized weights were assigned to the parameters using the analytical hierarchy process (AHP), constructed on their relative importance on waterlogging situations and incorporated in the GIS environment using the overlay method. Finally, the Potential Waterlogging Zone (PWZ) map was compared with the historical geo-hazard map of the concerned area for the validation of the work. The very high PWZ accounts for $66.69 \, \text{km}^2$ area of the district which is mostly the perennial marshy lands, whereas the majority of the portion ($1835.28 \, \text{km}^2$) comes under High PWZ mainly due to the inadequate drainage facilities during the flood discharge. Subsequently, the moderate PWZ comprises for about $1036.38 \, \text{km}^2$ area which is comparatively less vulnerable to the flood related health issues. Therefore, the overall approach of the study can be used to reduce the vulnerabilities related to water infectious diseases through comprehensive surveillance, emergency health facilities, early warning system, and well-coordinated collaborations.

Keywords Analytical hierarchy process (AHP) · Multi-criteria decision making (MCDM) · Potential waterlogging zone (PWZ) · Water-borne diseases · Vector-borne diseases

P. Paul · R. Sarkar (✉)
Department of Geography, Raiganj University, West Bengal, Uttar Dinajpur 733134, India
e-mail: itsrumki84@gmail.com

15.1 Introduction

Flood is among the common natural disaster occurring worldwide. Flood accounts for one-fifth of the global deaths [32]. In India, the Lower Ganga Basin is considered as one of the most flood-prone areas of the country. The main reasons for floods in the study area are higher intensity rainfall in short duration, insufficient drainage capacity of rivers, and primarily the geomorphic specification of the Rajmahal gap [24]. More importantly, Mahananda and other rivers of Uttar Dinajpur and Dakshin Dinajpur district get stagnated during the peak monsoon discharge as the downstream and upstream of Farakka barrage rules high and that does not allow to drain the flood discharge of the region [34]. Flood causes tremendous loss to the public property and healthcare system in that region. The adverse impact on health can be direct or indirect and immediate, mid-term or for long term as well [1, 4, 12, 21, 30]. The direct consequences happen at the time of floods which include drowning, hypothermia, snake bite, etc., and indirect effects are found after the floodwater has gone such health impacts are diarrhea, skin problems, cholera, etc. Flood also increases the mortality, morbidity, and economic distress of the society.

Since the natural occurrence of the flood cannot be prevented, the emphasis has moved from disaster response toward the risk management. Thereby, the mapping of flood zones has been turned out to be the most initial work in order to prepare the spatial management plans. The hydrological and hydrodynamic models were primarily used based on the magnitude, frequency, and extent of the flood [18, 25]. But such models have limited accuracy and hard to implement plans at administrative level [6].

In the last two decades, the application multi-criteria decision making analysis (MCDM) turned out as an evolutionary technique in terms of index-based zone identification of flood-affected region [2, 6, 13, 22]. The multi-criteria model incorporates those parameters that affect the output of the study most based on the provided normalized weight. The most of the study in MCDM models till date have relied on Analytical Hierarchy Process (AHP) for the input of weightage individually to those parameters [7]. The AHP is a mathematical model mainly prepared for multi-criteria assessment by Saaty [26, 27]. The validity of the work mostly relies on the selected parameters and subjective understanding and observation of the occurring phenomena. The number of parameters is not an issue for the multi-criteria models [6]. However, after the selection and weightage assignment to the parameters, Geographic Information System (GIS) facilitates the further analysis and interpretation [13]. The overlay result need to be further validated as sometime it can lead to uncertain results [7]. Historical flood map of the concerned region can be an important evidence to validate the output work [6].

For the concerned study, eight geo-hydrological parameters have been used for the identification of waterlogging zones, such as elevation, slope, relative relief, geomorphology, soil, groundwater depth, drainage density, and MNDWI. Further the overlaid map has been verified based on the historical geo-hazard map of the study area. The derived potential waterlogging map would be very much helpful for the proper planning and management of flood occurrence and associated outbreaks of diseases.

15.2 Study Area

The Uttar Dinajpur district is located between 25° 11′ N–26° 49′ N latitude and 87° 49′ E–88° 31′ E, with an area of 3140 km^2. In the eastern margin, it is bordered by the neighboring country Bangladesh, Bihar state in the West, Darjeeling and Jalpaiguri district in the north, and Malda district in the south (Fig. 15.1). The district comprises four statutory towns, five census towns and total nine Community Development blocks, namely Chopra, Islampur, Goalpokhar-I, Goalpokhar-II, Karandighi, Raiganj, Hemtabad, Itahar, and Kaliyaganj. The total population of Uttar Dinajpur district is over three million with a high density of 958 person/km^2 [8].

The general topography of the district is mainly flat in nature with the exception of highlands in the some parts of Chopra district. The elevation of the district varies from 23 to 101 m. The smooth undulating surface is ravined by numerous rivers flowing toward southward sloping direction. The major of them include Mahananda, Nagar, Kulik, Chiramati, Tangan, etc. The region is characterized by flood-prone zone and large numbers of marshes or beels are formed in the district due to the overflowing of the river [11]. The floods and associated waterlogging conditions are

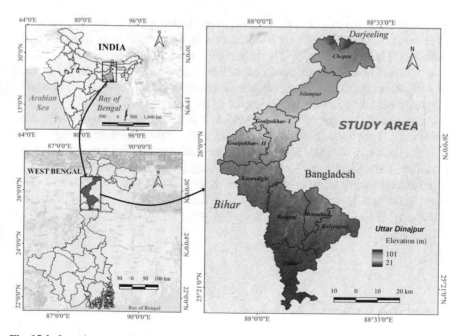

Fig. 15.1 Location map of the study area

Table 15.1 Flooding and communicable diseases (after [35])

Types of diseases	Sub-types
Water-borne	Cholera, Typhoid fever, leptospirosis and hepatitis A and E
Vector-borne diseases	Malaria, dengue and dengue hemorrhagic fever
contamination of groundwater	Contaminated drinking water induce stomach problems
Rodent: reptiles	Snake bite, mud contaminated with rodent and animal urine
Skin infection	Dermatitis, wound infections, conjunctivitis, ear, nose and throat infections

recurring event in the district which causes serious spreading of infectious diseases throughout the region. Flood can increase the probability of communicable diseases (Table 15.1), which includes both the water-borne and vector-borne diseases [35]. The water-borne diseases are mainly related with the contaminated water, whereas the vector-borne diseases mainly prevail when floodwater starts to recede. The spread of those diseases is very much prevalent during the period of June to December [3, 9, 15, 20]. Thus, it seems important to delineate those populated areas which have high potentiality of waterlogging.

15.3   Materials and Methods

The aim of the concerned study is to identify the waterlogged zones susceptible for water-borne diseases based on the multi-criteria decision making analysis through the Analytical Hierarchy Process (AHP) technique. The AHP method involves six major steps to reach the final output; statement of the unstructured problem, formulation of the AHP hierarchy, calculation of pairwise comparison matrix, assessment of relative weights, consistency check, and validation of the final output [19, 22].

The preparation of the considered thematic layers was followed by the classification of each layer into sub-classes on the basis of natural breaks clustering method [17]. This method is capable to minimize the variance within each class and can maximize inter-class variability [6, 17]. The integration of the thematic layers on the AHP-based relative weight assignment was made in the GIS environment to find out the potential waterlogged zones, which further was validated using the historical geohazard data. However, the detailed methodology of the work has been incorporated in Table 15.2.

Table 15.2 Source and details of the thematic layer used for the identification of Potential Waterlogging Zone

Parameters	Data source	Data details	References
Elevation	SRTM DEM	30 m × 30 m	https://earthexplorer.usgs.gov
Slope	SRTM DEM	30 m × 30 m	https://earthexplorer.usgs.gov
Relative relief	SRTM DEM	30 m × 30 m	https://earthexplorer.usgs.gov
Geomorphology	Geological Survey of India (GSI)	Vector layer, 1:1,091,958	https://www.gsi.gov.in
Groundwater depth	Geological Survey of India (GSI)	1:2,000,000	Hydrogeological Atlas of West Bengal (1988)
Soil	NBSS and LUP, Kolkata	1:250,000	USGS Soil Classification of West Bengal
Drainage density	HydroSHEDS	Vector Layer	https://www.hydrosheds.org
MNDWI	USGS	LANDSAT 8 OLI/ TIRS, 30 m × 30 m	https://earthexplorer.usgs.gov

15.3.1 Selection of Criteria

The identification of appropriate criteria is utmost important in the semi-quantitative MCDA-based models. Eight parameters were chosen under the sheds of physiography (elevation, slope, relative relief, and geomorphology), hydrology (groundwater depth and drainage density), and surface characteristics (MNDWI and soil) of the concerned area. All the parameters were projected into UTM Zone 45 (WGS 84 datum), with the spatial resolution of 30 m.

The elevation and the degree of sloping are of prime importance in terms of flow accumulation in the low lying floodplains [29]. Though most of the area comprises monotonous slope surface, different zones of relative relief have been found to be very much appropriate with the belonging depression zones. The thematic layers of elevation, slope, and relative relief have been prepared on the basis of SRTM DEM data (30 m × 30 m) processed in the ArcGis environment. The geomorphological data of the area is collected from the Geological Survey of India (GSI) web site in the vector layer format, which was further transformed into rasterized format in the ArcGIS environment. The groundwater depth and the drainage density data have been considered in terms of hydrological factors associated with the waterlogging situations. The retention of water in the depressed zones throughout the year heavily depends on the groundwater depth of an area. Thus, the average groundwater depth has been collected from the district resource map of Uttar Dinajpur district. The drainage network data is acquired from the *Hydrosheds* web site, and further processing was made for the drainage density map. Since the area is fluvially dynamic in nature, the low drainage density zones have been found to be most prevalent to the waterlogged zones. LANDSAT 8 OLI/TIRS data (30 m spatial resolution) has been acquired from the NASA's USGS Earth Explorer web site to prepare the MNDWI map using Xu's method (2006). The soil map is collected from the NBSS and LUP office, Kolkata, and categorization has been made based on the draining capabilities of them.

$$MNDWI = \frac{(Green - SWIR1)}{(Green + SWIR2)}$$

15.3.2 Weight Assignment and Normalization Using AHP Technique

Analytical Hierarchy Process (AHP) is among the most important techniques to perform the multi-criteria decision making models [22, 27]. It is a hierarchical structure to identify a problem where the weights are assigned for the substitutes based on the preference and judgments of the user. The relative importance scale of Saaty is used to compare the parameters in the pairwise comparison matrix (Table 15.3) and finally the normalization of the parameters was made using the Principle Eigen Vector measurement [23, 26].

Table 15.3 Pairwise comparison matrix and normalized weight for each selected layers

Parameters	GD	RR	GM	DD	EL	MNDWI	S	SL	Normalized weight
GD	1	1	1	3	3	1	5	7	0.21
RR	1	1	3	3	3	1	5	7	0.24
GM	1	0.33	1	1	3	1	3	3	0.14
DD	0.33	0.33	1	1	1	0.33	3	3	0.09
EL	0.33	0.33	0.33	1	1	1	3	5	0.1
MNDWI	1	1	1	3	1	1	1	3	0.14
S	0.20	0.20	0.33	0.33	0.33	1	1	3	0.05
SL	0.14	0.14	0.33	0.33	0.20	0.33	0.33	1	0.03

GD = Groundwater Depth, RR = Relative Relief, GM = Geomorphology, DD = Drainage Density, EL = Elevation, MNDWI = Modified Normalized Difference Water Index, S = Soil, SL = Slope

Table 15.4 Fundamental scale of Saaty for ranking (1–9) and RCI values [26, 27]

Intensity of importance	Definition	Random consistence index (RCI)
1	Equal importance	0
2	Weak	0
3	Moderate importance	0.58
4	Strong moderate	0.90
5	Strong importance	1.12
6	More than strong	1.24
7	Very strong importance	1.32
8	Very strong	1.41
9	Extreme importance	1.45

15.3.3   Consistency Check

The normalized weight of each of the parameters is further needed to be validated through the consistency check [26, 27].

$$CR = \frac{CI}{RI} \tag{15.1}$$

CR refers to the consistency ratio, and CI is the consistency index which is calculated using Eq. (15.2). The RI refers to the random index which differs with number of parameters taken and provided by Saaty (Table. 15.4). The value of RI is found as 1.41 for the study concerned. The λ_{max} is the maximum Eigen vector calculated based on Saaty's method (1990), and n is the number of parameter.

$$CI = \frac{(\lambda max - n)}{(n - 1)} \tag{15.2}$$

In the study, the values have been found as maximum eigenvalue $(\lambda max) = 8.646$, $CI = 0.092$, $n = 8$, and $CR = 0.065$. The value of CR is less than 0.1; thus, the assignment of weight to the parameters is consistent.

15.3.4   Classification of Parameters for Scoring

The parameters comprised both the qualitative and quantitative categories. The qualitative values are associated with geomorphology and soil type, and the rest of the parameters including elevation, slope, relative relief, MNDWI, groundwater depth, and drainage density have been sub-categorized using their quantitative values in the ArcGis environment based on the natural breaks clustering method. The scoring of the sub-categories was done using the fundamental scale (1–9) of [26, 27] based on the priorities and preferences for waterlogging conditions (Table 15.5).

15.3.5 Waterlogging Zone Mapping Using Weighted Overlay Analysis (WOA)

Weighted overlay analysis is a simulation method that constructs map using the composition of layers based on their given preferences and priorities [23]. It has been done using the following equation in the ArcGis environment.

$$\text{WOA} = \sum_{i=1 \text{ to } 8}^{j=1 \text{ to } 5} r_{ij}.w_i$$

where r_{ij} is the ranking of the ith parameter for jth category and w_i is the normalized weight of the parameter found using the AHP technique. After running the following algorithm, the final raster layer (waterlogging zones) is reclassified into five classes based on the natural breaks clustering to prepare the final zoning of Waterlogging zones in Uttar Dinajpur district.

15.3.6 Validation

The validation of the result is performed based on the historical geo-hazard map (1995) of the study area provided by Geological Survey of India (GSI). The map comprises characteristics of waterlogged areas, marshy lands, flood-prone areas, and water body of the concerned study area.

15.4 Results and Discussion

15.4.1 Parameters

15.4.1.1 Geomorphology

The Uttar Dinajpur district except the Chopra block in the northern-most portion falls within the entrenched low land valley of Mahananda [16] which is bounded by

Table 15.5 Weight assignment to the corresponding sub-classes for the desired waterlogging zone map

Parameters	Classes	AHP-normalized weights	Influence (%)	Weightage of sub-classes based on Saaty's scale
Relative relief	Very low (9–15 m)	0.24	24	9
	Low (16–19 m)			9
	Moderate (20–28 m)			7
	High (29–45 m)			5
	Very high (46–72 m)			3
Groundwater depth	<3 m	0.21	21	9
	3–4 m			7
	4–5 m			5
	5–6 m			3
	>6 m			1
Geomorphology	Water bodies	0.14	14	9
	Flood basin			9
	Younger alluvial plain			7
	Active flood plains			5
	Dissected alluvial deposits			3
MNDWI	Very high (−0.02 to 0.27)	0.14	14	9
	High (−0.11 to −0.03)			7
	Moderate (−0.16 to −0.12)			5
	Low (−0.20 to −0.17)			3
	Very low (−0.74 to −0.21)			1
Elevation	Very low (21–36 m)	0.10	10	9
	Low (37–46 m)			7
	Moderate (47–59 m)			5
	High (60–74 m)			3
	Very high (75–103 m)			1
Drainage density	Very low (0.02–0.31 km/km^2)	0.09	9	9
	Low (0.32–0.42 km/km^2)			7

(continued)

Table 15.5 (continued)

Parameters	Classes	AHP-normalized weights	Influence (%)	Weightage of sub-classes based on Saaty's scale
	Moderate (0.43–0.62 km/km^2)			5
	High (0.63–1.03 km/km^2)			3
	Very high (1.04–1.83 km/km^2)			3
Soil	Entisols	0.05	5	9
	Ustic			7
	Fluventic			5
	Ochric			3
	Inceptisols			3
Slope	Near level slope (0°–1.2°)	0.03	3	9
	Very gentle slope (1.3°–2.2°)			7
	Gentle slope (2.3°–3.4°)			5
	Moderate slope (3.5°–5°)			3
	Steep slope (5.1°–34.3°)			1

Teesta and Kosi mega-fan in its east and west stretches, respectively [5]. The entire portion of Mahananda basin in the concerned area is dominated by monotonous low slope surface often dissected by older alluvial channels [24, 31]. The nature of flat topography is sometimes disturbed and takes the form of depressions called *beels*. Morphologically, the area comprised five major types of landforms: dissected alluvial deposits (230.82 km^2; 7.47%), active flood plains (331.61 km^2; 10.74%), younger alluvial plain (2036.04 km^2; 65.92%), flood basin (197.32 km^2; 6.39%), and water bodies (292.69 km^2; 9.48%) comprising drainage network and marshy swamps (Fig. 15.3). The Piedmont alluvial fan in northern edge of the district and the older alluvium (*Barind*) found in the southern corner of study area is categorized under the dissected alluvial deposit zone. This zone is been given the least weightage to be waterlogged due to its higher elevation, and the water bodies are considered to be most vulnerable area to the waterlogged situations. The presence of numerous paleo-channels, cutoffs, swamps, etc., depicts that the concerned study area is highly dynamic in nature [5].

15.4.1.2 Groundwater Depth

The depth of groundwater is highly associated with the infiltration capacity of the surface water [6, 13]. Low groundwater depth does not allow the excess water to infiltrate sub-surface and thus are more reluctant to the waterlogged hegemony. The groundwater depth of the study area ranges from 2 to 7 m. For the waterlogged delineation purpose, five spatial classes (Fig. 15.3) have been prepared: very low (<3 m), low (3–4 m), moderate (4–5 m), high (5–6 m), and very high (>6 m). The groundwater depth of about 85% of the area ranges up to 5 m, and the depth of groundwater sharply declines moving southward. The southern zones are most susceptible to the waterlogged conditions.

15.4.1.3 Relative Relief

Relative relief is the change in the elevation per unit area. The areas with high relative relief zone are prone to higher surface run-off and lower infiltration and vice versa [33]. The relative relief map of concerned area has been prepared based on the SRTM DEM in the ArcGis environment. The relative relief zone (Fig. 15.3) has been divided into five major parts following Jenks's method (1967): very low (9–15 m), low (16–19 m), moderate (20–28 m), high (29–45 m), and very high (46–72 m). Since the four-fifth of the total area is classed under floodplain nature, thereby the very low relative relief zone covers an extensive area (>56%) and is the most potential zone of stagnation of water.

15.4.1.4 Drainage Density

Drainage density is considered as an important factor in terms of delineating the waterlogged zones. It is measured in unit length per square km area and directly related to the surface run-off and accumulation of flow [2]. The floodplain zones are comparatively low drainage density zone due to the flatness of the topography [13]. However, the total drainage density map was further divided into five classes (Fig. 15.3): very low (0.02–0.31 km/km^2), low (0.32–0.42 km/km^2), moderate (0.43–0.62 km/km^2), high (0.63–1.03 km/km^2), and very high (1.04–1.83 km/km^2). It is observed that the areas of low drainage density have higher trend of water depressions due to the poor drainage of excess water during the monsoon.

15.4.1.5 Slope

Slope plays an important role in terms of waterlogging as it controls the velocity of the surface run-off as well as the infiltration rate [6, 29]. The higher slope allows the water to move downslope, and the low slope area causes the water to stagnate. The slope of the Uttar Dinajpur district ranges from 0° to 35°; based on Jenks's method, (1967) they have been classified into five zones (Fig. 15.3): near level slope (0°–1.2°), very gentle slope (1.3°–2.2°), gentle slope (2.3°–3.4°), moderate slope (3.5°–5°), and steep slope (5.1°–34.3°). The low level slope is extensive throughout the district and the steep slope (1.63%) is confined to the dissected alluvial deposit in Chopra Block and in some parts of Itahar block.

15.4.1.6 Elevation

Elevation is considered to be an important factor and directly related to the water-logging of an area [13]. The areas of low elevation are comparatively more reluctant to the waterlogging conditions and vice versa [2]. The elevation map is prepared through the SRTM DEM (30 m × 30 m) in the ArcGis environment. The elevation contour successively declines in the southward directions, and the highest value is found on the topmost margin of the district. Five classes of elevations are (Fig. 15.3): very low (21–36 m), low (37–46 m), moderate (47–59 m), high (60–74 m), and very high (75–103 m). More than 62% of the study area comprises the low category elevation (Karandighi, Raiganj, Hemtabad, Kaliyaganj, and Itahar block) where the low land depression and associated waterlogged conditions are most prevalent.

15.4.1.7 MNDWI

Modified Normalized Difference Water Index (MNDWI) is more suitable than the NDWI to extract water content information of any region with the background of built-up area since it has the ability to avoid the built-up land noises over the NDWI [36]. The value of the MNDWI ranges from −1 to +1. The higher value indicates the presence of water bodies whereas the lower value suggests deficiency of water contention based on a certain threshold value [28]. The LANDSAT 8 OLI data has been used to apply the MNDWI method in the ArcGis environment. Based on the natural break method, five classes of MNDWI have been made (Fig. 15.3): very low (−0.74 to −0.21), low (−0.20 to −0.17), moderate (−0.16 to −0.12), high (−0.11 to −0.03), and very high (−0.02 to 0.27). The high and very high zones of MNDWI (>10%) are most prone to the waterlogging condition for longer period of time.

15.4.1.8 Soil

The soil types of Uttar Dinajpur district is mainly of fluvial origin. The region falls under the recent alluvial deposit zone and categorized under the USDA classification system [16]. Five major types of soil are found in that region: *Ustic* soil (5.90%) which is essentially dry in nature but contains enough water to support plant growth, *inceptisols* (23.88%) comprising fine and loamy structure with low content of clay and humus accumulation, *entisols* (45.12%) of fine and loamy structure and of more recent origin, *fluventic soil* (13.24%) that is found along the river banks and contains low organic matter, *and Ochric* (11.86%), poor in organic carbon. The inceptisols and entisols are most prevalent among the Chopra, Karandighi, Goalpokhar-I, Raiganj, Kaliyaganj, and Itahar blocks. The soil of Nagar and Kulik River (*inceptisols*) are very much efficient in terms of draining capabilities, whereas the *entisols* comprising in the Kaliyaganj, Itahar, Goalpokhar-I and II and parts of Chopra block is imperfectly drained and very much potential to the waterlogging situations (Fig. 15.3).

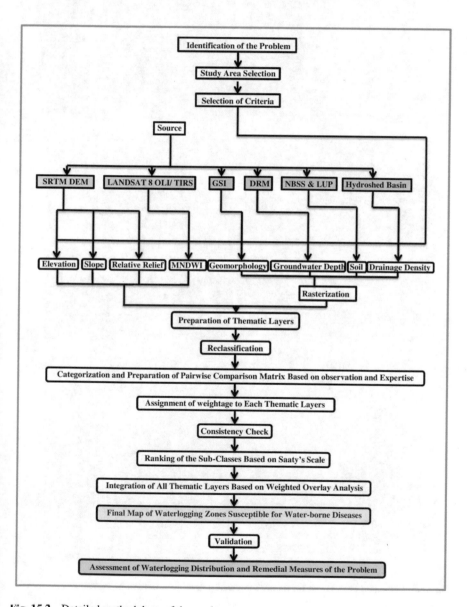

Fig. 15.2 Detailed methodology of the work

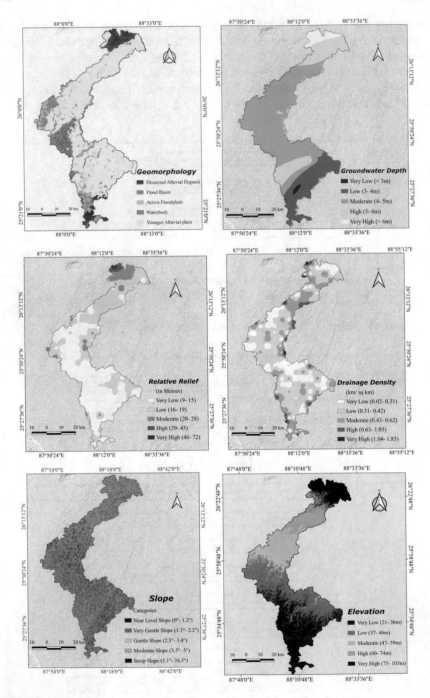

Fig. 15.3 Thematic layers used for identification of potential waterlogging zones

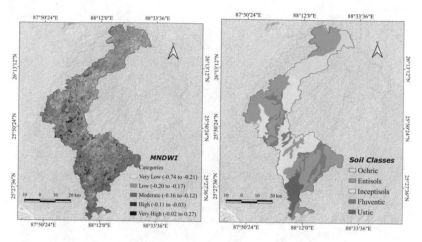

Fig. 15.3 (continued)

15.4.2 Estimation of Waterlogged Zones in Uttar Dinajpur District

The final map of potential waterlogged zones of the concerned area is produced by composing the selected parameters based on their normalized weights through the weighted overlay analysis. Further, the potential waterlogged zones have been categorized into five classes based on the natural breaks method (Fig. 15.4). The very high potential waterlogged zones (66.69 km^2; 2.24%) are the permanent wetlands of the district where water remains in those depressions throughout the year. The major extent of the district falls in the high potential waterlogged zones (1835.28 km^2; 61.60%) which get exposed during the peak monsoons in the blocks of Raiganj, Kaliyaganj, Itahar, Karandighi, and Goalpokhar-II. The moderate zones (1036.38 km^2; 34.79%) are found in major parts of Islampur, Chopra, Goalpokhar-I, and Hemtabad block. The low potential zones (40.74 km^2; 1.37%) are in the highlands of Chopra block and least susceptible to the waterlogged situations. The very low potential waterlogged zone comprises only 0.06 km^2 of area and has no chances of waterlogging (Table 15.6).

15.4.3 Result Validation

The potential waterlogging zone map has been validated based on the geo-hazard map of the district (1995, unpublished report of GSI) available on the District Resource Map (DRM) of Uttar Dinajpur district (Fig. 15.5). From the report, it has been found that more than four-fifth of the area comes under the potential flood zone. The same zone has been demarcated into three different parts (very high, high, and moderate

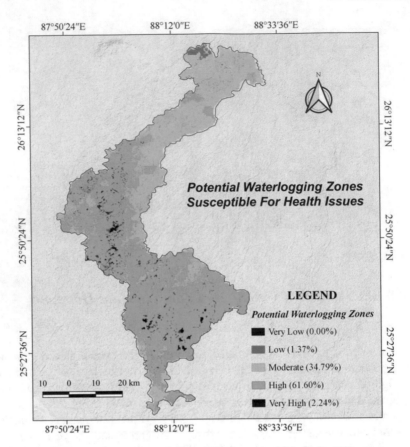

Fig. 15.4 Potential waterlogging zones of Uttar Dinajpur susceptible for health issues

Table 15.6 Distribution and spatial extension of potential waterlogging zone (PWZ) susceptible for water-borne disease in Uttar Dinajpur district

Potential waterlogging zone (PWZ)	Area (km²)	Area (%)
Very high	66.69	2.24
High	1835.28	61.60
Moderate	1036.38	34.79
Low	40.74	1.37
Very low	0.06	0.00

potential waterlogging zones) based on the AHP-MCDA model. The permanent waterlogging zones (very high zones) have been found in the very similar area to the marshy lands and waterlogged areas of the supported map. Therefore, it can be preferred that the concerned model has accurately made the prediction of potential waterlogging zones susceptible for the vector-borne diseases in the Uttar Dinajpur district.

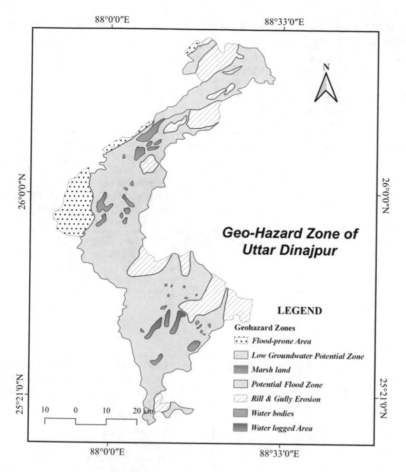

Fig. 15.5 Geo-hazard map of Uttar Dinajpur district (1995, unpublished report of GSI)

It has been a comprehensive AHP-MCDA-based micro level approach to identify the probable zones of spreading of waterlogging diseases in the Uttar Dinajpur
district (Table 15.7). However, the zone identification has been done mainly from the
perspective of physical setup of the region. Along with the geographical location,
the socioeconomic infrastructure of a society also plays important role in terms of
affecting the population with those serious epidemic diseases due to waterlogging.
Proximity to the hospital or health centers, societal awareness and economic condition of the individual also affect the quality of health and its deterioration of the
flood-affected people. Since the flooding causes the outbreaks of infectious diseases,
the identified Very High and High potential waterlogging zone need immediate attention to lessen the effect of flood consequences on the physical and mental health of
the population.

Table 15.7 Village-wise assessment of potential waterlogging zones (PWZ) to mitigate the spreading of water-borne diseases

Blocks	Potential waterlogging zones (PWZ) for spreading of water-borne disease	Area (km²)	Area (%)	Villages	Remarks
Itahar	Very high	12.00	23.05	Sibrampur, Patirajpur, Tilna, Guniakhanda, Ghughudanda, Mahamadpur, Bhagnail, Keotal, Sonapur, and Belul	Most vulnerable as the beels or marshes remain waterlogged more or less throughout the year
	High	248.34	73.40	Rest of the villages except the included villages of moderate and high PWZ	The spread of waterlogged diseases in this zone have high possibility mainly due to annual floods and associated inundation
	Moderate	77.99	3.55	Chhilampur, Nahanipur, Baidara, kaliganj, Thilbil, Khuniabari, Ghritatala, Tharais, Patinahar Parergram, and the villages of the southern part	Least vulnerable than the High and Very High PWZ
Kaliyaganj	Very high	6.86	2.27	Palihar, Bhuinhara, Chaipara, Maheshpur, PurbaGoalgaon, and some parts of Tarangapur and Mustafanagar	Waterlogging condition prevails throughout the year in those large marshes and *dighis*
	High	287.31	95.17	Rest of the villages except the villages of Very High and Moderate PWZ	Chances of spreading of waterlogged diseases remains during June to November

(continued)

Table 15.7 (continued)

Blocks	Potential waterlogging zones (PWZ) for spreading of water-borne disease	Area (km^2)	Area (%)	Villages	Remarks
	Moderate	7.72	2.56	Shursa, Chapair, Mirzzagar, and some parts of Bhelai	Vulnerable only during the major floods (e.g., 2017)
Hemtabad	Very high	0.59	0.32	Teghara, Binagram, Darimanpur, Pirojpur, Krishnabati, Ghagra, and Agapur	Very much reluctant to the waterlogged diseases due to perennial marshes and beels
	High	126.73	69.33	Rest of the Villages except the villages of Very High and Moderate PWZ	Vulnerable during the days of inundation of agricultural zones and built-up zones
	Moderate	55.48	30.35	Sunair, Malon, Bamair, Bishnupur, Kalua, Roshanpur, Turibar, Bhogram, and parts of Santara	Least chances of waterlogged due to local height of the region
Raiganj	Very high	11.91	2.51	Tenahari, Hathia, Mahishnagar, Jaynagar, Taherpur, Gaitar, Lakshmania, Jagadishpur, Maslandpur, and Malibari	Most vulnerable parts of the Raiganj block due to abundance of waterlogged zones
	High	419.71	88.56	Other parts of the block excluding the villages of Very High and Moderate PWZ	Mitigation measures need to be taken during the monsoon inundation
	Moderate	42.34	8.93	Adiar, Kantar, Kailadangi, Bindol, and Kaliadangi	Comparatively low-risk zone

(continued)

Table 15.7 (continued)

Blocks	Potential waterlogging zones (PWZ) for spreading of water-borne disease	Area (km²)	Area (%)	Villages	Remarks
Karandighi	Very high	20.51	5.40	Harasanda, Patanduba, Belua, Andharia, Bajargaon, Lahutara, Sabdhan, Dalmadh, Keshopur, Kararan, Baliamora, Dakshin Jadopur, Alampur, Rabanpur, and Sadipur	Most susceptible portion of waterlogging zone in the district
	High	289.97	76.40	Other villages excluding the included villages of the block	Vulnerable only during the annual floods
	Moderate	68.88	18.15	The villages covering the north-eastern and south-western corner of the block	Low damage concern
	Low	0.20	0.05	Pauti	Negligible risk
Goalpokhar-I	Very high	2.22	0.65	Jhiltalab, Dharampur, Churakuti, Pariharpur, and Raypur	Mitigation measures to be strictly followed
	High	157.67	46.22	Rest of the villages comprising the north-western and south-western part	Vulnerable due to the presence of numerous cutoffs and marshy lands
	Moderate	181.14	53.10	All the villages of north-eastern and south-eastern parts	Low chances of health hazard in terms of water-borne diseases
	Low	0.08	0.02	Goalin	Negligible risk

(continued)

Table 15.7 (continued)

Blocks	Potential waterlogging zones (PWZ) for spreading of water-borne disease	Area (km²)	Area (%)	Villages	Remarks
Goalpokhar-II	Very high	12.32	4.15	Arani, Gobindapur, Gandal, Dakshin Rampur, Dahuabari, Gochhra, Roshanganj, Kahalgaon, Udga, and Kaili	Most affiliated to waterlogging conditions throughout the year
	High	238.19	80.16	Rest of the villages except the Very high and moderate PWZ	Mitigation concern during monsoon
	Moderate	46.62	15.69	Bhelagachhi, Gerua, Bara Shikarpur, and the villages of middle and south-western corner	Low chances of waterlogging
Islampur	Very high	0.28	0.09	Kachna	Most reluctant and comes under the zone of spreading of waterlogged diseases
	High	47.84	14.92	Gaisal, Alipur, Rampur, Kuchlia, Goalgachh, Phulhara, Manikpur, and Bhagmahima village	Chances of inundation only during the annual flood events
	Moderate	272.47	84.99	Rest of the villages except the Very High and High PWZ	Prone to such waterlogged condition at mega flood events
	Low	0.02	0.01		Negligible risk
Chopra	High	19.52	5.68	Damodarkhuri, ChhotaDopaharu, Bara Panchmauz, and ChhotaMircha	Chances of inundation most probably at the times of flood

(continued)

Table 15.7 (continued)

Blocks	Potential waterlogging zones (PWZ) for spreading of water-borne disease	Area (km^2)	Area (%)	Villages	Remarks
	Moderate	283.75	82.54	Rest of the villages of the Chopra block except the including villages in other spatial zones	Moderate chance due to higher elevation of the region
	Low	40.44	11.76	Chitalghata, Jhajhri, Borobila, Bara Dangapara, and DangraDangri	Minimal chances of spreading of inundation-based waterlogged diseases
	Very low	0.06	0.02	Kuimari	Very negligible chance

15.5 Mitigation Measures

Mitigation refers to the act of reducing the adverse impact of particular events on health, society, or environment. However, the health consequences of flood can be categorized into two parts: direct and indirect [10]. The direct consequences are related to drowning, chemical contamination, hypothermia, etc. whereas the indirect consequences comprise infectious diseases, malnutrition, and poverty-related problems. Since the effects of natural event cannot be totally alleviated, rather could be minimized through enhanced disaster management plans. The mitigation measures are as follows:

- Improved disaster management plans for comprehensive surveillance of the flood-cum waterlogging zones.
- Establishment of early warning system prior to such waterlogging situations.
- Development of cell at all possible organization level to cope with the health hazard.
- Social awareness program should be promoted regarding the spread of infectious diseases, snake biting, etc., among the inhabitants and livestock animals.
- All possible medical infrastructures related to water-borne diseases, arrangement of anti-venoms should be there at village level for the endangered waterlogging zones.
- Furthermore, there should be arrangements for flood relief camp and emergency health treatment.

15.6 Conclusion

Flooding causes incredible hike in the number of cases of water and vector-borne diseases in the concerned study area. It was a necessity to identify the potential water-logging zones and associated settlements prone to such epidemic diseases. Therefore, an attempt has been made to categorize the district according to the susceptibility of those diseases using the analytical hierarchy process and GIS technique.

Eight geo-hydrological parameters (such as elevation, slope, relative relief, geomorphology, soil, groundwater depth, drainage density, and MNDWI) were selected to prepare the probable waterlogging areas. The study demonstrates that about 2.24% of total area comes under the very high zone, where waterlogging has been the most profound. The majority of the area belongs to the high waterlogging potential zone (61.60%) covering the southward portion of the district due to the abundance of cutoffs and marshes. The moderate waterlogging zone comprises about 35% of total area found mostly on the northern portion of the district due to the higher elevation. Furthermore, the validation of the work has been carried out by comparing the potential waterlogged zones with the geo-hazard (GSI report) map of the district. However, the accuracy of the present work would be increased if the socioeconomic status and infrastructure are incorporated as parameters.

The integration of AHP and GIS technique has delivered reliable result of potential waterlogging zones susceptible for water-borne diseases in the Uttar Dinajpur district and can be extensively used in other areas with such similar intensions. The identification of such vulnerable zones is important for the policymakers and local authorities to take necessary steps to mitigate the damages.

Acknowledgements First and foremost, we are extremely thankful to Prof. Dr. Nageshwar Prasad, Formerly Professor of Geography, University of Burdwan, for his guidance to the Research work. Furthermore, the authors would like to acknowledge NASA for DEM and LANDSAT images, GSI for geomorphology and groundwater depth data, and NBSS and LUP (Kolkata) for providing the soil data. Without those datasets, the research work could not be possible.

Funding The authors did not receive financial support from any kind of institution regarding the research work.

Declarations

Conflict of Interest Authors state that they have no known conflict of interest.

References

1. Alderman, K., Turner, L. R., & Tong, S. (2012). Floods and human health: A systematic review. *Environment International, 47*, 37–47. https://doi.org/10.1016/j.envint.2012.06.003
2. Allafta, H., & Opp, C. (2021). GIS-based multi-criteria analysis for flood prone areas mapping in the trans-boundary Shatt Al-Arab basin, Iraq-Iran. *Geomatics, Natural Hazards and Risk, 12*(1), 2087–2116. https://doi.org/10.1080/19475705.2021.1955755

3. Anandabazar. (2022). *Outbreak of dengue in North Dinajpur, Some are hospitalised,* 1–3. https://www.anandabazar.com/west-bengal/north-bengal/outbreak-of-dengue-in-north-din ajpur-some-are-hospitalised-dgtld/cid/1310479

4. Bich, T. H., Quang, L. N., Thi, L., Ha, T., Thi, T., Hanh, D., & Guha-sapir, D. (2011). Impacts of flood on health: Epidemiologic evidence from Hanoi. *Vietnam, 1,* 1–8. https://doi.org/10.3402/gha.v4i0.6356

5. Chakraborty, T., & Ghosh, P. (2010). The geomorphology and sedimentology of the Tista megafan, Darjeeling Himalaya: Implications for megafan building processes. *Geomorphology, 115*(3–4), 252–266. https://doi.org/10.1016/j.geomorph.2009.06.035

6. Dash, P., & Sar, J. (2020). Identification and validation of potential flood hazard area using GIS-based multi-criteria analysis and satellite data-derived water index. *Journal of Flood Risk Management, 13*(3), 1–14. https://doi.org/10.1111/jfr3.12620

7. De Brito, M. M., & Evers, M. (2016). Multi-criteria decision-making for flood risk management: A survey of the current state of the art. *Natural Hazards and Earth System Sciences, 16*(4), 1019–1033. https://doi.org/10.5194/nhess-16-1019-2016

8. District Census Handbook. (2011). *Census of India 2011 West Bengal Series-20 Part XII-B District Census Handbook Uttar Dinajpur.* https://censusindia.gov.in/2011census/dchb/1904_PART_B_DCHB_UTTARDINAJPUR.pdf

9. District Disaster Management Plans of Uttar Dinajpur District 2020–2021. (2021). http://wbdmd.gov.in/pages/district_dm_plan.aspx

10. Du, W., Fitzgerald, G. J., & Clark, M. (2010). *Health impacts of floods.*

11. Strong, F. W. (1912). *Eastern Bengal District Gazetteers : Dinajpur.* The Pioneer Press.

12. Few, R., Ahern, M., Matthies, F., Kovats, S., Few, R., & Ahern, M. (2004). *Floods, health and climate change : A strategic review.*

13. Hammami, S., Zouhri, L., Souissi, D., Souei, A., Zghibi, A., Marzougui, A., & Dlala, M. (2019). Application of the GIS based multi-criteria decision analysis and analytical hierarchy process (AHP) in the flood susceptibility mapping (Tunisia). *Arabian Journal of Geosciences, 12*(21). https://doi.org/10.1007/s12517-019-4754-9

14. Health Effects of Climate Change in the UK. (n.d.).

15. Hindu, T. (2021). *6 children dead, hundreds in hospital in Bengal,* 1–3. https://www.thehindu.com/news/national/other-states/hundreds-of-children-hospitalised-for-viral-infection-in-ben gal/article36502718.ece

16. Human Development Report. (2010).

17. Jenks, G. F. (1967). *The data model concept in statistical mapping.*

18. Kuldeep, Garg, P. K., & Garg, R. D. (2016). Geospatial techniques for flood inundation mapping. *International Geoscience and Remote Sensing Symposium (IGARSS), 2016-Novem (July),* 4387–4390. https://doi.org/10.1109/IGARSS.2016.7730143

19. Lee, A. H. I., Chen, W. C., & Chang, C. J. (2008). A fuzzy AHP and BSC approach for evaluating performance of IT department in the manufacturing industry in Taiwan. *Expert Systems with Applications, 34*(1), 96–107. https://doi.org/10.1016/j.eswa.2006.08.022

20. News, A. (2021). Several babies & children admitted to hospitals with fever, flu like symptoms, respiratory distress in West Bengal. 2021. https://newsonair.com/2021/09/15/several-babies-children-admitted-to-hospitals-with-fever-flu-like-symptoms-respiratory-distress-in-west-ben gal/

21. Okaka, F. O., & Odhiambo, B. D. O. (2018). *Relationship between flooding and out break of infectious diseasesin Kenya : A review of the literature.*

22. Papaioannou, G., Vasiliades, L., & Loukas, A. (2015). Multi-criteria analysis framework for potential flood prone areas mapping. *Water Resources Management, 29*(2), 399–418. https://doi.org/10.1007/s11269-014-0817-6

23. Roy, S., Bose, A., & Mandal, G. (2021). Modeling and mapping geospatial distribution of groundwater potential zones in Darjeeling Himalayan region of India using analytical hierarchy process and GIS technique. *Modeling Earth Systems and Environment, 0123456789.*https://doi.org/10.1007/s40808-021-01174-9

24. Rudra, K. (2018). *Rivers of the Ganga Brahmaputra Meghna Delta: A fluvial account of Bengal*.https://doi.org/10.1007/978-3-319-76544-0_8
25. Ruji, E. M. (2007). Floodplain inundation simulation using 2d hydrodynamic modelling approach. *Simulation*.
26. Saaty, T. L. (1980). *The analytic hierarchy process : Planning, priority setting, resource allocation*. McGraw-Hill International Book Co.
27. Saaty, T. L. (1990). How to make a decision: The analytic hierarchy process. *European Journal of Operational Research, 48*(1), 9–26. https://doi.org/10.1016/0377-2217(90)90057-I
28. Sar, N., Chatterjee, S., & Das Adhikari, M. (2015). Integrated remote sensing and GIS based spatial modelling through analytical hierarchy process (AHP) for water logging hazard, vulnerability and risk assessment in Keleghai River Basin, India. *Modeling Earth Systems and Environment, 1*(4), 1–21.https://doi.org/10.1007/s40808-015-0039-9
29. Souissi, D., Zouhri, L., Hammami, S., Msaddek, M. H., Zghibi, A., & Dlala, M. (2020). GIS-based MCDM–AHP modeling for flood susceptibility mapping of arid areas, southeastern Tunisia. *Geocarto International, 35*(9), 991–1017. https://doi.org/10.1080/10106049.2019.156 6405
30. Taylor, P., Brown, L., Murray, V., Brown, L., & Murray, V. (2013). Examining the relationship between infectious diseases and flooding in Europe: A systematic literature review and summary of possible public health interventions, 37–41. https://doi.org/10.4161/dish.25216
31. Thakur, B., & Singh, D. Sen. (2018). *The Indian Rivers*, 165–175. https://doi.org/10.1007/978-981-10-2984-4
32. Jha, V., & Bairagya, H. (2011). Environmental impact of flood and their sustainable management in deltaic region of west bengal, India. *Caminhos de Geografia*, 283–296.
33. Vijith, H., & Dodge-Wan, D. (2019). Modelling terrain erosion susceptibility of logged and regenerated forested region in northern Borneo through the Analytical Hierarchy Process (AHP) and GIS techniques. *Geoenvironmental Disasters, 6*(1). https://doi.org/10.1186/s40677-019-0124-x
34. West Bengal Disaster Management. (2021). http://wbdmd.gov.in/pages/flood2.aspx
35. WHO. (2000). Flooding and communicable diseases fact sheet Risk assessment and preventive measures. *Most*, March, 1–9.
36. Xu, H. (2006). Modification of normalised difference water index (NDWI) to enhance open water features in remotely sensed imagery. *International Journal of Remote Sensing, 27*(14), 3025–3033. https://doi.org/10.1080/01431160600589179

Chapter 16
An Automated Geoprocessing Model for Accuracy Assessment in Various Interpolation Methods for Groundwater Quality

Baskaran Venkatesh and **M. A. M. Mannar Thippu Sulthan**

Abstract The qualities of different groundwater parameters rely on the sources of groundwater recharge, industrial effluent, urban interactions, agriculture, aquifer pumping, and waste disposal. Domination of any of these factors alters the groundwater quality. GIS enables the tracking of changes throughout time within an area/watershed and correlating changes in water quality with these alterations. Choosing an appropriate GIS-spatial interpolation technique is a critical success element in surface analysis since different interpolation methods might result in different surfaces and, eventually, different outcomes. Though no particular interpolation technique is completely optimum, the best interpolation method for a given circumstance can only be determined by comparing their results. An automated tool has been developed to assess popular interpolation techniques like inverse distance weightage, radial basis function, ordinary kriging, universal kriging, empirical Bayesian kriging, and kernel interpolation in terms of root mean square error and mean absolute error. The model provides researchers and official groundwater quality monitoring and assessment teams with the help to pick the optimal algorithm.

Keywords Groundwater quality · GIS · Spatial interpolation · Geoprocessing model · Automated tool · Accuracy assessment

B. Venkatesh (✉) · M. A. M. Mannar Thippu Sulthan
Department of Civil Engineering, Government College of Engineering, Tirunelveli 627007, India
e-mail: venkatesh@gcetly.ac.in

16.1 Introduction

The total freshwater available on earth is less than 3%, in which 2.15% of water is trapped in the polar region as a glacier, which is not usable for human consumption. In the earth water resources, less than one percent of water is used for domestic purposes in which 0.32% of water is present in deep below as groundwater, and 0.33% of water is readily available as surface water like in a river, lake, and pond. Groundwater is widely used for domestic, irrigational, and industrial purposes [1, 2]. Urban and industrial activities disturb groundwater's natural infiltration by interventions the surface water movement, which results in sub-standard quality. Groundwater's chemical composition decides the suitability for necessary function [3]. The contamination leads to a decrease in the groundwater quality, which results in potential health problems, avoidance of water withdrawal, and higher remedial costs [4].

Groundwater quality can be improved by monitoring and controlling the external factors which cause risk to it [5]. A prompt monitoring program may identify possible restoration work and determine the harmful external factors [6, 7]. With the interpolation technique in geographic information system (GIS), monitoring and controlling are straightforward to achieve [8]. GIS helps in hydrologic and hydrogeologic-based decision-making through mapping and analyzing appropriate spatial data [9].

Interpolation is the technique to estimate unknown data from known data. These techniques were initially used in mining and further extended to all fields [10]. Spatial interpolation is a scientific way to tell about groundwater's status around the extent, where samples were collected [11, 12]. Deterministic and geostatistical methods are the two interpolation methods based on smoothness and certainty [13, 14]. Generally, geostatistical methods perform better than deterministic methods [15].

Geostatistical interpolation from the GIS technique provides spatial variation within the prescribed environmental setting [16]. Hoover et al. [17] used the GIS technology for dynamic visualization of interpreted water quality detail for unregulated drinking water.

The GIS-based statistical study is efficient in groundwater quality assessment compared to the conventional method. It is also very essential to estimate the accuracy of various geostatistical modeling techniques [18]. The selection of a valid interpolation method is considered a necessary factor for producing an environmental-based spatial variability map [19, 20]. Developing an advanced statistical strategy for spatial analysis in a GIS platform would be challenging [9].

Global polynomial interpolation, Bayesian interpolation, kernel interpolation, radial basis function interpolation, inverse distance weighted interpolation, simple kriging interpolation, ordinary kriging interpolation, and tension spline interpolation are popular geostatistical interpolation methods [21]. The performance of each interpolation technique can be evaluated by cross-validations methods [22] like root mean square, mean absolute error, and mean bias error [18, 23–25].

The geoprocessing model helps in automatic spatial analysis with better data management. ModelBuilder in the GIS platform is a drag-and-drop tool for automating and reusing workflows. In ModelBuilders, the geoprocessing tools are sequenced to set the predecessor output as an input for a successor [26]. This extensive decision support tool solves a knotty problem in a pretty simple way. The framed model can be customized and run multiple times with a single click on the mouse [27].

16.2 Methods

16.2.1 Geostatistical Interpolation

Interpolation is preferred when data availability is limited and operates based on the principle of autocorrection by assuming the objects nearby behave similarly to the thing apart [28]. Geostatistical techniques apply mathematical and stochastic models which perform autocorrelation, while deterministic techniques apply only mathematical models [29]. As the number of steps needs to be followed while executing the different interpolation methods cross-validation, creating a model makes the process easier with performing automatic tasks in a defined sequence [30]. The research work reported using geostatistical modeling techniques in water quality assessment has clinched to the positive trend line since 2000. The effectiveness of the study on modeling the water quality should be measured [31]. As we cannot measure everything to assess the suitability of the techniques, the model with minimal predictive error should be given preference for that particular dataset [32, 33]. Various authors have identified different modeling techniques as the best ones based on cross-validation. The predictive performance depends on sampling design, sample distribution, data quality, and all [34, 35]. The workflow for this automated tool is briefed in Fig. 16.1.

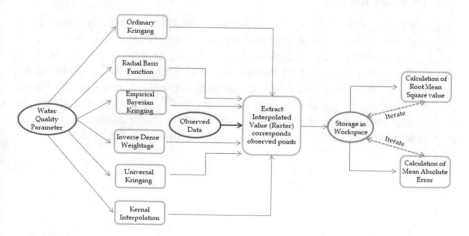

Fig. 16.1 Workflow of automated tool for assessment of different interpolation method

(a) Inverse Distance Weightage

This method creates unknown attribute values by considering certain inverse weightage to known attribute values [36, 37]. The attribute value that is only available at sampled location creates a spatial surface by connecting determined values. The influence of sampled location decreases with greater distance [38].

$$z = \sum_{i=1}^{n} \frac{Z_i}{(d_i)^p} / \sum_{i=1}^{n} \frac{1}{(d_i)^p} \tag{16.1}$$

where

Z the estimated value of the unknown prediction point
Z_i the measured value of know sample point
d_i the distance between prediction point and sample point
p the power weightage parameter, and
n the number of sample points.

(b) Radial Basis Function

RBFs are preferred for a large number of data points to create smooth surfaces. The radial basis function (RBF) predicts values from the generated surface, passing through all the measured sample points. This function minimizing the total curvature of the surface from measured values results in gentle variation in surface [19]. The predicted value shall either be greater than the maximum or lesser than the minimum measured value. This geostatistical method contains five different basis functions: thin-plate spline, spline with tension, completely regularized spline, multi-quadric spline, and inverse multi-quadric spline.

(c) Ordinary Kriging

Ordinary kriging (OK) is a popular kriging model used for the study of water quality mapping. The estimated neighborhood location through variogram estimates the value around any point in the region [39]. Assuming constant mean as unknown, this kriging model is considered flexible and straightforward. OK is preferred for its exactness, where interpolated values or local averages coexist with the known sampled locations [40]. It is an advantage for integrating diverse predicted datasets and used for data with decent trends. Accommodation of modification and changes in the mean value of the surface is possible [41]. OK assumes the model,

$$Z(s) = \mu + \varepsilon(s) \tag{16.2}$$

where

$Z(s)$ the variable of interest
μ unknown constant mean
$\varepsilon(s)$ an error with constant mean.

(d) Universal Kriging

The universal kriging model is a generalized version of ordinary kriging [42]. Estimation and prediction are two-step procedures indulged here [43], and it uses either semivariograms or covariances forms for autocorrection and measurement of error [44]. Here, the error $\varepsilon(s)$ is random and modeled as autocorrelated. The model assumed here is based on a deterministic function and a variation (error) as

$$Z(s) = \mu(s) + \varepsilon(s) \tag{16.3}$$

where

$Z(s)$ the variable of interest
$\mu(s)$ some deterministic function
$\varepsilon(s)$ random variation or error.

(e) Kernel Interpolation

Kernel interpolation is a first-order local polynomial interpolation variant in which instability in the calculations is avoided by utilizing an approach similar to that used to estimate the regression coefficients in ridge regression. By introducing a slight degree of bias to the equations, the ridge parameter in the kernel smoothing model corrects the problem of substantial prediction standard errors and dubious predictions [45]. For the output surface type, this interpolation only supports prediction and prediction standard error. The shortest distance between points is used in this model to connect locations on both sides of the given nontransparent (absolute) barrier with a sequence of straight lines.

(f) Empirical Bayesian Kriging

EBK is one of the reliable automatic interpolator models classified under geospatial interpolation. It differs from classical kriging methods because the semivariogram model estimation reduces the introduced error [46]. The process is attained by considering various iterative semivariogram models rather than a single semivariogram model. This model is preferred for its fast and reliability in a large set of spatial data interpolation [47, 48]. By not considering uncertain semivariogram estimates, this kriging does not underestimate prediction's standard error.

16.2.2 Qualitative Assessment

The various spatial interpolation techniques use different models and assumptions. The selection of appropriate interpolation depends on accuracy, applicability to large datasets, robustness, and flexibility in defining multiple phenomena, smoothening noisy data, running time, and ease of use [49]. Currently, selecting the suitable interpolation method by considering all these attributes is not an easier task. The selection of inaccurate methods and unsuitable parameters ends with a distorted spatial inter-polation model, leading to wrong decisions from this misleading spatial information [19]. The observed and estimated values for different interpolation techniques are compared for various water quality parameters. There are various methods to calcu-late interpolation error. We have selected root mean square error (RMSE) and mean absolute error (MAE) for this study. Both express average model prediction errors and are widely reported in environmental and climate change literature [50]. The inconsistency between RMSE and MAE shall be due to differing error variance of associated sets [51]. RMSE is sensitive to extreme outliers than MAE. Because of its simplicity in the calculation, these two methods are often used to assess the model's performance [18, 52, 53]. RMSE and MAE are calculated as follows:

$$\text{RMSE} = \sqrt{\frac{1}{n} \sum_{i=1}^{n} (Z_i - E_i)} \tag{16.4}$$

$$\text{MAE} = \frac{\sum_{i=1}^{n} |(Z_i - E_i)|}{n} \tag{16.5}$$

Where Z_i and E_i are the observed and estimated value of any water quality parameter, respectively, at a point i within our study area.

16.3 Result and Discussion

16.3.1 Automated Geoprocessing Model

A model tool can be built with the set tools arranged in sequence and connected adequately in Arcmap ModelBuilder. The automated tool is accomplished by blending a large processing workflow in ModelBuilder that includes various tools, variables, parameters, and preconditions. This setup automates the tool's perfor-mance as per the structure and ends with a single raster or vector file as workflow's output. The ModelBuilder also allows to create, edit, and manage the models [54]. It again allows the user to modify the model as per the reforms in the objective. Gener-ally, the elements in the model are classified into input data, geoprocessing tool, and output data. The shape and color of elements display their functional uniqueness, as shown in Fig. 16.2.

Input (blue-oval shape) is generally geographic data in either raster or vector format before the model is executed. Tools (yellow-rectangular) are the tasks to be performed on the input. A process would be one of the various ready-to-use system tools offered in ArcGIS, which can be easily grabbed from the ArcToolbox and positioned onto the ModelBuilder interface. Output (green–oval shape) is data created when the model runs by the tools or scripts, which does not exist until the

Fig. 16.2 Diagram property and symbology of model elements

process is over [55]. They can be used in another process once they have been created. Raster, vector, and non-geographical values are the possible derived elements created while running a tool [56]. The geoprocessing tools shall be connected in parallel or series, but there should be proper input data to validate the model. Since manually selecting the inputs for the number of tools in the process workflow consumes solid time, the first advantage of any automated geoprocessing model is time-saving [57].

16.3.2 Interpolation Accuracy Assessment Model

The model begins by interpolating the acquired data on water quality. The input attribute table of collected data on water quality must include specifics such as parameter names and their values. By making the parameter name the head of the associated column, the tool's user can select which parameter to examine for accuracy. All six interpolation algorithms mentioned here are parallelized to receive water quality parameter data as a common input vector layer. All of these processing tools produce a raster dataset as their output. The user can specify any folder or geodatabase as the destination for all derived data. The required environmental settings for the entire model, such as processing extent, output coordinates, cell size, mask layer, and compression, can be overridden by the user via the model properties menu bar.

The final goal of this automated model is to determine the accuracy of a widely adopted geostatistical interpolation technique in water quality assessments by comparing the predicted value to observed data within the study area [58]. The observed data should be spatially consistent but contained inside the research area. An additional two empty columns with the names MAE and RMSE should be added to the attribute table of this vector layer to minimize the model steps as shown in Fig. 16.3. The next step is to extract predicted water quality data from an interpolated raster surface at the exact location, where the observation was made.

Specific outputs generated during model processing are required to generate the final output only and are of no value once the model is complete. Such outputs are marked as intermediate data [27]. All six resulting extracted vector datasets will be saved as intermediate data in a specified location. Since MAE and RMSE require only the difference between predicted and actual values, almost all the modeling work is done.

Following that, the model does a basic mathematical computation using MAE and RMSE formulas. Compounded by the fact that Python lacks a direct mathematical

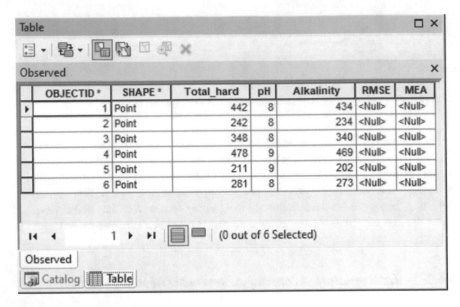

Fig. 16.3 Attribute table format for observed data

function for MAE and RMSE, the computing steps are a bit long. A single tool cannot do operations such as difference, square, square root, summation, and average in ArcMap. Calculate field and summary statistics are the tools that this model uses to execute simple calculations to furnish the required outcome. If these tools are linked with the output of all six extract values from the point tool as their input, the model may appear to be a cluster of similar tools. Iterating the procedure while evaluating the prediction error will result in a pleasing and straightforward model.

Iteration, commonly called looping, involves repeating a procedure with a certain degree of automation. It is significant since repeated task automation lowers the time and effort necessary to carry out the job. In each iteration, a process can be repeated over and over with changing settings or data. ModelBuilder also gives flexibility in iteration, as it is possible to repeatedly run an entire model or merely a specific tool or process [59]. In this model, the iterate feature class, which iterates across a workspace's or feature dataset's feature classes, is adopted.

Though the similar series of operations to compute MAE and RMSE is curtailed by the iterator tool, the model looks too lengthy, and modifying the settings as per user requirements will be difficult. The entire model is customized to reduce complexity. The first part of the automated tool till extraction of interpolated value to the observed point is modeled as a sub-model.

A sub-model, nested model, or model within another model is an approach that involves embedding and executing one model tool within another model. This model hierarchy is used for two major reasons: simplifying a large, complicated model and enabling more extensive model iterators usages. Whenever an error occurs in one sub-model, just that model should be corrected, and then the single model should be repeated rather than the complete process. The sub-model for the automated tool is shown in Fig. 16.4. The collect values tools are intended to collect iterator output values or combine a list of multi-values into a single input, which helps connect the sub-model to the main model.

Setting precondition in combining sub-model to the main model is mandatory for this automated tool. Preconditions can be used to regulate the sequence of operations in a model deliberately. By making the output of the first process a precondition for

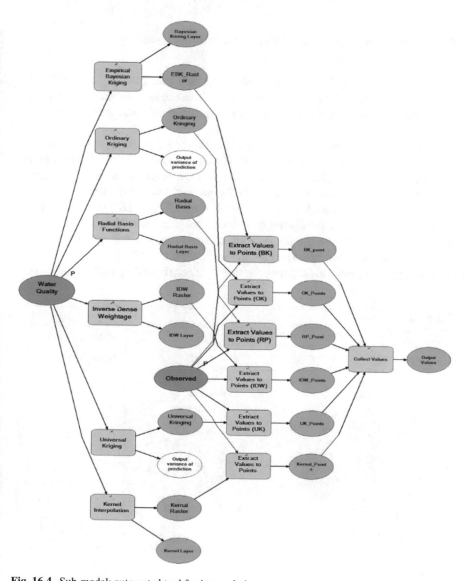

Fig. 16.4 Sub-model: automated tool for interpolation accuracy assessment

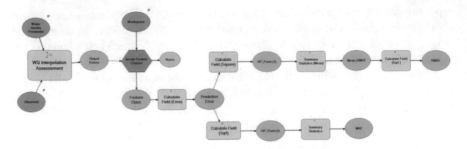

Fig. 16.5 Main model: automated tool for interpolation accuracy assessment

the second process, a process may be made to run only after another. Any variable can be used as a precondition for executing a tool, which can have several preconditions. The complete main model configured with sub-model preconditioning is shown in Fig. 16.5.

Every model variable can be converted into a model parameter. The value of a variable can be provided in the model tool dialog box when it is turned into a model parameter. The letter P displays next to the variable in the model once it is set as a parameter. Creating model parameters are preferred when the model user must specify variable values each time the model is run. Figure 16.6 depicts the automated tool's dialog box for assessing various interpolation techniques. Workspace or geodatabase, water quality parameter for interpolation, and observed data to assess six interpolation techniques are marked as model parameters for this model. The user shall select the appropriate file for all the three variable tabs from the tool dialog box.

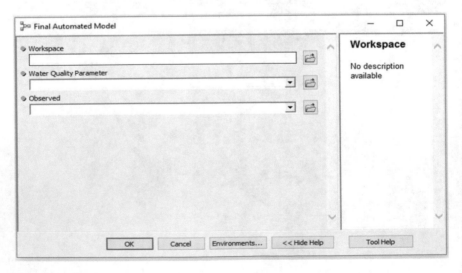

Fig. 16.6 Automated model tool dialog box

The user can run the entire model from the tool dialog box or the run button on the ModelBuilder edit screen. Once the model is finished, a running tab will alert the user that the model is completed, and this automated tool will notify them within a few minutes. If a user manually executes the equivalent set of operations with a clear objective, it will take more than 30 min. Again the performance depends on the processing efficiency of hardware. Remembering the number of tools, the sequence of tools to be used, selecting the correct file as input, storing the output in the desired location, and adjusting the environmental settings for all tools will be challenging in manual operation. After successfully running this automated tool, each six output tables of MAE and RMSE corresponds to six geostatistical interpolations as shown in Fig. 16.7.

A single-rowed table containing either the RMSE or MAE value of these six interpolations will be saved in the desired location as specified in the tool dialog box before executing the model. Combining all the RMSE and MAE values in a single table may make the model intricate and complicate. Finally, the table of any

Fig. 16.7 Output table of RMSE and MAE for six interpolations stored in a geodatabase

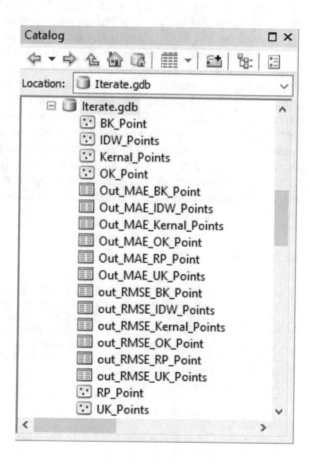

technique with the most negligible value of compounded error shall be preferred to study water quality parameters.

16.4 Conclusion

The model frameworks and applications outlined in this study demonstrate the potential benefits of utilizing GIS models in evaluating various interpolation mechanisms for water quality spatial mapping. These enable us to save time, develop reusable and shareable tools, depict workflows in an easy-to-understand diagram, and create sophisticated models as per our project objectives, all of which will help us save time and energy. The significant barriers were identified when a user prefers manual operation to accomplish the objectives of determining the best are spatial interpolation techniques. A few are performing workflow in the correct sequence, selecting valid input for all tools, modifying the storage directory for output of any tool, and repeatedly using the calculate field and summary statistics tool.

The recharge and discharge activities shall influence the quality of the groundwater parameter. Modeling any parameter of groundwater quality by taking account of typical external forces is a challenging task. It is challenging to model any groundwater quality parameter accurately. No technique will ever be able to match the genuine reality on the ground. Since each interpolation algorithm's assumptions and expressions are unique, the automated tool can help reduce uncertainty by recommending the best technique with the least prediction error for any water quality parameter. This automated tool helps select the suitable interpolation technique for any research related to the study on a parcel's water quality. The developed tool will allow policymakers and city planners working on water quality restoration and sustainable groundwater management to make evidence-based decisions promptly.

References

1. Asadi, E., Isazadeh, M., Samadianfard, S., Ramli, M. F., Mosavi, A., Nabipour, N., Shamshirband, S., Hajnal, E., & Chau, K. W. (2020). Groundwater quality assessment for sustainable drinking and irrigation. *Sustainability, 12*, 1–13. https://doi.org/10.3390/su12010177
2. Shigut, D. A., Liknew, G., Irge, D. D., & Ahmad, T. (2017). Assessment of physico-chemical quality of borehole and spring water sources supplied to Robe Town, Oromia region, Ethopia. *Applied Water Science, 7*, 155–164. https://doi.org/10.1007/s13201-016-0502-4
3. Babiker, I. S., & Mohamed, M. A. A. (2007). Assessing groundwater quality using GIS. *Water Resources Management, 21*, 699–715. https://doi.org/10.1007/s11269-006-9059-6
4. Nas, B., & Berktay, A. (2010). Groundwater quality mapping in urban groundwater using GIS. *Environmental Monitoring and Assessment, 160*, 215–227. https://doi.org/10.1007/s10661-008-0689-4
5. Mohamed, A.K., Liu, D., Mohamed, M.A.A., & Song, K. (2018). Groundwater quality assessment of the quaternary unconsolidated sedimentary basin near the Pi River using fuzzy evaluation technique. *Applied Water Science, 8*. https://doi.org/10.1007/s13201-018-0711-0.

6. Wiersma, G. B. (1990). Conceptual basis for environmental monitoring programs. *Toxicological and Environmental Chemistry, 27*, 241–249. https://doi.org/10.1080/02772249009357578

7. Babaei Semirom, F., Hassan, A. H., Torabia, A., Karbass, A. R., & Hosseinzadeh Lotf, F. (2011). Water quality index development using fuzzy logic: A case study of the Karoon River of Iran. *African J. Biotechnol., 10*, 10125–10133. https://doi.org/10.5897/ajb11.1608

8. Yao, L., Huo, Z., Feng, S., Mao, X., Kang, S., Chen, J., Xu, J., & Steenhuis, T. S. (2014). Evaluation of spatial interpolation methods for groundwater level in an arid inland oasis, northwest China. *Environment and Earth Science, 71*, 1911–1924. https://doi.org/10.1007/s12665-013-2595-5

9. Machiwal, D., Cloutier, V., Güler, C., & Kazakis, N. (2018). A review of GIS-integrated statistical techniques for groundwater quality evaluation and protection. *Environment and Earth Science*. https://doi.org/10.1007/s12665-018-7872-x

10. Zimmerman, D., Pavlik, C., Ruggles, A., & Armstrong, M. P. (1999). An experimental comparison of ordinary and universal kriging and inverse distance weighting. *Mathematical Geology, 31*, 375–390. https://doi.org/10.1023/A:1007586507433

11. Mitasova, L. M. A. H. (2005). Spatial interpolation. In *Geographic information systems: Principles, techniques, management and applications* (pp. 481–492). Springer International Publishing. https://doi.org/10.1007/s40808-017-0355-3

12. Chen, L., & Feng, Q. (2013). Geostatistical analysis of temporal and spatial variations in groundwater levels and quality in the Minqin oasis, Northwest China. *Environmental Earth Sciences, 70*, 1367–1378. https://doi.org/10.1007/s12665-013-2220-7

13. Shahmohammadi-Kalalagh, S., & Taran, F. (2020). Evaluation of the classical statistical, deterministic and geostatistical interpolation methods for estimating the groundwater level. *International Journal of Energy and Water Resources*. https://doi.org/10.1007/s42108-020-00094-1

14. Tanjung, M., Syahreza, S., & Rusdi, M. (2020). Comparison of interpolation methods based on geographic information system (GIS) in the spatial distribution of seawater intrusion. *Journal Natural, 20*, 24–30. https://doi.org/10.24815/jn.v20i2.16440

15. Ohmer, M., Liesch, T., Goeppert, N., & Goldscheider, N. (2017). On the optimal selection of interpolation methods for groundwater contouring: An example of propagation of uncertainty regarding inter-aquifer exchange. *Advances in Water Resources, 109*, 121–132. https://doi.org/10.1016/j.advwatres.2017.08.016

16. Javari, M. (2016). Comparison of interpolation methods for modeling spatial variations of Precipitation in Iran. *International Journal of Environmental and Science Education, 11*, 349–358. https://doi.org/10.12973/ijese.2016.322a

17. Hoover, J. H., Sutton, P. C., Anderson, S. J., & Keller, A. C. (2014). Designing and evaluating a groundwater quality Internet GIS. *Applied Geography, 53*, 55–65. https://doi.org/10.1016/j.apgeog.2014.06.005

18. Salekin, S., Burgess, J. H., Morgenroth, J., Mason, E. G., & Meason, D. F. (2018). A comparative study of three non-geostatistical methods for optimising digital elevation model interpolation. *ISPRS International Journal of Geo-Information, 7*, 1–15. https://doi.org/10.3390/ijgi7080300

19. Ikechukwu, M. N., Ebinne, E., Idorenyin, U., & Raphael, N. I. (2017). Accuracy assessment and comparative analysis of IDW, Spline and Kriging in spatial interpolation of landform (topography): An experimental study. *Journal of Geographic Information System, 09*, 354–371. https://doi.org/10.4236/jgis.2017.93022

20. Kyriakidis, P. C., & Goodchild, M. F. (2006). On the prediction error variance of three common spatial interpolation schemes. *International Journal of Geographical Information Science, 20*, 823–855. https://doi.org/10.1080/13658810600711279

21. ESRI (Environmental Systems Research Institute): Using ArcGIS geostatistical analyst. (2001)

22. Murphy, R. R., Curriero, F. C., & Ball, W. P. (2010). Comparison of spatial interpolation methods for water quality evaluation in the Chesapeake Bay. *Journal of Environmental Engineering, 136*, 160–171. https://doi.org/10.1061/(asce)ee.1943-7870.0000121

23. Fazeli Sangani, M., Namdar Khojasteh, D., & Owens, G.(2019). Dataset characteristics influence the performance of different interpolation methods for soil salinity spatial mapping. *Environmental Monitoring and Assessment, 191*. https://doi.org/10.1007/s10661-019-7844-y
24. Faraj, M., & Megrahi, E. L. (2019). Comparison of spatial methods to determining the best interpolation method for estimation of groundwater quality parameters (Elzawia City—Libya, A Case Study). In: *International Conference on Technical Sciences (ICST2019)* (pp. 04–06).
25. Arslan, H., & Ayyildiz Turan, N. (2015) Estimation of spatial distribution of heavy metals in groundwater using interpolation methods and multivariate statistical techniques; its suitability for drinking and irrigation purposes in the Middle Black Sea Region of Turkey. *Environmental Monitoring and Assessment, 187*. https://doi.org/10.1007/s10661-015-4725-x
26. AbdelRahman, M. A. E., & Tahoun, S. (2019). GIS model-builder based on comprehensive geostatistical approach to assess soil quality. *Remote Sensing Applications: Society and Environment, 13*, 204–214. https://doi.org/10.1016/j.rsase.2018.10.012
27. ESRI (Environmental Systems Research Institute): ModelBuilder for ArcView Spatial Analyst 2 (2000)
28. Setianto, A., & Triandini, T. (2015). Comparison of kriging and inverse distance weighted (Idw) interpolation methods in lineament extraction and analysis. *Journal of Applied Geology, 5*, 21–29. https://doi.org/10.22146/jag.7204
29. Wu, C. Y., Mossa, J., Mao, L., & Almulla, M. (2019). Comparison of different spatial interpolation methods for historical hydrographic data of the lowermost Mississippi River. *Annals of GIS, 25*, 133–151. https://doi.org/10.1080/19475683.2019.1588781
30. Nguyen, T. T., Ngo, H. H., Guo, W., Nguyen, H. Q., Luu, C., Dang, K. B., Liu, Y., & Zhang, X. (2020). New approach of water quantity vulnerability assessment using satellite images and GIS-based model: An application to a case study in Vietnam. *Science of the Total Environment, 737*, 139784. https://doi.org/10.1016/j.scitotenv.2020.139784
31. Shirazi, S. M., Imran, H. M., & Akib, S. (2012). GIS-based DRASTIC method for groundwater vulnerability assessment: A review. *Journal of Risk Research, 15*, 991–1011. https://doi.org/10.1080/13669877.2012.686053
32. Bennett, N. D., Croke, B. F. W., Guariso, G., Guillaume, J. H. A., Hamilton, S. H., Jakeman, A. J., Marsili-Libelli, S., Newham, L. T. H., Norton, J. P., Perrin, C., Pierce, S. A., Robson, B., Seppelt, R., Voinov, A. A., Fath, B. D., & Andreassian, V. (2013). Characterising performance of environmental models. *Environmental Modelling and Software, 40*, 1–20. https://doi.org/10.1016/j.envsoft.2012.09.011
33. Elumalai, V., Brindha, K., Sithole, B., & Lakshmanan, E. (2017). Spatial interpolation methods and geostatistics for mapping groundwater contamination in a coastal area. *Environmental Science and Pollution Research, 24*, 11601–11617. https://doi.org/10.1007/s11356-017-8681-6
34. Li, J., & Heap, A. D. (2014). Spatial interpolation methods applied in the environmental sciences: A review. *Environmental Modelling and Software, 53*, 173–189. https://doi.org/10.1016/j.envsoft.2013.12.008
35. Kumari, M. K. N., Sakai, K., Kimura, S., Nakamura, S., Yuge, K., Gunarathna, M. H. J. P., Ranagalage, M., & Duminda, D. M. S. (2018). Interpolation methods for groundwater quality assessment in tank cascade landscape: A study of ulagalla cascade, Sri Lanka. *Applied Ecology and Environmental Research, 16*, 5359–5380. https://doi.org/10.15666/aeer/1605_53595380
36. Lu, G. Y., & Wong, D. W. (2008). An adaptive inverse-distance weighting spatial interpolation technique. *Computers & Geosciences, 34*, 1044–1055. https://doi.org/10.1016/j.cageo.2007.07.010
37. Stachelek, J., & Madden, C. J. (2015). Application of inverse path distance weighting for high-density spatial mapping of coastal water quality patterns. *International Journal of Geographical Information Science, 29*, 1240–1250. https://doi.org/10.1080/13658816.2015.1018833
38. Achilleos, G. (2008). Errors within the inverse distance weighted (IDW) interpolation procedure. *Geocarto International, 23*, 429–449. https://doi.org/10.1080/10106040801966704
39. Wackernagel, H.(1995). Ordinary kriging. In: *Multivariate geostatistics* (pp. 74–76).
40. Chabala, L. M., Mulolwa, A., & Lungu, O. (2017). Application of ordinary kriging in mapping soil organic carbon in Zambia. *Pedosphere, 27*, 338–343. https://doi.org/10.1016/S1002-0160(17)60321-7

41. Kostopoulou, E. (2021). Applicability of ordinary Kriging modeling techniques for filling satellite data gaps in support of coastal management. *Modeling Earth Systems and Environment, 7*, 1145–1158. https://doi.org/10.1007/s40808-020-00940-5

42. Kumar, V. (2007). Optimal contour mapping of groundwater levels using universal kriging—A case study. *Hydrological Sciences Journal, 52*, 1038–1050. https://doi.org/10.1623/hysj.52.5.1038

43. Tonkin, M. J., Kennel, J., Huber, W., & Lambie, J. M. (2016). Multi-event universal kriging (MEUK). *Advances in Water Resources, 87*, 92–105. https://doi.org/10.1016/j.advwatres.2015.11.001

44. Gundogdu, K. S., & Guney, I. (2007). Spatial analyses of groundwater levels using universal kriging. *Journal of Earth System Science, 116*, 49–55. https://doi.org/10.1007/s12040-007-0006-6

45. Mühlenstädt, T., & Kuhnt, S. (2011). Kernel interpolation. *Computational Statistics and Data Analysis, 55*, 2962–2974. https://doi.org/10.1016/j.csda.2011.05.001

46. Krivoruchko, K., & Gribov, A. (2019). Evaluation of empirical Bayesian kriging. *Spat Stat, 32*, 100368. https://doi.org/10.1016/j.spasta.2019.100368

47. Gribov, A., & Krivoruchko, K. (2020). Empirical Bayesian kriging implementation and usage. *Science of the Total Environment, 722*, 137290. https://doi.org/10.1016/j.scitotenv.2020.137290

48. Le, N. D., & Zidek, J. V. (1992). Interpolation with uncertain spatial covariances: A Bayesian alternative to Kriging. *Journal of Multivariate Analysis, 43*, 351–374. https://doi.org/10.1016/0047-259X(92)90040-M

49. Mitas, L., & Mitasova, H. (2005) Spatial interpolation. In: *Geographic information systems: Principles, techniques, management and applications* (pp. 481–492). https://doi.org/10.4324/9781351243858-7

50. Rodríguez-Amigo, M. C., Díez-Mediavilla, M., González-Peña, D., Pérez-Burgos, A., & Alonso-Tristán, C. (2017). Mathematical interpolation methods for spatial estimation of global horizontal irradiation in Castilla-León, Spain: A case study. *Solar Energy, 151*, 14–21.

51. Willmott, C. J., & Matsuura, K. (2005). Advantages of the mean absolute error (MAE) over the root mean square error (RMSE) in assessing average model performance. *Climate Reserach, 30*, 79–82.

52. Filik, Ü.B., & Filik, T. (2017).Wind speed prediction using artificial neural networks based on multiple local measurements in Eskisehir. In: *3rd International Conference on Energy and Environment Research* (pp. 264–269). https://doi.org/10.1016/j.egypro.2016.12.147

53. Kazemi, E., Karyab, H., & Emamjome, M. M. (2017). Optimization of interpolation method for nitrate pollution in groundwater and assessing vulnerability with IPNOA and IPNOC method in Qazvin plain. *Journal of Environmental Health Science and Engineering, 15*, 1–10. https://doi.org/10.1186/s40201-017-0287-x

54. Hemakumara, G. (2015). Gis based analysis for suitability location finding in the residential development areas of greater Matara Region. *International Journal of Scientific and Technology Research, 4*, 96–105.

55. Schaller, J., & Mattos, C. (2010). ArcGIS modelBuilder applications for landscape development planning in the region of Munich, Bavaria. In: *Digital landscape architecture* (pp. 42–52).

56. Schaller, J., & Mattos, C. (2009). GIS model applications for sustainable development and environmental planning at the regional level. In: *GeoSpatial visual analytics. NATO science for peace and security series C: Environmental security* (pp. 45–57).

57. Schröder, D., Omran, A., & Bastidas, M. (2015). Automated geoprocessing workflow for watershed delineation and classification for flash flood assessment. *International Journal of Geoinformatics, XI*, 31–38.

58. Adhikary, P. P., & Dash, C. J. (2017). Comparison of deterministic and stochastic methods to predict spatial variation of groundwater depth. *Applied Water Science, 7*, 339–348. https://doi.org/10.1007/s13201-014-0249-8

59. Ghadirian, P., & Bishop, I. D. (2008). Integration of augmented reality and GIS: A new approach to realistic landscape visualisation. *Landscape and Urban Planning, 86*, 226–232. https://doi.org/10.1016/j.landurbplan.2008.03.004

Printed in the United States
by Baker & Taylor Publisher Services